Michael Blake

The Origins of
Agriculture in the
Lowland Neotropics

The Origins of Agriculture in the Lowland Neotropics

by

DOLORES R. PIPERNO
Smithsonian Tropical Research Institute

DEBORAH M. PEARSALL
University of Missouri, Columbia

ACADEMIC PRESS

San Diego   London   New York   Boston   Sydney   Tokyo   Toronto

Front cover photograph: A sample of some fruits currently grown in the Neotropics. (For more details, see Plate 3.8.)
Background art: Map of the tropical evergreen forest from Fort Sherman, Panama. (For more details, see Plate 2.3.)

This book is printed on acid-free paper. ∞

Academic Press
a division of Harcourt Brace & Company
525 B Street, Suite 1900, San Diego, California 92101-4495, USA
http://www.apnet.com

Academic Press Limited
24-28 Oval Road, London NW1 7DX, UK
http://www.hbuk.co.uk/ap/

Library of Congress Card Catalog Number: 98-84003

International Standard Book Number: 0-12-557180-1

PRINTED IN THE UNITED STATES OF AMERICA
98 99 00 01 02 03 QW 9 8 7 6 5 4 3 2 1

To Jenny,
who makes all of my days bright and fills them with love.
—Dolores

To Mike,
for all his loving support.
—Debby

Contents

2   The Neotropical Ecosystem in the Present and the Past

3   The Phytogeography of Neotropical Crops and Their Putative Wild Ancestors

4 The Evolution of Foraging and Food Production

5 From Small-Scale Horticulture to the Formative Period: The Development of Agriculture

Preface

This volume focuses on the origins and development of agriculture in the lowland New World, or Neotropical forest. It is the outcome of the authors' individual, career-long fascination with and research into the topic. Tropical forests have always had an aura of mystery about them. Occupying great land areas in the Northern and Southern Hemispheres of the New and Old Worlds, and containing most of the world's plant and animal species, they have long been a source of fascination, as well as frustration, to investigators from the natural and social sciences. The difficulties of carrying out research in the tropical forest biome are legion and legendary and sometimes exaggerated. During the past 20 years, archeologists, botanists, paleoethnobotanists, ecologists, paleoecologists, and molecular biologists employing an arsenal of techniques have elevated both the quality and the quantity of field and laboratory investigations to impressive levels. This volume is a reassessment and resynthesis of early human adaptations and agricultural history in the Neotropical forest in light of evidence from these disciplines obtained largely during the past two decades.

In 1952, the cultural geographer Carl O. Sauer published a book that argued for a single origin of agriculture in the tropical forests of Southeast Asia. In a series of papers written before and after that tome, he offered his views of American plant domestication that also emphasized the early importance of the tropical biome, especially the seasonally dry forests where annual precipitation was punctuated. Inspired by Sauer, the archeologist Donald W. Lathrap sought New World agricultural origins and other important cultural innovations in the lowland Neotropical forest in books and papers published during the 1970s. The information that we present in this volume leads us to believe that many of the ideas developed by Sauer and Lathrap were essentially correct and that the humid and lowland tropical forests were important, independent, and possibly the earliest centers of plant cultivation and domestication in the New World. We also conclude that the Neotropical forest supported other major cultural developments, such as the emergence and spread of truly effective and productive agricultural systems and ceramic production.

Our theoretical approach to the problem is ecological and evolutionary. The time has long since past when archeologists could afford to ignore these disciplines in considerations of agricultural origins. The alliance of formal economic logic with models from the field of evolutionary or behavioral ecology, leading to what is known as optimal foraging theory, is a particularly powerful theoretical approach to the study of agricultural origins. Such models are not intended to be perfectly realistic representations of human subsistence decisions in all their complexity. Instead, they are heuristic tools used to generate and test hypotheses about specific questions and to build robust explanations. We believe that foraging theory has the potential to elucidate many aspects of the transition from hunting and gathering to food production in the Neotropics and elsewhere.

A variety of audiences at different levels should find interest in this book. We sought to make it user-friendly to the beginning student in Latin American archeology and paleoethnobotany by providing a review of some basic ecological, climatological, and geographic concepts. We also targeted the higher-level scholar in anthropology and the biological sciences with detailed discussions of some major questions relating to tropical agriculture, ecology, and evolution. We avoided lengthy reviews of some problems (e.g., Amazonian ecology and prehistoric human adaptations) that have seen prolonged controversy and that quite likely have relatively little bearing on agricultural origins. We refer the reader to key papers and syntheses on topics such as these.

Although we review many archeological sequences from the lowland Neotropics, this is not a book on lowland Neotropical prehistory. We apologize in advance for excluding what may be someone's favorite archeological site or study region. We focused on those sites and regions most pertinent for examining the central topic of the volume and those where macro- and microbotanical remains were systematically recovered. Also, the relevant literature is so huge and comprises so many disciplines that a pruning of sources and citation of a relatively few basic references were sometimes necessitated.

We hope this volume will help to stimulate much more research in the humid and lowland tropics and that readers will find the tropical forest a more forgiving and rewarding place for long-term study than has been commonly supposed.

Dolores Piperno developed the theoretical perspective of the work and is the primary author of all chapters but Chapter 3. Deborah Pearsall contributed sections of Chapters 1, 2, and 5 and is the primary author of Chapter 3.

Acknowledgments

In a work of this scope it is naturally impossible to acknowledge more than a few of the intellectual debts owed to colleagues and collaborators we have worked with during our careers. Dolores Piperno has benefitted enormously from interactions with colleagues at the Smithsonian Tropical Research Institute (STRI) concerned with evolutionary biology and tropical ecology. Particularly, Neal Smith, Egbert G. Leigh, Jr., Stephen Hubbell, Jeremy Jackson, S. Joseph Wright, Robin Foster, and Annette Aiello are acknowledged. She is grateful to Richard Cooke of STRI and Anthony J. Ranere, Temple University, for the long and fruitful collaborations and numerous stimulating conversations about tropical archeology.

Anthony J. Ranere and Richard Cooke also deserve special thanks for spending long hours reading drafts of the manuscript. Their comments and suggestions greatly improved the final product. Cristóbal Gnecco, Carlos López, Inéz Cavelier, Luisa Fernanda Herrera, Paulo De Oliveira, Thomas Andres, Karen Stothert, John Jones, Jack Rossen, and Tom Dillehay read and commented on sections of the manuscript. The STRI library filled numerous interlibrary loan requests. Donna Conlon prepared all the figures except Figs. 1.1, 3.18, 3.19, 5.1, 5.5, and the Obelisk Tello plant depictions, and she drew the plant and animal illustrations. The STRI digital imaging laboratory provided facilities for Conlon's work. Irene Holst organized the photo sessions of the tropical fruits and provided invaluable assistance when the final manuscript was being assembled. Roberto Ibañez carried out the statistical analysis of *Cucurbita* phytoliths. María Isabel Alfaro typed the tables and the Reference section.

Martin H. Moynihan, the founding director of the STRI and an eminent animal behavorist, died on December 3, 1996. More than anyone, Martin was an inspiration and source of support for the senior author, beginning from her first trip to Panama in 1979 as a Master's thesis student. She does not know what she would have accomplished without his encouragement and loving friendship.

Piperno's research was supported by the Smithsonian Tropical Research Institute, grants to the STRI from the Andrew W. Mellon Foundation, and grants from the National Science Foundation and the National Geographic Society.

Deborah Pearsall thanks Robert Benfer, Tom and Sheila Pozorski, and Edward Buckler for reading and commenting on sections of the book. Despite such expert help from these and researchers mentioned previously, any errors of fact or judgment are entirely the authors'. Brigitte Holt coordinated the computer literature search on crop phytogeography and arranged for numerous interlibrary loans. Research in the Jama River Valley, Ecuador, was supported by National Science Foundation grants to James A. Zeidler and Deborah M. Pearsall. Pearsall's research is conducted at the facilities of the American Archaeology Division of the Department of Anthropology, University of Missouri, Columbia. Pearsall thanks all the archeologists who have entrusted their botanical remains to her over the years for analysis.

The following figures and photographs are included in the text with permission of the authors and photographers: Figure 4.10, Anthony J. Ranere; Figure 5.4, Mora *et al.* 1991; Figures 5.12 and 5.13, John Jones; Plates 2.1 and 2.2, Neal Smith and the Smithsonian Tropical Research Institute; Plates 2.3, 2.4, 2.6, 3.1, 3.2, 3.8, and 5.3, Marcos Guerra and the Smithsonian Tropical Research Institute; Plate 2.5, George Angehr; Plates 2.7, 3.9, and 4.16, Carl Hansen and the Smithsonian Tropical Research Institute; Plates 3.3, 3.4, 3.5, and 3.6, Thomas Andres and the New York Botanical Garden. (Plate 3.6 was originally taken by T. Plowman and J. Alcorn.); Plate 3.7, Karen Stothert and the Muséo del Banco General, Guayaquil, Ecuador; Plates 4.1, 4.9, 4.10, and 4.11, Richard Cooke and the Smithsonian Tropical Research Institute; Plates 4.2, 4.3a, and 4.7, Karen Stothert; Plate 4.3b, Anthony J. Ranere; Plate 4.8, Warwick Bray and Richard Cooke; Plates 5.1 and 5.2, James Zeidler.

CHAPTER 1

Background of Tropical Agricultural Origins

INTRODUCTION

Native Americans domesticated more than 100 species of plants before the arrival of Europeans in the 15th century, which is an impressive achievement in plant breeding. Many of the species, such as maize, the white and sweet potato, squash, beans, and tomatoes, are familiar foodstuffs. Others, such as the North American domesticate "sumpweed" (*Iva annua*), a relative of sunflower, have probably seldom been eaten by any modern human being.

An interesting fact is that of the total repertoire of American crop plants, more than half, including many of the staple foods that supported indigenous populations at the European contact, are known or thought to have been originally taken under cultivation and domesticated in the warm and humid tropical lands of Central and South America (Sauer, 1950; Brücher, 1989; Sauer, 1993). Even maize, whose origins were once attributed to the high and arid Mexican valleys, is now considered to have come from the tropical deciduous forest of low elevations in Mexico (Doebley, 1990).

Some geographers have long held that the American lowland tropics were an important early center of domestication, but until recently scholars interested in the problem have not had the techniques and the data necessary to examine this proposition in-depth. Limited archeological survey and excavation had been undertaken in the humid tropics. Poor preservation of botanical materials in archeological sites hampered interpretations of early subsistence. The development of effective approaches for doing tropical archeology and new methods for recovering, identifying, and dating plant remains during the past 20 years, together with the use of interdisciplinary research strategies, has advanced us to the point where we can now do more than just speculate on the age and nature of early plant cultivation in the tropics.

In the low-lying regions between southwestern Mexico and the southern rim of the Amazon Basin, rains are bountiful and predictable, soils are fertile, lands teem with plant life never at risk to killing frosts, and many of the wild ancestors of the major crop domesticates still reproduce successfully without a human hand. We believe that in these tropical lands, the origins of New World agriculture are to be found.

Students of the problem know that our thesis is not original. Two prominent investigators of agricultural origins, the cultural geographer Carl Sauer and archeologist Donald Lathrap, posited the primacy of agriculture in the forested lands of the humid tropics (e.g., Sauer, 1952; Lathrap, 1970, 1973a,b, 1977a). For the most part, they did not have the benefit of modern techniques for recovery and dating of botanical remains from archeological sites or genetic studies to precisely determine the relationships between wild and domesticated plants. Therefore, they built their hypotheses on what evidence they could muster from modern distributions of crop plant progenitors, the ecological requirements of these plants, and linguistic and other cultural affiliations of tropical people. Still, some of their hypotheses were elegant and, as new evidence shows, some were remarkably accurate.

Initial reactions to Sauer's and Lathrap's claims for agricultural origins in the tropical forest were skeptical, and they continue to elicit comment, both positive and negative (Mangelsdorf, 1953; Meggers and Evans, 1957; Mangelsdorf et al., 1964; Harlan, 1992; Stahl, 1998; Maloney, 1994). The tropical forest was a long-neglected area of study and one that was commonly depicted as inimical to cultural innovation and development (e.g., Steward, 1948, 1949; Mangelsdorf et al., 1964; Meggers, 1954, 1971). In his major review article for the then-young journal American Antiquity, Mangelsdorf (1953, p. 90) remarked that Sauer's (1952) hypothesis "was almost completely lacking in factual basis" and probably could not be made to be more untestable.

In these early debates, there were also more valid conceptual questions about the ability of the tropical ecosystem to support sedentary, agriculturally based societies (e.g., Stewart, 1949; Meggers and Evans, 1957). If, as commonly thought, the soils were so poor, settlement densities so low, and populations so mobile, even at the European contact, how could the tropical forest have been a hearth of food production? These questions still resonate and have been broadened to include the suitability of the tropical forest habitat even for low-density, foraging lifeways (e.g., Bailey *et al.,* 1989; Headland, 1987).

However, doubts concerning the agricultural potential of the tropical forest were based on examples drawn largely from the Brazilian Amazonian *terra firme* or nonfloodplain forest, whose soils and overall resource base for foragers, we contend, are atypically poor in comparison to other Neotropical regions and probably have little relevance to the question of agricultural origins. On the other hand, the *varzea* or floodplain forest of the Amazon River offers an atypically rich resource base for Neotropical foragers and would have given early people little incentive to grow plants. Also, neither of these zones were likely to have housed the wild ancestors of most of the major tuber and seed crops. For all these reasons, the interior of the Brazilian Amazon probably had little to do with the origins of Neotropical food production.

We will also attempt to show that the tropical forest habitat generally is neither as hostile (e.g., Richards, 1952; Bailey *et al.,* 1989) nor as benevolent for human occupation and plant experimentation (e.g., Colinvaux and Bush, 1991; MacNeish, 1991) as researchers have suggested and, in these kinds of environments, neither too demanding nor too benign for plants and people, are to be found the origins of New World food production.

Another notable reason for the lack of attention to the lowland Neotropics in discussions of agricultural origins has been the success of investigators in studying the problem in the Near East. Ecologically, this region could not be more unlike the tropical forest, and it precisely fits many researchers' notions of the type of environment (semi-arid) suited for agricultural origins and the fostering of cultural development in general (e.g., Fritz, 1994; Mangelsdorf *et al.,* 1964; Smith, 1995a,b; Steward, 1949).

Decades of research in the Near East have yielded important results bearing on the fundamental questions of the how, when, where, and why of food production (e.g., Bar-Yosef and Belfer-Cohen, 1989, 1992; Byrd, 1992; Harris, 1996; Henry, 1989; McCorriston and Hole, 1991; Miller, 1992; Moore and Hillman, 1992; Price and Gebauer, 1995; Russell, 1988; Wright, 1993). (The quality and quantity of effort devoted to the question in the Near East has reached such high levels that it is possible to cite only a few of the important studies). These studies have also securely identified the wild progenitors of most of the initial crop domesticates, including wheat, barley, and lentil, and clarified their current and past geographical ranges and ecological habitats (e.g., Wright, 1993; Zohary, 1989, 1992; Zohary and Hopf, 1988).

In the absence of substantive data on the timing and nature of early cultivation in the tropics, people understandably extrapolated from the well-studied Near East. For example, early plant domestication in the Americas was thought to (i) have occurred first in arid or semi-arid zones and to have primarily involved manipulation of the seed-bearing structures of plants (Fritz, 1994; MacNeish, 1991; Mangelsdorf *et al.,* 1964; Smith, 1995a,b) and (ii) have been preceded by fully sedentary life and some considerable social complexity (Fritz, 1994, 1995; Gebauer and Price, 1992a; Hayden, 1995; Smith, 1995a,b).

The tropical biome must be considered on its own terms and on the basis of empirical evidence derived from within its boundaries. Reed (1977) hit the nail on the head when he noted:

> The kind of culture in which plant cultivation in the tropical forest could originate and be nurtured sucessfully would probably look simple and primitive to the person who, familiar with the Near Eastern Natufian, thinks automatically of that culture as the adaptive plateau . . . necessary for any agricultural origin . . . but obviously different standards prevailed elsewhere. Any attempt to rank such preagricultural adaptive plateaux from different parts of the world would be a basic error. I believe firmly that each such area [of independent agricultural origin] *should be studied from the viewpoint of the sequence of environmental and cultural stages that preceded the initiation of agriculture in that region* (p. 885, italics added).

The following is a summary of our current understanding of the evidence:

1. Tropical forest food production emerged at approximately the same time as it did in the Near East and earlier than currently demonstrated in highland Mexico and Peru.[1]

2. The lowland tropics witnessed climatic and vegetational changes between 11,000 and 10,000 B.P. that were no less profound than those experienced at higher latitudes, and which led to major shifts in resource densities and distributions and necessitated significant cultural responses relating to the food supply.

3. Systematic cultivation of small plots adjacent to residential structures (gardens) was under way during the 10th and 9th millennia B.P. in the humid, tropical lowlands of Panama, Peru, Ecuador, and Colombia. By at least 9000–8000 B.P. evidence of morphological and other changes (larger seed size and phytolith size) associated with systematic cultivation and probably indicating domestication is apparent in some economic plants.

4. By 7000 B.P. larger scale food production characterized by the preparation of substantial areas away from house-side locations (fields) had emerged. With the extension of cultivated plots into the forest, the felling and/or killing of trees to admit sunlight to the seed and tuber beds became compulsory and the effects of what is referred to now as slash-and-burn agriculture become apparent in paleoecological records. By the beginning of the Christian era, these methods of cultivation intensified and expanded. By this time, they included most of the cultivated species witnessed 15 centuries later by the first Europeans, and many tropical people were living in nucleated and sedentary villages.

The purpose of this book is to present the evidence for this series of developments and to offer some explanations for why and when they occurred. We are particularly

[1] An article published when this volume was in press demonstrates that the *Cucurbita pepo* squash remains from Guilá Naquitz and, therefore, plant domestication in Mesoamerica, data close to 9000 B.P. (Smith, 1997). Because a wild ancestor for this domesticate is unknown and was probably not native to the arid, highland regions of south and central Mexico, we see little reason to revise this statement at this time. All dates in this book are in uncalibrated radiocarbon years.

concerned with the earlier parts of the process—when hunters and gatherers in the tropical forest decided to intensify their use of the wild flora, turned their attention to the propagation of certain plants, and subsequently developed systems of food production known today as slash-and-burn or swidden cultivation. We also consider, in less scope and detail, the evidence for the time when people in possession of a suite of fully domesticated and productive crop plants began to live in villages and developed more complex forms of social structures, technologies, and exchange systems.

Although we are mainly concerned with the humid forested lands of the tropics at low and lower mid-elevations (sea level to approximately 1200 m) our discussion will sometimes range into higher zones. We consider, for example, the origins of a number of crops (beans, squashes, and chiles) that were probably domesticated and grown in the low to mid-elevation Andes (1000–2000 m, depending on the latitude) and that became important components of lowland agricultural systems. We also consider the north and central coast of Peru, the southern-most limit of early agriculture along the dry western coast, where many species of crops domesticated in the lowland tropical forest were grown. We will not discuss in any detail the high-elevation Andes and the crops that were domesticated there, such as the white potato.

The decision not to formally incorporate the high-elevation Andes into our model for the emergence of agriculture in the Neotropics was made for both theoretical and practical reasons. The Andean mountains provide a complex array of environments for hunter–gatherers, with ready access in many regions to diverse environmental zones created by local variations in effective rainfall, slope, aspect, and elevation. This complexity provides the potential for very different people–plant interrelationships than those we describe for the lowlands. The presence of large game animals, and their importance in the early Holocene occupation of high-elevation grasslands following the retreat of alpine glaciers (sites on the Junin puna in Peru, for instance), argues for a different balance of the importance of plant and animal resources in decisions concerning settlement location and subsistence strategies during the immediate post-Pleistocene period.

Domestication of plants favoring disturbed habitats, such as quinoa (*Chenopodium quinoa*) and maca (*Lepidium meyenii*), a root crop of the Junin region, is perhaps intertwined with the process of camellid domestication (Pearsall, 1989a). To explore the processes of plant and animal domestication in the Andes and their relationships to developments in the lowlands, which we believe have primacy in food production origins, would require a book-length treatment.

From the practical point of view, very few paleoethnobotanical data sets are available from early to mid-Holocene sites in the mid- to high-elevation Andes; these have previously been reviewed (Pearsall, 1992).

Interpretation of two of these early data sets—Guitarerro Cave and a series of sites in the Ayacucho region, Peru—is complicated by potential disturbance of dry sediments, dating ambiguities, and, in the case of the Ayacucho materials, incomplete publication. In the higher elevations of the Ecuadorian Andes, vulcanism has a severe impact on site discovery. The few Formative period botanical data sets available are from sites buried under meters of volcanic ash, and no record exists for plant use in earlier periods. Lake coring, which has proven valuable in expanding the data on early agriculture in the lowlands (see below and Chapters 4 and 5), holds equal potential for tracing the introduction of maize into the northern Andes (i.e., Athens, 1990) and for tracking the trajectory of plant domestication and agricultural intensification there. Applying the perspective we present here on recovering and interpreting data to the issue of plant and animal domestication in the Andean mountains might prove informative, but we leave this task to someone else.

A TAXONOMY OF TROPICAL FOOD PRODUCTION

All too often the terms *cultivation* and *domestication,* and *food production, horticulture,* and *agriculture* have been used synonomously, resulting in a lack of clarity about the kind of plant propagation system actually being described for any point in time. Especially in regions where food production independently arose and subsequently intensified into larger scale entities that supported larger and more complex settlements, some taxonomy of food-producing behavior seems necessary to distinguish among the social, demographic, and environmental correlates of the various systems present through time.

We use the terms referred to previously in the following ways. Our usage of the terms largely follows that proposed by Harlan (1992), Ford (1985), and Harris (1989). Cultivation in the broadest sense refers to all human activities involved with caring for plants. We limit this term, however, to activities surrounding the preparation of plots specified for plant propagation and repeated planting and harvesting in these plots. It is these types of cultivation activities that lead to marked genetic and morphological changes in the plants being cared for—the process of domestication. We use the terms crop and cultivated plant synonomously to refer to plants that are planted and harvested, regardless of their domesticated status.

Domesticated species are those that have been genetically altered from the wild form through human (artificial) selection and normally are dependent on human actions for reproduction. The genetic changes leading to complete domestication are cumulative and difficult to identify archeologically. Because some plants may be less altered than others even after persistent human selection, we sometimes discuss species thought by tropical botanists to have been "semidomesticated." Food production is used in the most general sense to envelop all scales of plot preparation and planting behavior.

We use the terms horticulture and agriculture in an evolutionary continuum to denote, respectively, small-scale plantings (house gardens), which typically contain a range of plants from morphologically "wild" to clearly domesticated, and larger scale field systems, in which domesticated plants are common and come to dominate as staple crops. Although we envision an evolutionary continuum for these types of food production, they probably coexisted in the tropics after agriculture developed, depending on local ecology.

The term protocultivation has been commonly used to indicate early tropical food production, but it has also denoted such manipulations of wild plants by foragers as replanting the head of tubers into the hole left by the harvested plant or use of fire to ensure the renewal of exploited wild plant resources (e.g., Jones, 1969; Hallam, 1989; Bahuchet *et al.,* 1991). These activities fall outside of our definition of cultivation. To avoid confusion we prefer the term horticulture for early systems of tropical food production.

We believe that, for the most part, the process of domestication is strictly linked to the cycle of planting and harvesting (as argued by Harlan, 1992). Genetic changes in certain tropical cultigens such as tree crops might have occurred prior to cultivation (as argued by Rindos, 1984). Nevertheless, as Pearsall (1995a) has demonstrated, testing Rindos's "coevolutionary" model (see discussion below) archeologically or through study of paleoenvironmental records is extremely difficult. It is more useful to focus on those practices that involve moving plants out of their natural habitats and clearing vegetation for food plots because they clearly lead to genetic changes in crops, and they can more easily be documented in the prehistoric records.

Some students of tropical food production may take issue with our focus on cultivation practices that involve food plot preparation. In the tropics, particularly, there are many intermediate stages between the manipulation of plants in their wild states and their movement to prepared plots (Levi-Strauss, 1950; Harlan, 1992). Sophisticated forms of plant management have been observed in the Neotropical forest and other tropical regions today. Even if they do not fit the more limited definition of cultivation offered previously, they come close to doing so and may promote some genetic and morphological modification of the plants (e.g., Anderson and Posey, 1989; Balée, 1988, 1989; Casas *et al.,* 1996; Clement, 1997; Groube, 1989).

These practices—protection, selective pruning, and planting of perennial tree species, particularly palms, in their natural settings in the forest (rather than in prepared plots), plus the replacement of tubers in the ground—undoubtedly were characteristic of prehistoric forest peoples, and may be among the more ancient interactions between humans and plants in the forest setting (Groube, 1989). By themselves, however, they are unlikely to have led to the settlement, demographic, and environmental changes associated with the establishment and spread of food-producing behavior.

We believe, as do many others (e.g., Anderson, 1952; Lathrap, 1970, 1977a; Harris, 1989), that tropical food production began in small-scale house gardens that served as the laboratories of plant domestication. In these gardens, morphologically wild plants may have been under cultivation and all early cultivated plants were not necessarily domesticated, either in part or in full (Harris, 1989; Harlan, 1992). Even after a protracted period of time, tropical horticulture probably involved the exploitation of almost as many "wild" as "domesticated" species (Harris, 1989, p. 20). In fact, we will document a number of cases in which (i) cultivated plants that were important food items during one time period later decline or disappear from the record altogether, and (ii) plants do not develop characteristics of genuine domesticates despite a prolonged association with humans.

We will argue that in some areas swidden cultivation (slash and burn and slash and mulch) developed out of dooryard horticulture, and that this was an intermediate stage in the evolutionary continuum between horticulture and settled village agriculture. In other areas, more sedentary populations may have grown their crops on rich alluvial lands of small rivers for many millennia without cutting and burning vegetation, with swidden cultivation coming into play only after more fertile lands were fully utilized.

We will also argue that many lowland tropical societies practiced food production for at least 5000 years before the emergence of village life, and that this "late" emergence (Flannery, 1986a) should not be seen as an anomaly but rather as a necessary and logical outcome of the ecology and demography of food production in the Neotropics. Where natural abundances of resources permitted settled life based on wild food resources, for example, in some riverine and coastal environments, food production appears later in the record, as does dependence on domesticated plants. It is important, therefore, that we separate the issue of the origins of food production from the origins of agriculture; these phenomena may not have completely congruent explanations.

To review, in defining food production and seeking to identify when and where it started in the lowland Neotropics, we require that it (i) be systematic (i.e., practiced on a regular basis), (ii) involve preparation of plots and the planting and harvesting of plants within those plots, and (iii) provide some real contribution to the diet. Creating small plots and sowing plants in them is the first step in the creation of an agricultural system; as such, we feel it reflects concerns about food plants rather than simply spices, medicinal plants, or containers (although we expect to see such plants represented as well). All the previous criteria presume a social structure permitting at least semisedentary living, scheduling of food-producing activities, and returns to the same or nearby parcels of ground from year to year.

Such early food-producing societies should leave evidence in the archeological record in the form of small but fairly durable settlements located on circumscribed parcels of land of high fertility from which can be recovered artifacts and, perhaps, well-defined activity areas related to cooking and plant processing and, with luck, remains of the plants themselves. Paleoenvironmental data should reveal relatively small-scale environmental modification; groups not preparing food plots or returning routinely to the same region would be largely invisible in the record.

POSTULATED CHARACTERISTICS OF TROPICAL FOOD PRODUCTION COMPARED TO NEAR EAST FOOD PRODUCTION

Because we have argued that tropical food production should be studied on its own terms, it is informative to compare our views on the nature of early food production in the Neotropics to characterizations of this process in the Near East. We do not require that many of the first cultivated plants eventually be domesticated or even extensively cultivated later in prehistory. We do not require that early domesticates be domesticated rapidly, i.e., within several to 25 human generations of continuous cultivation. These two characteristics distinguish early Neotropical food production from the Near Eastern example, in which many of the earliest cultivated plants, such as wheat, barley, lentils, and chick peas, were apparently quickly domesticated (a time period of between 30 and 200 years is thought to be a reasonable estimate for wheat and barley domestication) and continue even today to be staple foods (e.g., Hillman and Davies, 1990; Zohary, 1989).

Furthermore, we do not require that the first plants were cultivated by sedentary people living in villages with fairly complex social organizations. Full-time living by foragers in multi-house communities, which characterized Near Eastern settlement on the eve of the Neolithic, was made possible by a fairly stable and dense wild resource supply that included many plant carbohydrates. An extended period of exploitation of some of the wild resources that came to be cultivated in the Near East led to an increase in regional human population densities and a decrease in hunter–gatherer foraging ranges.

Plant resource densities are structured very differently in tropical forest settings, in which especially plant carbohydrates are generally not present in substantial numbers. It is also important to remember that many of the earliest locales of Neotropical food production were sparsely populated 10,000 years ago. Population pressure, in the sense used by Cohen (1977a), or a decrease in available foraging territories arising from human population growth (Bar-Yosef and Belfer-Cohen, 1992) probably had little to do with the decision to start and maintain garden plots in the New World. They may have been involved in the subsequent intensification of early food production into intermediate and fully agricultural systems.

Finally, we suggest that the wild ancestors of some of the earliest crop plants, such as maize, squash, manioc, and other tubers, were not subject to a lengthy period of collection before they were taken under cultivation. We acknowledge that there are relatively few data on plant use in the Neotropics prior to the appearance of cultivated plants during the early Holocene, but available paleoenvironmental evidence leads us to conclude that prior to 10,000 years ago, many groups were emphasizing different resources on very different late-glacial landscapes, a point we discuss later in this chapter and in Chapter 2.

There are also significant parallels, which should not be minimized, between our model of early food production and those proposed for the Near East. We will argue that the transition from foraging to farming was part and parcel of the profound environmental changes that were associated with the end of the last Ice Age, between 11,000 and 10,000 years ago. The same arguments are being advanced by researchers in the Near East (e.g., Henry, 1989; McCorriston and Hole, 1991; H. E. Wright, 1993; K. I. Wright, 1994). Our model also assumes that a nomothetic explanation should be sought for the transition from foraging to food production occurring, as it did, within 1500 years after the end of the Pleistocene in at least three or four regions of the world.

We now discuss the explanations and theories that have been advanced to account for the origins of food production and offer in more detail our perspective on the matter.

EXPLANATIONS OF NEW WORLD FOOD PRODUCTION ORIGINS

> *To do science is to search for repeated patterns, not simply to accumulate facts. . . . Doing science is not such a barrier to feeling or such a dehumanizing influence as is often made out. It does not take the beauty from nature.*
>
> Robert H. MacArthur (1972)

Explanations of the origins of food production in the New World, as in the Old World, have run the gamut from those relying on social causality to those using Darwinian theory to, most recently, the optimality models of modern behavioral ecology (e.g., Bender, 1978, 1985; Layton, *et al.*, 1991; Piperno, 1997a; Russell, 1988; Smith, 1987; Watson, 1991). The differences among the various models are not trivial and involve real paradigmatic choices as to whether Darwinian processes may account for shifts in the subsistence behavior of the modern human species or if human decision making is isolated from ecological/economic influences and mediated mainly by the social and political matrix.

We will not follow the time-honored tradition of reviewing in detail the explanations that have been advanced for the transition from hunting and gathering to farming. Several excellent reviews have been undertaken over the years (e.g., Flannery, 1973, 1986a; Redding, 1988; Reed, 1977; Rindos, 1984; Stark, 1986; Watson, 1991, 1995) and the issue would not benefit from yet another extended treatment here. Rather, we briefly describe the kinds of explanations most frequently proposed and then evaluate them as they contrast with our own explanation, which relies on modern ecological and evolutionary theory.

After Redding (1988), explanations for the origins of food production are divided into the following seven types:

1. No cause is given, and food production is seen to be the result of human innovation (e.g., Sauer, 1952; Braidwood, 1960).

2. Climatic change is the prime mover. Childe's (1952) oasis model is the first and most famous example of this explanation. Climatic change arguments are being refined by Wright (1993) and others with the use of robust empirical data from paleoecological studies, which now include information on the past distribution of some wild ancestors of Old World crop plants on the landscape.

3. Demographic stress factors, especially population growth, are the primary mechanism (Smith and Young, 1972; Cohen, 1977a). People turn to food production as a last resort when imbalances between the number of people and the food supply begin to reach critical levels. An important assumption of these explanations is that early farmers experienced diminishing returns to labor as they were forced to increasingly rely on lower quality, less preferred foodstuffs (seeds, tubers, etc.).

4. An interaction of population growth and climate change led to food production (e.g., Binford, 1968; Hassan, 1977, 1981).

5. Food production is seen as the end result of long, mutualistic associations between people and plants within an evolutionary (Darwinian) framework (Rindos, 1984). In this view, close, fitness-enhancing relationships existed for thousands of years between people and the plants that came to be domesticated, and they resulted in behavioral changes in human populations that were largely unconsciously motivated. An important part of Rindos's model is that harvesting, incidental dispersal, and protection of wild plants by humans begins the process of genetic change in some species before systematic cultivation takes place.

6. Food production is seen as the outcome of a set of subsistence decisions made on the basis of relative return rates (usually caloric yield/time or yield/unit area invested) of exploitable resources in the environment (e.g., Gremillion, 1996; Hawkes and O'Connell, 1992; Kaplan and Hill, 1992; Keegan, 1986; Piperno, 1997a; Russell, 1988; Winterhalder and Goland, 1993). Significant changes in food procurement practices may come about through alterations in the abundance of the most highly valued (least costly) resources, which may result from such factors as environmental change or human population growth. The outcomes are the inclusion into the diet of new or little used foods that had borne unacceptably high costs before the disappearance or depletion of the previously exploited resources and, possibly, the intensification of use of these items leading to their production. These explanations, grouped under the framework of "optimal foraging theory," also see the human–plant association in Darwinian terms and are developments in a field called behavioral or evolutionary ecology. We find them to be the most compelling and will discuss them in detail later in the chapter.

7. Various social factors (e.g., desire for prestige, competitive feasting, and cosmology) are seen as the prime forcers of subsistence changes leading to food production (e.g., Bender, 1978, 1985; Hayden, 1990, 1992, 1995; Price, 1995).

How can these explanations be evaluated in light of what we currently know about the New World archeological and paleoecological records and considering developments in modern ecological and evolutionary theory?

Population Pressure

We do not think that population pressure was a significant factor in the New World because the area was settled by so few people on the eve of food production. Rather, increasing numbers of people on the landscape after horticulture was established and started to spread probably led to the intensification of horticulture into swidden cultivation.

Moreover, written into all the demographic pressure arguments is the assumption that at some point shortly before the origin of food production human populations had exceeded their carrying capacity, usually understood by anthropologists to be the upper limit of population growth that can occur in a given habitat without the degradation of the resource base (e.g., Glassow, 1978; Hayden, 1975). This point of view posits that as the carrying capacity ceiling was reached, food production became necessary to replace the depleted wild plant and animal populations.

Framed as such, the carrying capacity concept was subject to all kinds of difficulty when it was used as an analytical tool for evaluating foraging behaviors. For example, how did one go about calculating the carrying capacity of any biome for humans at any technological level? How could one be sure that calculations made on the basis of modern-day landscapes and resources have relevance to past conditions?

As Winterhalder et al. (1988) explain, the anthropological concept of carrying capacity has been significantly altered from the one commonly used in ecology and weakened as an analytical construct because it does not consider the dynamic relationships that exist between predators and their prey. In contrast to the use of the concept in anthropology, the resource base of ecological predator/prey theory is never static. As resources are exploited by foragers, their population sizes will vary, and changing prey availability, in turn, will have significant consequences for both the forager population size and subsequent resource selection. This latter point is important because if foraging and food production represent two distinct subsistence strategies, then a selection among alternatives of resource availability and procurement must have occurred to bring about the transformation we observe in the archeological record.

The anthropological carrying capacity approach has also neglected the economics of foraging decisions (e.g., optimal foraging theory, category 6 in the previous list, and see below), so no insight into how foragers may respond to changing resource densities and select among the alternatives is possible. It now appears that the most productive and realistic approach to investigating the relationships between human foragers and their resources is to combine population biology and foraging models (e.g., Winterhalder *et al.,* 1988; Winterhalder and Goland, 1993). The former may elucidate the demographic trends associated with the emergence of food production, and the latter may robustly explain the course of resource selection and intensification that resulted in the onset of the Neolithic period.

One last but exceedingly important point relating to population pressure arguments is their implicit or explicit assumption that the shift to food production carried with it declining returns to labor. If true, this would reenforce the notion that turning to cultivation was a last resort for foragers, initiated only when populations had grown and/or were territorially constrained to the point that wild resources were in danger of deterioration and people needed to increase their yield of food per unit area exploited. In reality, very few estimates have been made of the relative costs and returns of wild and cultivated resources.

In Chapter 2, we present evidence that the returns to labor from tropical horticulture are likely to be greater, not less, than the returns from foraging in a tropical forest. Under these conditions, the "correct" or most energetically efficient selection among the alternatives of resource availability between 10,000 and 9000 years ago would have proceeded along the path from foraging to food production.

Climate Change

Climate change as an explanation of food production has been subjected to much vilification, which has largely missed the point. Proponents of most well-reasoned objections dislike using a physical phenomenon to determine the direction of complex human behavior (e.g., Wagner, 1977). However, climate change could not have "caused" food production any more than an especially rainy day could force a human family to eat rice instead of beans—if both are available in the kitchen cupboard. What has been largely missed is climate's role in influencing the selection among alternatives of resource availability.

Helped by improving paleoecological techniques, which are making reconstructions of environments on the eve of food production more robust (e.g., Wright, 1993; Leyden *et al.,* 1994), the role of climate change in the transition to the Neolithic period has been restored to respectability (Harris, 1996; Henry, 1989; McCorriston and Hole, 1991; Watson, 1995), although it is still controversial, especially among those that posit social causation (Hayden, 1995). We view climatic change not as a "prime mover," which we submit was human decision making under selection pressures, but as the key physical element of the process because it triggered shifts in resource density and abundance in Neotropical environments that compelled the human adjustments that led to cultivation behavior (see below and Chapter 2).

Perhaps the best way to evaluate the role of climate, and one's own feeling about the role of climate, is to ask the following question: Would food production have started in Mesoamerica, South America, the southern Levant, and southern

China when it did if the Pleistocene had not ended when it did (about 10,000 years ago)? Our answer is decidedly no. If this conforms to standard definitions of environmental determinism then so be it, but we enter a plea that our brand of determinism is at least dependent on human decision making in the context of evolutionary relationships for the outcome to be effected.

Social Causation

We generally find explanations of food production rooted in social theories to be the least satisfying of all. Some posit necessary changes in social relationships that are often difficult to identify using empirical data from archeological sites. They also lack a unifying explanation for a phenomenon that arose in several regions of the world at approximately the same time, closely followed major, global environmental shifts, and hence, we believe, was begun by people largely for the same (ecological and evolutionary) reasons and not because of historical accidents in widely dispersed and different social systems.

We will undoubtedly face objections from our colleagues in the anthropological community who do not want to see human culture "reduced" to broad, scientific "laws" of explanation. It is worth remembering that when faced with similar worries from naturalists who were busy recording the astounding array of life forms from the earth's different zones, the wise ecologist Robert MacArthur (1972, p. 1) urged them not to "take refuge in nature's complexity as a justification to oppose any search in patterns." Also, he added that the application of the scientific method to determine and explain the shape of biogeographical distributions did not "have any power to take away nature's beauty" (p. 1).

Similarly, human cultural complexity should not prevent a search for pattern in the evolution of food production and other signal human developments or lead to denial of a pattern should it become evident upon empirical study using the scientific method. A finding that uniformitarian processes associated with environmental and biological factors shaped subsistence decisions and accounted for major changes in human behavior would not make the actors in question any less a part of a unique social sphere.

For the Neotropical cases of agricultural development that we discuss, social correlates integral to theories of social causation for food production occur after the appearance of domesticated plants (e.g., Cooke and Ranere, 1992b; Pearsall, 1995a; Pohl et al., 1996). Emerging empirical data indicate that the social units we identify in Chapter 4 as practicing early food production were small and simple and do not meet the requirements of the social models, which have status-conscious big men accumulating agricultural surpluses and labor-intensive foods enhancing the power of individuals or groups (Hayden, 1995; Price, 1995). The level of settlement organization among early Neotropical food producers appears to have been something like the modern, tropical hamlet and hamlet cluster, in which no more than one to a few nuclear families shared a residential community and any "commitment" to the community was nonbinding. The nuclear family was probably the main unit of production and consumption. In the absence of superordinate political structures and social commitments beyond the household, food procurement decisions were free of social constraints and even more likely to follow opportunistic and economically rational strategies.

Evolutionary Theory and the Origins of Food Production

There are increasingly common applications—or calls for applications—of evolutionary (Darwinian) theory in archeology (e.g., Dunnell, 1989; Ladefoged, 1995; O'Brien and Holland, 1990; Winterhalder and Smith, 1992; Teltser, 1995). Bettinger (1991, p. 151) explains as follows: "the term *Darwinian* applies to theories that explain macrolevel phenomena as the cumulative consequence of explicitly defined processes (e.g., selection and others) acting on a microlevel, specifically on reproductive individuals." "Darwinian theories move from process to consequence" (p. 151), in contrast to other approaches to explaining human behavior (Marxism, neofunctionalism, and social causality) in which the study of process first emphasizes "macrolevel" phenonema, which then become subjects for generalization. However, such generalizations cannot be seen as explanations for the phenomena (Bettinger, 1991). We believe that the origins of food production and the subsequent development of agricultural systems are fundamentally evolutionary problems and thus use Darwinian theory in framing a hypothesis for Neotropical food production and deriving testable expectations from it.

The development of food production as a form of coevolution, an approach first articulated by Rindos (1984), has been postulated by several investigators using Darwinian theory to examine the process (Gremillion, 1989; Smith, 1992). We find merit in the coevolutionary thinking of Rindos and others, especially as it concerns the importance of the creation and maintenance of an anthropogenic landscape near settlements before people actually engaged in systematic cultivation and with regard to its role in creating opportunities for predomestication semisedentary living in various environments. As we will see, these factors may have been especially important in the tropical forest. However, we do not believe that a protracted period of mutualistic interactions between people and the plants taken under cultivation preceded tropical food production.

A major point of divergence between our model and coevolutionary models is that the latter focus on the "how" of domestication because, if cultivation behaviors and domestication are the inevitable outcome of certain kinds of human–plant interactions, it makes little sense to ask "why" people started to cultivate plants. We disagree with Rindos (1984, p. 141) that to ask why humans began close associations with certain plants is a question "without real meaning." Rather, the why question is at the forefront of a type of evolutionary theory increasingly being applied to human behavior called evolutionary or behavioral ecology (Smith and Winterhalder, 1992).

Behavioral ecology seems to us to be the most appropriate way to explain the transition from human foraging to food production. Developed by biologists during the past several decades, this framework comprises the ways in which behavior contributes to survival and reproduction of organisms in relation to the ecology of the organisms (Hawkes *et al.*, 1997; Krebs and Davies, 1993). In biology, it has provided fresh insights into a spectrum of complex issues, including the origins of group living, primate social organization, and parental care and mating systems (Krebs and Davies, 1993).

Models drawn from behavioral ecology differ from coevolutionary and other types of evolutionary theory used in anthropology in a number of important ways. Behavioral ecology emphasizes decision making by animals capable of flexible and learned behavior who have the capacity to adjust quickly via the phenotype to varying ecological circumstances (this capacity having evolved because it offerred a competitive advantage). The outcomes of these decisions are screened through the filter of reproductive fitness and, thus, are differentially replicated in subsequent generations (Smith and Winterhalder, 1992). Behavioral ecological models attempt to identify the underlying processes, the selective pressures, that must have been operating to favor the establishment of food producing and other important behavioral changes (Russell, 1988).

Distinct from sociobiological explanations, behavioral ecology does not posit that particular behaviors emanating from the phenotype are coded in particular genes, but emphasizes that their expression is multicausal and heavily dependent on the environment. In its broadest sense, the environment can be defined "as everything external to an organism that impinges upon its probability of survival and reproduction" (Winterhalder and Smith, 1992, p. 8). Importantly, with the use of behavioral ecology the "intentionality" question, which has long been a thorn in the side of attempts to apply evolutionary theory to humans and has engendered much confusion (Dunnell, 1980; Flannery, 1986a; Rindos, 1984; Sahlins, 1972; Watson, 1995), becomes much less problematic because it is allowed that human behavior has a strong component of motivation (if no less subject to the action of natural selection than other behavioral attributes discussed). Again, the cognitive abilities underlying conscious decision making by individuals are assumed to have evolved largely under natural selection (Smith and Winterhalder, 1992).

Obviously, no proponent of evolutionary theory would allow that intentionality can include a long-term goal in mind for any action; i.e., that the earliest teosinte cultivators foresaw the production of a monstrous ear with many rows of seeds. In this longer term sense, evolution has indeed had little to do with "intentionality."

One of the theories developed within the framework of evolutionary ecology that we most heavily rely on (optimal foraging theory) is derived from formal economics, specifically microeconomics. Humans are seen to be rational actors in environments in which resources are limited and needs must be continually met. Human actors have the ability to "assess payoffs and choose or learn the best alternative under any given set of circumstances" (Smith and Winterhalder, 1992, p. 33). They will try different foraging strategies and repeat and copy those that are most successful (Hawkes and O'Connell, 1992).

As discussed previously, foraging theory predicts actual subsistence choices by actors, and the choices are seen to be based largely on the relative return rates of exploitable resources in the environment. The "currency" used to measure the rate of subsistence returns per unit time, usually energy (calories), is assumed to be highly correlated with the fitness of the actor and that, all things being equal, natural selection should have favored more efficient strategies at the expense of less efficient strategies.

This seems to be a fair assumption because in many circumstances more food, less exposure to risks through shorter foraging time, and more time spent in activities other than the food quest (such as caring for children) should have been associated with increased fitness (Kaplan and Hill, 1992). Conversely, factors other than actual food shortages are likely to select for efficient foraging strategies, and foragers do not have to absolutely maximize the total amount collected to benefit from efficient food procurement practices (Smith, 1983).

Formal testing of these and other predictions of optimal foraging theory with modern hunters and gatherers and horticulturalists have supported the propositions that energy is a useful currency to use in foraging models and that energetic concerns are major constraints on foraging decisions (e.g., Alvard, 1993, 1995; Beckerman, 1993; Gragson, 1993; Hames and Vickers, 1982; Hawkes et al., 1982; Keegan, 1986; O'Connell and Hawkes, 1981, 1984; Hill et al., 1987; Kaplan and Hill, 1992; Smith, 1981).

The particular approach in foraging theory that we rely on most heavily is the "diet breadth model" (Hawkes and O"Connell, 1992; Hill et al., 1987; Kaplan and Hill, 1992; Winterhalder, 1981), which, in addition to being a "paragon of robustness" (Winterhalder, 1986, p. 372), makes a number of valuable, and sometimes counterintuitive, predictions concerning food choice and subsistence change that are highly relevant to food production origins. They can be summarized as follows: (i) resources will enter the diet as a function not of their own abundance but of the abundance of higher ranked (least costly) resources; (ii) as the abundance of higher ranked resources on the landscape declines, foragers begin to do better by investing less search time in them and more time in handling lower ranked resources; (iii) foragers will now choose a broader diet because it results in a higher return rate than could be achieved by more searching for food; (iv) the reduction of search time will permit greater investments in storage and food processing, adding to the nutritional quality of what is eaten and extending the use life of food items; (v) broader diets and decreased search time will also lead to smaller foraging radii and, possibly, increases in residential stability; and (vi) changes in diet breadth may result in human demographic change, whose direction (increase, decrease, or no change at all) is dependent on the characteristics of the resources newly incorporated into diets (this prediction is from a more complex model developed by Winterhalder and Goland, 1993, and is discussed in more detail in Chapters 2 and 6).

It is obvious that the diet breadth model demonstrates close affinities to hypothesized and archeologically documented processes linked to the emergence of food production, including the "broad spectrum revolution" (Flannery, 1969). As several behavioral ecologists concerned with human behavior have indicated (O'Connell and Hawkes, 1984; Hawkes *et al.,* 1997; Hill *et al.,* 1987; Kaplan and Hill, 1992), the diet breadth model is particularly well-suited for studying major directional changes in human subsistence over time because of its ability to make robust, qualitative predictions of prey choice and dietary diversity, especially where paleoecological data are robust across the Pleistocene/Holocene boundary and can serve as proxies for changing resource distribution and foraging costs in the absence of detailed archeological records on subsistence.

In summary, we believe that the fundamental transition of the human lifeway considered in this volume was driven by changing selection pressures on hunter–gatherer resource procurement and, ultimately, their search for successful adaptations in changing environments. We agree with Winterhalder and Smith (1992, p. 4) that modern evolutionary theory has much to offer to the social sciences and that "any comprehensive explanation of human behavior requires evolutionary forces."

EARLY CONSIDERATIONS OF TROPICAL FOREST AGRICULTURE: CARL ORTWIN SAUER, DAVID HARRIS, AND DONALD LATHRAP

Many years ago, three prominent scholars, Carl Ortwin Sauer, David Harris, and Donald Lathrap, considered the question of how plant domestication arose by identifying some of the ecological and social circumstances that likely were correlated with early food-producing strategies in the tropics. Many of their ideas continue to enlighten tropical agricultural studies today. Among their most enduring contributions was that each viewed the tropical forest as being of immense complexity but also opportunity and not as the inhospitable and inhabitable "Green Hell," which was the dominant view of the tropical forest during their early scholarly years.

Carl Ortwin Sauer

Sauer's (1952) most often cited work, *Agricultural Origins and Dispersals,* deals with his contention that food production first arose among sedentary, affluent riverine people of the tropical zone who had the necessary absence of resource stress and hence leisure time to experiment with plants and invent agriculture. He felt that in the tropics the necessary biological and physical factors favoring food production, such as marked plant diversity (creating a large pool of genes to be experimented with), benign temperature, good soils, and adequate rainfall, were to be found.

Also, Sauer (1936, 1952) noted that in wooded lands cleared, fertile areas for planting could easily be achieved by ringing or deadening trees (felling was unnecessary), which then permitted sunlight to enter the forest floor where a litter rich in nutrients and mulch for the crop was waiting. Only an implement to break the soft cambium of the outer tree trunk and a digging stick were required to complete the process. On the other hand, grassy lands were unlikely to have been exploited by primitive cultivators because of the unyielding character of the stoloniferous grasses and underlying sod.

Most of all, Sauer (1952:20) felt that food production "did not originate from a growing or chronic shortage of food" and, thus, he probably would not have been favorably inclined toward the various population pressure arguments proposed during the past 25 years. "People living in the shadow of famine do not have the means or time to undertake the slow and leisurely experimental steps out of which a better and different food supply is to develop in a somewhat distant future" (Sauer, 1952, p. 21). Rather, sedentary folk blessed with abundant and secure plant carbohydrate and animal protein supplies, as are typically found at the edges of fresh water, would have been most likely to invent agriculture. Sauer also proposed that the earliest cultivators combined a number of multipurpose plants, which provided starch food, poisons, necessary equipment for fish nets and lines, and other necessities of daily life. Because rich wild resources were available to these people, he suggested that food production was perhaps not the most important reason that people grew the first plants.

Sauer believed that physical and biotic characteristics of Southeastern Asia best met all the requirements for early food production and that in this region could be found the peoples who practiced agriculture the earliest. Thus, the cultivation of root and tree crops indigenous to this area arose before the seed crop agricultural systems based on rice native to the subtropical climes to the north. Agriculture diffused from this single center in Southeast Asia to all over the world.

Sauer's hypothesis met with much criticism that still resonates today. His extreme diffusionist views have found few friends, particularly because increasing empirical evidence shows that food production emerged independently in several far-flung and temperate regions of the globe. Also, for many of his critics, the tropical forest was (is) a uniform, uncomfortable, ever-wet zone of bewildering diversity that must have no less discomfited and bewildered early humans. There was (and still is) a strong feeling that the cultivation of maize, beans, and other seed crops must somehow have occurred prior to the tropical forest root crop complex (e.g., Fritz, 1994; Mangelsdorf et al., 1964; Smith, 1995a,b) simply because the latter complex was indigenous to the uninspiring tropical forest.

Lost amid these biases and preconceptions has been the essential, crystal-clear logic of Sauer's ideas about plant husbandry in the tropical biome and his many wonderful insights into the early ecological relationships between humans and tropical forest plants (Sauer, 1936, 1947, 1958).

For example, Sauer (1936) argued that the origins of tropical forest food production were to be sought not in the ever-wet tropical rain forest but in the seasonally dry zone of semi-evergreen and deciduous woods, where the annual punctuation of rainfall spurs the production of seeds and tubers. Here, there is a well-defined season when maturity takes place en masse and bursts of harvestable resources occur during a short period. These are conditions that may lead to scheduling of resource acquisition and storage, which are important prerequisites for intensification of resource use and semisedentary living. As shown in Chapter 2, these "seasonal" types of forests once occupied large land areas throughout the Neotropics before they were cut for agriculture.

Sauer's comments on the importance of the seasonal cycle of plant production in the tropical forest long anticipated the recognition of the importance of these phenomena by botanists and ecologists who now closely study how they affect the nonhuman animals that feed on the flora (e.g., Leigh et al., 1982). Perhaps most important, the seasonal tropical forest would also offer less heavily leached and more highly fertile soils for agriculture in addition to being a generally more hospitable environment in which to live because it receives far less annual rainfall.

Sauer (1936) recognized that food production must be conceived as an integral part of the ecological conditions under which it first originated. He noted how the cultivated plant assemblage of the New World, including maize, was largely "mesophytic;" that is, growth began under warm temperatures and rain and continued through the rainiest parts of the year. Maturity then occurred under a marked dry season. Few of the primitive crop plants in the Americas, including maize, were tolerant of low, uncertain, and irregular rainfall or of cold temperatures; these adaptations would come later in time for some of them (Sauer, 1936).

The natural plant associations of many of the cultigens and, presumably, their areas of origin were to be found in the seasonal, especially deciduous, tropical forest habitat (Sauer, 1936). That the mesophytic character of the American plants is in marked contrast to the character of the western Asian plant complex is a point frequently lost on students of the problem, some of whom continue to believe that the first crop plants originated in, or quickly disseminated into, arid, highland zones.

Sauer (1947, 1958) shed much light on the character of the "tropical forest" in relation to its exploitation by humans. He noted that descriptions of the biome were subject to "distortion" and "oversimplication." For example, the first modern book dedicated to the subject of tropical forests (Richards, 1952) subsumed large areas of seasonally drier forest in America into the tropical rain forest category. Richards (1952) also believed that the forests in the very large area he defined as tropical rain forest contained very poor soils and had been little occupied and altered by prehistoric populations, especially by people cultivating plants.

In contrast, Sauer (1958) stressed the great variation in forest and soil types in the American tropics and noted how the more optimal environments for humans, including soils for agriculture, were found in Central and northern South America and, generally, outside of the vast interfluve region of Amazonia. Again, the seasonal forests, especially where the ancestors of the modern crop plants still flourished, were the areas that had seen the earliest human penetration and plant propagation. On a local scale, he proposed that the edges of water bodies (lakes, rivers, and streams) witnessed the earliest, permanent settlements of humans who entered low latitudes. In these disturbed, sunny habitats within the forest were to be found many nutritionally and technologically important plants as well as abundant and stable supplies of animal food.

Sauer (1947, 1958) also emphasized how humans using a simple and well-known technology, fire, could have fundamentally altered the seasonal, pristine forest and made it a more useful habitat by encouraging the reproduction of heliophytic (sun-loving) successional plants. These plants are most able to provide the fuel that human stomachs can digest because they invest less in lignified and chemically and/or mechanically defended tissues. He also noted that "The mastery of the forest by man requires no axe" (1958, p. 189) because simple ringing and fire could kill and fell trees.

It will be seen in Chapters 4 and 5 how many of Sauer's propositions concerning early settlement, resource use, and forest disturbance are being borne out by archeological and paleobotanical data. Many of the sterile debates concerning the suitability of the tropical forest habitat for human exploitation (Chapter 2) could have been avoided by careful attention to his views on the tropical biome.

David Harris

David Harris (1969, 1972, 1973, 1977a,b), like Sauer, offerred a realistic assessment of the tropical forest as home to prehistoric human populations by carefully distinguishing those zones of optimum potential for early hunters and gatherers. He, too, felt that the beginnings of cultivation were to be found in ecosystems of high diversity in which many wild plant species were available for experimentation. Harris (1972, p. 185) added that the conditions leading to plant cultivation and domestication existed "among those forager bands who established relatively permanent settlements along forest and woodland margins and who created open habitats within the territories they occupied." Here, collecting wild plants and harvesting animal protein would have been easier.

Also like Sauer, Harris viewed the "intermediate dry zone" or the seasonal tropical forest with dry seasons of intermediate length (3–7 months) as the home of the tuber crops because such plants are adapted to survive long dry seasons and then quickly mature upon onset of the rains. They do so by accumulating starch in their roots or stems during the growing period or rainy season.

Harris (1969, 1972, 1977a,b) believed that what he called "protocultivation" or fixed-plot horticulture close to dwellings was humankind's earliest system of food production in the tropics, and that stress, or kinds of "push factors," mainly those arising from human demographic changes, were instrumental in the emergence of food-producing societies. He placed considerable importance on four fundamental features of food production systems: their structure, function, degree of stability, and evolution through time. Harris made the distinction between "seed culture," systems based on the reproduction of seeds, such as maize and beans, and native to Mesoamerica, and "vegeculture," the propagation of mostly roots and tubers common at European contact in South America. He noted that each possessed quite different inherent stabilities and demographic implications.

Seed culture is much more demanding of nutrients in the soil, generally involves fewer plants, and may be expected to require more frequent shifts from one clearing to another. Vegeculture, on the other hand, more nearly duplicated the natural ecosystem with its typically large variety of cultivated plants in any plot, was less demanding of soil nutrients, and because it was a more ecologically stable system it was unlikely to spread as rapidly as seed culture. We will see in Chapters 4 and 5 that the development and spread of slash-and-burn agriculture is often accompanied by the presence of maize.

Harris also laid the basis for explaining the apparent "delay" in the development of integrated and specialized agricultural systems and full-fledged village life in the New World (Reed, 1977; Flannery, 1986a). Because tropical food production duplicated the high-diversity/low number of individual species example from the natural environment and first operated under conditions of low human population density and little social complexity, it was likely to have lasted for a considerable period of time before it evolved into swidden and other more intensive agricultural systems and spread.

Other factors also contributed to the slow evolution of agriculture in the New World, and we digress here to consider some of them. Some crops, including maize, underwent profound morphological change before they became productive food plants (e.g., Iltis, 1987). The complex genetics of maize, which along with teosinte have seen considerable study (e.g., Doebley, 1992, 1994a,b; Buckler and Holtsford, 1996), made the fruiting structure of the plant yield its wild morphology only grudgingly to the human hand. Because the effective population size of maize is very large, a protracted period of cultivation was required—at least an estimated 2000–3000 years—to produce cobs similar to those recovered from archeological sites in the Tehuacan Valley (Buckler et al., 1995). Doebley (1990, pp. 16, 24) characterized the maize domestication process as a "series of improbable mutations," adding that "the conversion of teosinte into maize is so improbable that it is difficult to imagine that it happened several times."

Genetic data also suggest that some major morphological traits may have been independently acquired in different geographic regions (Goloubinoff *et al.*, 1993). This does not require several independent domestications of maize but only that the "improbable series of mutations" responsible for the unique morphological characters of modern maize may have been completed by cultivators at different times and in different places on cobs that were in various stages of domestication. Subsequent pooling of traits eventually led to specialized maize forms from Mexico and South America well-known from later prehistoric times.

Other Neotropical cultigens whose development was prolonged may include lima and common beans. They contain components that range from poisonous to difficult to digest and may have been used as "snap beans" (as green vegetables before seed development) for a long period until selection and changes in cooking technologies allowed their use as a dried pulse (see Chapter 3).

The ecology of some important crop plants may have also contributed to the slow evolution of agriculture in the New World. For example, the wild ancestors of squash, beans, and the various tuber crops are not found at particularly high densities or in large clumps on the landscape and certainly not in densities remotely comparable to the wild cereal stands of the Near East. This, together with the fact that New World cultivators were dealing with plants that they tended, planted, and harvested one by one (Sauer, 1936), precluded mass harvesting, storing, and, thus, mass selection of genetic traits that may have contributed to slower genetic change under cultivation.

Donald Lathrap

For many years, Donald Lathrap was the primary spokesperson in archeology for the preeminence of the humid lowland tropical forest in New World cultural development, particularly the origins (or origin, as he saw it) of agriculture. He inspired a generation of students to pursue archeological and botanical research in an environment then largely seen as impervious to study and unimportant.

His approach to the problem of agricultural origins and dispersals was always creative, never dull, and frequently brilliant. A central feature of his view of this process was that agriculture emerged once. Lathrap's (1977a) unitary model drew inspiration from Sauer's views of the primacy of the lowland tropics for agricultural origins—although Lathrap chose tropical Africa, not Southeast Asia, as the hearth—and from Spinden's (1917) proposition that all New World civilizations rested on a single Neolithic foundation. Lathrap differed from Spinden, however, in choosing manioc, not maize, as the crop that gave the initial impetus to the emergence of food production. In Lathrap's view, the New World Neolithic was characterized by genetic modification in crops for increased yields, rescheduling of human activities around food production, and demographic upsets caused by increasing food supplies. The Neolithic began with the intensification of bitter manioc cultivation. He believed that this process was centered in the alluvial flood plains of northern South America and the Amazon, and that its roots were in the house garden.

Lathrap's conception of the house garden as experimental plot was another key feature of his views on the domestication process. Whether or not one accepts his notion that the arrival of bottle gourd from Africa initiated the process of domestication in the New World tropics (Lathrap 1977a), the idea that semisedentary foragers would move useful plants to cleared areas around their dwellings, thus placing the plants under new selective pressures and under direct human control, provided a mechanism for crop evolution. In Lathrap's words (1977, p. 719), "the artificial propagation of bottle gourd and certain other technologically significant crops such as cotton and fish poisons imposed particular disciplines on man and in the context of these behavioral patterns all of the other nutritionally significant agricultural systems arose." Lathrap believed that the emergence of a new behavior—bringing plants to a controlled space around the house—was fundamental to the domestication process.

Much of Lathrap's influence lies in his approach to the topic of agricultural origins. This approach had four essential tenets: (i) an economic pattern behaves like a radiating species, adapting to its environment and spreading throughout the range where its subsistence system can be practiced (Lathrap, 1976, originally written as a class paper in 1956); (ii) there is no contradiction between historical (age area) and ecological (evolutionary) approaches to understanding culture change: The state of a system at any point in time can be understood only in terms of its state at all prior points (Lathrap, 1984); (iii) agriculture is not an event or an invention because behavior is rooted in goals (e.g., the best manioc for beer or the best for flour) and this, on some level, leads to intentional behavior (Lathrap, 1984); and (iv) it is essential not to become fixated on the available data in hand (Lathrap, 1984).

As discussed previously, Lathrap felt that the assumption that manioc and other tropical roots had their own hearth, and that maize/beans/squash was another system widely separated from the former and a prime mover in the development of New World civilization, was incorrect. Inspired by the diffusionism of Sauer and Spinden, Lathrap envisioned a uniform Neolithic cultural base from Mesoamerica to Peru and a uniform agricultural substrate, tropical forest agriculture, that only diverged at ca. 3000 B.P. in Mesoamerica and the Andes (Lathrap, 1970, 1973a, 1974, 1977a,b, 1984, 1987).

The nature and area of origin of this uniform cultural base also came from the influence of Sauer: lowland fisher folk, in this case along the Amazon and Orinoco [the location being derived from historical (age area) data, i.e., the distributions of related lowland languages]. The model for the spread of these people and the system they developed was "adaptive radiation," a concept borrowed from ecological and evolutionary theory. Lathrap's laboratory for the emergence of domesticated plants—the house garden—also stemmed from the work of Sauer (1952) and

Anderson (1952) and Lathrap's own observations of tropical forest peoples (the Shipibo).

An important point to be made is that the bringing together of historical and ecological approaches underlies Lathrap's use of linguistic data to model population movements in Amazonia. The key to his model is an outward movement of peoples along floodplains from the central Amazon (Lathrap, 1970) and northern Colombia (Lathrap, 1977a) in search of new agricultural lands. The most general description of this model, economic systems as radiating organisms (Lathrap, 1976), makes the point that if we view society as an organism adapting to its environment, then any change in environment or subsistence influences population densities. An expanding society will spread throughout the range in which its subsistence system can be practiced. A society that reaches nearly the maximum adaptation in a highly limited geographic range can be called specialized. In areas where existing subsistence is not well adapted, a broader range of options will be used, and such societies can be called generalized.

Lathrap (1976) felt that agriculture was more likely to emerge in a generalized society, because experimentation would be common. When agriculture developed to the point that increased food and increased population resulted, then it spread into zones in which it was better adapted than previous subsistence systems. Spread would slow or stop when societies were encountered that were better or as well adapted, until a change in efficiency of the agricultural system changed the balance (increased yield of crops, loss of day length limitations, and so on).

Similar ideas were also developed by Rindos (1984), who marshalled theoretical data to discuss a similar tendency for agricultural systems to expand. In Rindos' view, agriculture is, on the one hand, expansive, and, on the other hand, inherently unstable. Populations will also expand in search of new lands after crop failures.

Another key concept from Lathrap's general model concerns substitution of crops. As agriculture spreads into zones in which it is less adaptive, new plants may be brought into the system that allow its extension into new zones. For example, Lathrap considered potato and quinoa as being substitutes for manioc and maize at high elevations. On this score, he was influenced by Sauer's arguments for the priority of lowland vegeculture from the beginning.

The final contribution noted at the beginning of this discussion, the importance of not becoming fixated on the available data, may strike some as frivolous and others as a weakness in Lathrap's approach. To us, this aspect of Lathrapian thinking is the most appealing. If one is party to the idea that data relevant to the origins of agriculture should be found in the lowland tropics, where preservation of botanical remains is very poor, rather than in the arid uplands, where it is excellent, then one must look for relevant data in the lowlands (and not become fixated on data from dry caves). Those who have looked have found, as we will discuss in Chapters 4 and 5.

There is another very important implication, however. If one must look where preservation is very poor, then methodologies must be developed to find the data one needs. This point was brought home to Pearsall during her dissertation research at the Real Alto site in southwestern Ecuador. It is no accident that a way to identify maize using inorganic residues (phytoliths) was the result. Lathrap had read a little-known manuscript (Matsutani, 1972) about the attempt to identify maize using silica skeletons (articulated phytoliths) at the Kotosh site in Peru, and he suggested trying it in the humid lowlands. He is the "godfather" of phytolith analysis in the Neotropics.

THOUGHTS ON THE DEVELOPMENT OF TROPICAL FOOD PRODUCTION

Although Sauer, Harris, and Lathrap have had a great deal of influence on our thoughts and discussions of New World agriculture, our views also differ from theirs in some important respects, particularly in relation to the earliest parts of the process. For example, we doubt that the interior of the Amazon Basin, including the middle to lower stretches of the Amazon River itself—envisioned by Lathrap as the primary cradle of agriculture—had much to do with the early cultivation and domestication of plants. We believe this for two reasons. First, the distributions of many crop plant progenitors appear to have been outside or on the margins of the basin (Chapter 3).

Second, major river valleys, with their endless supply of deep alluvium subject to dramatically fluctuating water levels and lengthy floods were probably not favored by the low-density and socially simple earliest food producers, who were interested mainly in feeding their families. Rather, the early and middle Holocene cultures of the lower Amazon River documented by Roosevelt et al. (1991, 1996) may stand as one of the few examples of "complex hunting and gathering" that was permitted by the tropical forest habitat. On the other hand, much of the terra firme Amazonian forest had such poor soils and poor resource availability as to have generally made it very marginal both for hunter–gatherers and early food producers (Chapter 2).

We believe that the ecological settings of early plant cultivation were more anxiety producing than Sauer envisioned. Sauer believed that the tropics had remained essentially stable during the Ice Ages and that the post-Pleistocene occupants were under no imperative to alter or improve their food supply. They had the affluence to experiment casually with and to manipulate plants, and they could afford to fail. Although we believe that the tropical forest is far from the implacably hostile place for human beings that some have suggested (e.g., Bailey et al., 1989; Headland and Bailey, 1991), we propose that the earliest habitats of food production in the tropics were costly to exploit using only wild plants and animals, resource (especially carbohydrate) poor, and unpredictable compared with the ecological circumstances that had immediately preceded them.

Thus, we differ from Lathrap in believing that the earliest forms of food production had as their primary goal the production of food, although a diversity of utilitarian plants, including bottle gourd and cotton, were, with little doubt, also taken under cultivation. The dynamic ecological circumstances to which we refer were those that accompanied the last retreat of the glaciers at higher latitudes—the end of the Pleistocene and the advent of the modern climate.

These were profound ecological perturbations, probably of greater magnitude and impact than anything seen in the 120,000 years that came before and certainly unlike anything that occurred later in the Holocene (Chapter 2). In this book, we repeatedly stress that in order to understand Neotropical food production, one must grasp how different the tropical world during the Late Pleistocene was from the one we know today. Many currently well-watered areas now under tropical forest held habitats that were much drier, cooler, and more open. The drier types of tropical forest were replaced by thorny scrub, grassy, and other kinds of open-land vegetation where herds of big game roamed and fed and where there was an abundance of edible plants for humans. These resources provided a much higher return per unit of effort of foraging than tropical forest plants and animals and would have been heavily favored for exploitation by Late Pleistocene hunters and gatherers.

Beginning 10,500–10,000 B.P., when the Pleistocene ended, the forests claimed the open lands and assumed their modern distributions. Consequently, the abundance of "higher ranked" (in optimal foraging schemes) animal and plant resources was decreased, whereas full-time living in a tropical forest and exploitation of its resources was demanded of many human populations. However, these resources offerred much lower return rates from foraging than had the big, extinct game and thorny scrub plants, making the production of some tropical forest plants a strategy whose energetic efficiency was more acceptable and, ultimately, beneficial to human populations than their collection. Thus, unlike Harris, we favor factors that "pulled" rather than "pushed" people into food production.

In short, the most important effect of the post-Pleistocene changes on the selective pressures favoring food production may have been to lower the overall foraging return rate dramatically in some areas beginning about 10,000 years ago, leading to the development of alternative means of food acquisition in order to maintain the previous "standard of living." It is important to point out that we do not view people as anywhere near starvation at this juncture. Subsistence needs almost certainly could have been met as they were met elsewhere in the world during previous, similar times before 100,000 years ago—by adjustments in mobility, settlement densities, and wild resource procurement until some kind of demographic balance was achieved and populations became fairly stablized on the landscape again, if initially at lower levels.

What made this period different everywhere in the world is that for the first time people had the cognitive capacity to respond in a fully economically rational manner to dramatic perturbations in the environment. This is our answer to the query of why humans never responded with systematic food production during the multiple glacial/interglacial perturbations that had previously occurred during the Pleistocene, an oft-raised objection to associating post-Pleistocene environmental changes with the emergence of food production (e.g., Hayden, 1992). We argue that this was probably the first time that a fully modern human species capable of modern cognition and behavior was confronted with such a perturbation (Reed, 1977; Klein, 1992, 1995).

There is an increasing unwillingness on the part of archeologists and paleoanthropologists to accept that the behavior of "hunters and gatherers" prior to the emergence of anatomically modern humans about 120,000 years ago was qualitatively the same as that of people practicing foraging on the eve of food production. Marked differences in subsistence practices and social organization may instead have been present (e.g., Foley, 1988; Mellars, 1996; Mithen, 1996). It has been argued persuasively that the behavioral adaptations leading to the expression of fully modern human culture did not occur until a final reorganization of the human brain took place between 50,000 and 40,000 years ago (Klein, 1992, 1995). Only after this time do innovations such as net sinkers, fish hooks, bows and arrows, basketry, microliths, sleds and canoes, and plant-grinding implements regularly and widely occur in the archeological record [recent evidence suggests that some of these phenomena may have first appeared in Africa ca. 120,000 years ago (Brooks *et al.*, 1995) but still in association with morphologically modern humans].

In this view, the human line has not been practicing hunting and gathering for more than 99% of its time during the past two and one-half million years. Rather, hunting and gathering as we usually conceptualize it—with large home ranges, central-place foraging, an emphasis on herds of big game utilizing communal hunts with complex technology, and an ability to exploit aquatic and plant resources with increasing energetic efficiency—possibly occupied only approximately 5% of the hominid time span before the emergence of food production (Foley, 1988). Why food production emerged shortly after 10,000 years ago becomes much easier to understand as it becomes increasingly clear that the necessary cultural and ecological prerequisites first converged at this time.

Also of relevance to the timing of plant cultivation and domestication worldwide is the fact that this last retreat of the glaciers may have represented the most extreme perturbation of climate, vegetation, and, hence, resource adjustments in several parts of the world, including the Neotropical lowlands, that had been seen during the past 120,000 years (Lister and Sher, 1995). The few pollen and phytolith spectra that cover most or all of the last entire transglacial period in the Neotropics plus the substantial evidence from deep sea and ice cores suggest that the magnitude of the change associated with the Pleistocene–Holocene transition was much greater than that during the several stadial–interstadial perturbations that had occurred since approximately 120,000 years ago (Bush and Colinvaux, 1990; Colinvaux *et al.,* 1996a; DeOliveira, 1992; Haberle, 1997; Piperno, 1997b; Shackleton,

1987). The fact that extinctions of large game did not occur during earlier interstadials supports this model (Lister and Sher, 1995).

When, where, and how many times did the propagation of plants begin in the New World? In Chapter 4, we discuss botanical evidence from four regions where it appears that systems of plant cultivation that had already developed or that incorporated such domesticates as squash (*Cucurbita* spp.) and *leren* (*Calathea allouia*) were established by 9000–8000 years ago: coastal Ecuador, northwestern Peru, the Colombian Amazon, and central Pacific Panama. Two of these regions (Peru and Panama) occupy the lowland, deciduous forest habitat, and a third (Ecuador) is in a thorn-scrub/deciduous forest ecotone. The fourth is in an ever-wet forest. The Ecuadorian, Peruvian, and Panamanian areas also share in having been in close juxtaposition to several different environmental types and resource zones. The Balsas Valley of southwestern Mexico was also a center of early Holocene food production based on molecular evidence and the appearance of maize much further south by 7000 years ago.

The presence of horticultural societies in several rather geographically separated regions of the New World during the early Holocene obviously does not mean that food production developed independently in all of them. Separating a truly independent origin from rapid spread of people or a plant or an idea is a difficult task, particularly when distributions of some major crop ancestors have not been delineated and archeological data are not abundant from some important regions, and even totally lacking from others. Unlike Sauer and Lathrap, we believe that domestication began independently in one or more areas of the New World, not as a result of diffusion from the Old. We also do not accept Lathrap's notion that new crops were necessarily brought under domestication as a result of the substitution of species as agricultural peoples spread into new environments. While this model may appear to fit some crop distribution data (see the discussions of chile peppers and squashes, for example, in Chapter 3), independent domestication over a widespread geographic area fits equally well; we cannot currently distinguish between these explanations for some crops. However, given that we have been in a state of no knowledge for so long and that we have tremendously increased the quality and quantity of our botanical database for the early Holocene period in a number of regions, we find this an exhilerating challenge.

By combining the current archeological, botanical, and molecular evidence, we are inclined to see independent origins of plant cultivation and domestication in the low-lying regions of at least three areas: southwestern Ecuador/northern Peru, northern South America (Colombia/Venezuela/the Guianas/northern Brazil), and southwestern Mexico. Central Pacific Panama is a strong candidate for an independent origin of "nondomestication cultivation" (Hillman and Davies, 1992). Molecular work in progress may reveal whether it was home to the wild ancestor of the major domesticated lowland squash, *Cucurbita moschata*. The facts that agriculture involved so many different species in the New World, often two or more species in the same genus from different hemispheres were domesticated, and that the environmental impact of the termination of the Pleistocene seemed to have been felt throughout the low and mid-elevational areas of Central and South America at about the same time, also lead us to believe that there were several independent origins of plant cultivation and domestication.

In the Old World, the origins of plant cultivation and the origins of major domesticated species are seen to be largely congruent in time and in space because the earliest crop plants are relatively few in number, became the staples, seem to have been domesticated rapidly, and probably derived from a single domestication event (domesticated strains were developed in a very localized area) (Hillman and Davies, 1990, 1992; Zohary, 1989). Note that these factors do not exclude the possibility that cereal cultivation, perhaps practiced on a smaller scale, was independently developed throughout a broader area of the Near East during the ca. 12,000–10,000 B.P. period, when major environmental changes were occurring, and that only a few populations in the southern Jordan Valley eventually combined the requisite harvesting and planting methods that selected for superior domesticated strains (Unger-Hamilton, 1989; Hillman and Davies, 1992; Hillman, 1996).

In the New World, nondomestication cultivation may have synchronously started in various places, especially because the influential environmental processes impacted large areas of Central and South America at virtually the same time. The rates of domestication in many New World crop plants are currently unknown. Current molecular and botanical evidence indicates that the initial development of domesticated forms of some crops, such as maize, manioc, squash, beans, and cotton, probably took place in localized areas, but that hybridization with related species and/or substantial experimentation leading to morphological change occurred over a protracted period in some of these and other crops subsequent to their initial dispersals (Chapters 3–5).

These processes are related to Harlan's (1971) concept that South America was a "domestication noncenter," whereby peoples over a wide geographic area were engaging in early cultivation and domesticatory relationships with plants and came to influence the early development of some domesticates after these plants left their domestication cradles. In tropical America, we may eventually have to draw a broader distinction between the origins of plant cultivation, which we view as the most crucial issue, and the appearance of certain plant domesticates, at least in the forms that are most familiar to us today.

STUDYING THE PROBLEM OF THE ORIGINS OF TROPICAL FOOD PRODUCTION

The only rules of scientific method are honest observations and accurate logic. To be great science it must also be guided by a judgment, almost an instinct, for what is worth studying.

Robert H. MacArthur (1972)

What Kinds of Data Should We Study?

The question of what is worth studying is currently an important one in the problem of food production origins. Some investigators are placing disproportionate reliance on macrobotanical remains (e.g., Fritz, 1994; Smith, 1995a,b), an approach we strongly disfavor because, by not taking advantage of new developments in the field and interdisciplinary skills, it provides a very limited and incomplete perspective on prehistoric plant use. Others are seeking to improve and diversify paleoethnobotanical techniques for gathering evidence from archeological sites.

It is becoming clear that important data may often be obtained from sites where people never lived, such as lakes and swamps. Sediments from these "off-site" contexts contain identifiable microscopic prints of human impact on the vegetation that speak of former land clearance and the presence of agricultural plots. Often, they contain the pollen grains and phytoliths of crop plants themselves. In this book, we use a multifaceted approach to the evolution of food production because, simply, multiple lines of evidence provide more data and robust insights into the question, especially in areas such as the tropics where species diversity is high, subsistence alternatives are many, and plant remains may be preferentially destroyed by climatic conditions.

The evidence we cite for the evolution of tropical food production is primarily of four types: (i) botanical remains from archeological sites, both macrofossil (seeds, tubers, wood, corn, and cob fragments) and microfossil (phytoliths, pollen, and starch grains); (ii) the vegetational (primarily pollen and phytolith) records obtained from perennially wet areas, mainly lakes and large swamps, near sites where people lived; (iii) the settlement characteristics and lithic inventories of these sites; and (iv) the molecular prints of living crop plants and their wild ancestors. The latter reveal in very detailed terms the relationships between domesticated and wild plants and often provide strong hints as to where a crop plant was originally taken under cultivation.

From time to time, we also consider other forms of evidence, such as the geographic distributions of crop plant wild ancestors on the eve of food production, as suggested by molecular and paleoecological data, and the timing and nature of major environmental changes brought about by climatic perturbations as revealed from the sequence of vegetational changes in lake and swamp sediments. This latter evidence may also offer a guide to the most likely geographic areas of crop domestication and define the kinds of environments that the last foragers and incipient cultivators occupied. We argue that the records of vegetational change in the Neotropics at the close of the Pleistocene may even serve as proxies for resource density and distribution through time and permit estimates of relative foraging return rates during the late Pleistocene and early Holocene periods.

What the Different Classes of Botanical Remains Can Tell Us

Because we have reviewed techniques for recovering botanical data that we have found effective for the lowland tropics elsewhere (Pearsall, 1995b; Piperno, 1995a), we will not consider them in detail here. Rather, our goal is to discuss how differences in the character of the different classes of plant remains discussed in Chapters 4 and 5 (as they relate mainly to deposition and preservation) affect how we use them and their value as indicators of early agriculture. It will be seen that what may appear to be contradictions among indicators for the antiquity of agriculture in the Neotropics are, in large part, simply reflections of the nature and relative strength of those indicators.

Desiccated macroremains discussed here that are relevant to the origins and early development of agriculture in the lowland tropics come from sites along the desert coast of western South America and from dry caves in central Mexico. As has been recognized for quite some time, these sites are outside the areas of origin of many crops they contain. These data are important, however, for establishing minimum ages for crops not well documented in their areas of origin (e.g., manioc, lima bean, and peanut) in sites on the western coast, for documenting the types of combinations of cultivated and wild plants used at different points in prehistory (e.g., suites of wild seeds, domesticated tubers, and semidomesticated tree fruits from the western coast), and, in some cases, for documenting changes within crops from human selection (e.g., changes in cob morphology from materials at Coxcotlan Cave, Mexico). The main weaknesses of these data are recovery bias and dating ambiguities.

Recovery bias refers to the fact that the excellent preservation of organic materials in dry caves and desert settings led to much archeological research being carried out before the importance of recovering all size classes of biological materials was realized (and in some cases before archeologists were interested in subsistence issues). Many sites along the Peruvian coast, for example, were excavated without screening matrix or with use of only wide aperture screens ($\frac{1}{2}$ in. or $\frac{1}{4}$ in.)

Caches of smaller remains were recovered *in situ* in some cases, however, and desiccated human fecal material (coprolites) was sometimes saved (gut contents of mummies from sites in Peru are also available for study). The use of fine sieves to recover all size classes of materials from a subset of matrix (the same principle as flotation in nondesiccated sites) is becoming more common. Luckily, several important early Holocene sites, Paloma on the central Peruvian coast and sites in the Zaña valley on the northern coast (discussed in Chapters 4 and 5), were excavated using modern botanical recovery techniques, as were a number of sites from the late preceramic and early ceramic periods. To examine the transition from the appearance of domesticated plants to dependence on agriculture on the coast of Peru (Chapter 5), we selected data that were less impacted by recovery bias, and we focused on how common domesticated plants were for each time period (simple presence, by site).

Examining the relative importance of wild and domesticated resources using ratios or percentage presence within sites or time periods, approaches commonly used in other regions (e.g., ratios of starchy seeds:nuts and corn:starchy seeds and percentage presence of corn used in the midwestern United States), is difficult for coastal Peruvian data because many wild foods leave inconspicuous remains and are therefore underrepresented at many sites.

The Tehuacán sequence is an example of the second major problem with desiccated materials: uncharred food remains are potential foods for burrowing or commensal animals and are thus subject to repositioning in deposits. AMS dating represents a significant advance for our understanding of the age of these and other plant materials, but direct dating of desiccated material is not without its own problems (Rossen *et al.,* 1996). As with conventional dates, what is dated must be carefully considered beforehand, and the results should then be evaluated with respect to other cultural materials. This is as true for other plant remains that yield suitable substrates for dating (carbonized macrofossils and phytoliths) as it is for desiccated specimens.

Preservation of botanical macroremains at the vast majority of archeological sites in the Neotropics is through accidental charring. This mode of preservation characterizes all the macroremain data from Brazil, Ecuador, Colombia, Venezuela, and Central America discussed in Chapters 4 and 5. Although desiccation does not produce a perfect record of past foodways (i.e., some foods are eaten away from the site or completely consumed, leaving no seeds or husks), preservation by charring introduces a new suite of biases: Only foods that come into contact with fire are preserved. Only "tougher" charred remains, such as palm kernels and other hard fruit fragments, may survive burial and recovery, and only material that survives with distinctive features intact can be identified. Tuberous parts of plants are notorious for their failure to enter the record of carbonized materials.

If site soils are high in dense clays, as is usual in the humid tropics, shrinking and swelling may break up materials, leaving fewer to recover and further confounding identification. Deeply buried materials at multiple component sites can also be broken up by soil compaction and pressure. Figure 1.1 illustrates how abundance of charred material can drop off with depth and age in sites in alluvial soils with high clay content. This example, from a site in the Jama River valley, Ecuador, represents approximately 3000 years of deposition (E. Engwall, personal communication). Although some of the decline in charcoal abundance correlates with a change in intensity of site occupation, density of charred material begins to decline (40– 60 cm level) while artifact densities are still high (through the 120-cm level).

The pattern is thus produced by a combination of depositional and taphonomic factors. The interpretive problem arises when one tries to compare the presence and abundance of plant species among assemblages with very different combinations of these two factors. For example, in Fig. 1.1, where declining abundance of charred material and decreasing species richness are both correlated with increasing depth, is the presence of charred maize in the upper levels, but not in the lower levels, interpretable? Are relative abundances of foods in the lower levels comparable to ratios from the upper levels?

To avoid the interpretive dilemma outlined previously, we take the conservative stance of not attempting to quantify charred macroremain data from early assem-

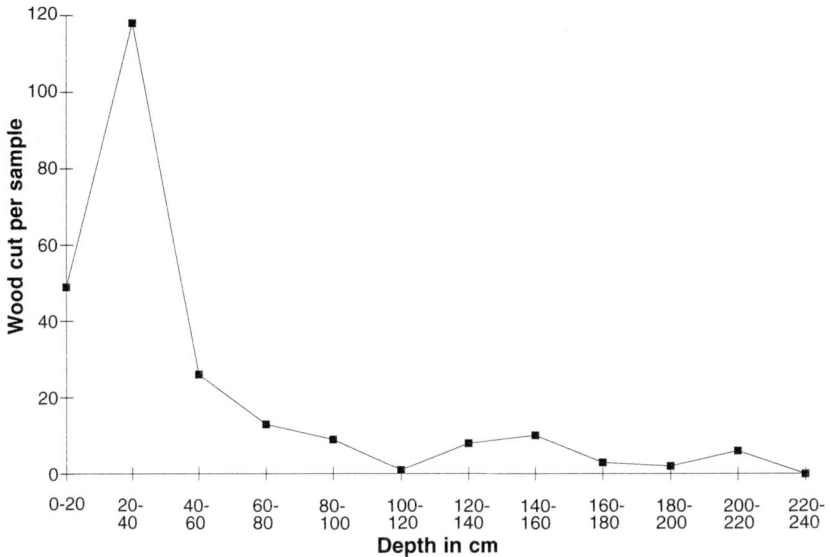

FIGURE 1.1   Declining wood preservation with depth at a site from the humid, lowland tropics.

blages. We discuss these data using simple presence and do not attempt to interpret absence. In many cases, foods that are absent in the macroremain record are present in the phytolith record. This is the case for the site illustrated in Fig. 1.1. Maize phytoliths occur in both the upper and the lower levels (the lower part of the graph is the Chorrera period, when charred maize is present at other sites in the valley). While one of us (DMP) has long preached the importance of carrying out systematic flotation at lowland sites (and we would know much less about the shift to dependence on agriculture without flotation-derived macroremain data), the plain truth is that sometimes the charred remains are just not there, no matter how much soil is floated.

A number of archeological sites relevant to the origins and spread of agriculture in the Neotropics have been sampled for microfossil remains of plants. These data include pollen extracted from sediments and desiccated coprolites, phytoliths from site sediments (coprolites can also contain phytoliths, but no studies of this type are available for the region), and starch grains. Although we use several coprolite studies in examining changes in plant use along the Peruvian coast, the majority of site microfossil data we discuss are phytoliths from sites in Ecuador, Panama, Colombia, Belize, and Mexico (pollen data are generally more important in off-site cores, as discussed below, but are also available from several early occupation sites).

Starch grain analysis is not a new technique in Neotropical archeology (e.g., Rossen *et al.*, 1996; Ugent *et al.*, 1984, 1986), but as far as we are aware, we present the first data from the humid tropics in the form of grains recovered from the edges of plant grinding stones and the surfaces of grinding bases. Because starch grain analysis has been less commonly used in the humid tropical areas of the New World, we offer a brief assessment of the technique here. For a complete discussion, see Piperno and Holst, 1997.

Starch grains are microscopic granules that serve as the principal food (energy) storage mechanism of plants. They are found mainly in rhizomes, tubers, and seeds (Loy, 1994). The fact that starch grains of different plants possess a large variety of forms has been recognized for some time (Reichert, 1913). The morphology of the grains can be diagnostic to individual genera and even species of plants. Particularly, when aggregates of starch grains can be isolated from sampled contexts, species-specific identification may be possible. When present on tools, they provide direct evidence for the species of plant that was processed. Thus, they have the potential to elucidate a major lacuna in our understanding of tropical plant domestication—the origin and history of the tuber crop complex of the humid lowlands.

Although the potential exists to recover starch grains from permanently inundated and anoxic environments such as lake sediments, we confine our discussions to the results obtained from the analysis of grinding stones and grinding bases from archeological sites. Apparently, the grains survived on these tools because they became embedded in tiny surficial cracks and crevices, where they were protected from the effects of the humid climate through time. For starch grain identification, the modern reference collection of tropical economic species housed in Piperno's laboratory was used (Piperno and Holst, 1996a; Piperno and Holst, 1997). Piperno and Holst also relied on the large monograph published by Reichert (1913), which provides descriptions and photographs of more than 300 species and varieties of important economic plants from around the world, including many New World tropical specimens.

Phytoliths become deposited in archeological deposits from the *in situ* decay (or burning) of plants used and discarded and from the deposition of soil. Because people use plants for much more than food, typical soil samples contain a diversity of phytoliths: from the decay of roofing thatch, from the ashes of wood or grasses burned as fuel, and so on. A major strength of phytolith analysis, of course, is the resistance of these microfossils to dissolution and decay over long periods of time. Because phytolith preservation is generally held constant through time, we examine changing presences and frequencies of phytoliths in various sites and at various times to decipher changes in human plant use.

Identification criteria for many of the lowland crops that can be identified by phytoliths (maize, squash, achira, arrowroot, *Calathea,* and palm) have been published (Bozarth, 1987; Pearsall, 1989b; Piperno, 1985a, 1988a, 1989a) and will not be discussed here. However, we do employ the results of a new method for using size of phytoliths produced in the rinds of *Cucurbita* to evaluate whether plants leaving the remains were wild or domesticated species (Piperno and Holst, 1996b) (see discussion of the Vegas site in Chapter 4). We also briefly summarize identification criteria for some crop plants, such as *Calathea allouia* and bottle gourd, that are not yet in print.

The antiquity of crops identified by phytoliths for the studies we discuss has been established, in some cases, through dating of associated charcoal, in others by artifact association, and in several important cases by direct AMS dating of carbon extracted from phytoliths (Mulholland and Prior, 1993).

Direct dating of phytoliths by the AMS technique represents a significant advance for our understanding of early tropical food production. Such an approach is possible because when phytoliths form in living plant cells some of the organic material of the cell becomes trapped inside the phytolith, and it remains there over long periods of time. Also, because the carbon is locked within silica bodies, it remains protected from the various modes of postdepositional contamination over the life of the phytolith. Because individual phytoliths obviously cannot be dated, phytoliths extracted from a single soil sample are evaluated as an assemblage. Usually, one handful of dirt is all that is needed for a radiocarbon determination on phytoliths. The integrity of direct ^{14}C dates on phytoliths can be evaluated by examining the error ranges and stratigraphic positioning of a series of dates as well as their relationship to ^{14}C determinations derived from other cultural materials.

In terms of phytolith assemblages dated by association with standard ^{14}C dates or artifacts, there is no evidence that downward movement of phytoliths poses a special problem in archeological sediments (Piperno, 1988a; Pearsall and Piperno, 1993). In situations in which soil is mixed (e.g., by earthworms in upper strata, by burrowing animals, and by construction of later features), phytoliths obviously will be transported along with other associated materials. In tropical soils phytoliths are chemically bound up in humic colloids (semidecayed organic matter) and clays; indeed, most of the process of extracting phytoliths from soil involves releasing them from these bonds—phytoliths are by no means "loose" in soil.

The incorporation of paleoecological sequences as a primary source of data for considerations of the emergence and spread of food production has not been a common undertaking in the New World. We realize that for those not familiar with pollen or phytolith analysis and the tropical flora following the sequence of vegetational changes in paleoecological diagrams can be heavy going. Whenever possible, we use simplified pollen and phytolith diagrams to illustrate the results of the paleoecological data discussed in this volume. However, we emphasize that empirical paleoecology in the humid lowland tropics based on these techniques is viable. Because relatively few people are carrying out the work and many of them have, at one time or another, collaborated with each other, the techniques are mostly standardized. This means that data sets from different regions can be mean-ingfully compared (see Piperno, 1995a, for a review of this issue).

The near explosion of pollen and phytolith paleoecological records from humid, lowland sites (Chapters 2, 4, and 5) testifies to the mounting maturity of these two disciplines in the tropics. Contrary to earlier concerns that the lowland tropical forest would leave little in the way of a useful pollen record because of poor production, preservation, and taxonomic specificity (e.g., Faegri, 1966), palynolo-gists are finding that pollen influx into lakes is comparable to that of the temperate zone (e.g., Bush and Colinvaux, 1988, 1990; Bush *et al.*, 1992; Colinvaux *et al.*, 1996 a,b). Pollen preservation in perennially wet environments is excellent, and concerted efforts to build reference collections and to study the pollen rain of extant forests, including those where botanists have identified, mapped, and censused the vegetation, have resulted in the identification of a large number of families and genera (more than 160 taxa are now routinely being identified in lake sediments from the lowlands), with concurrent reduction in the proportion of unknown pollen types (e.g., Bush, 1991, 1992; De Oliveira, 1992; Colinvaux *et al.*, 1996a,b; Jones, 1994).

Parallel developments have taken place in phytolith analysis (Piperno, 1995a). Increasingly, phytoliths and pollen are being analyzed in tandem, an approach that significantly improves the robustness and the precision of climatic and vegetational reconstruction because the limitations of one technique in way of production or taxonomic specificity are offset by the strengths of the other (Piperno 1993, 1995a).

In an important paper on Neotropical palynology, Bush (1995) explains why lake sediments in the lowland tropics contain much pollen. Entomophilous (insect-pollinated) species are well represented in lake pollen records because outcrossing is a common reproductive strategy in this group and will produce more flowers and pollen than once thought. Bush (1995, pp. 601–602) notes: "Plants that outcross using these small generalist insect pollinators and sexual dimorphism can be viewed as the reproductive equivalents of the temperate anemophilous [wind pollinated] . . . taxa, and they are similarly richly represented in the pollen record."

We also emphasize that the factors accounting for the complex and diverse plant and animal associations of tropical forests are not as undecipherable as people have commonly thought. During the past 15 years, a group of ecologists have dedicated their lifes' work, and sometimes their lives, to understanding the distribution of biotic variability and biota in the humid tropics. Predictable patterns of plant distributions and succession in the Neotropical lowlands, as well as patterns of plant responses to climatic change, are being elucidated (Chapter 2).

For an increasing number of regions, including those where archeological and paleoecological data testify to the early development and growth of food production, we know what associations of families and genera are likely to represent cool or warm habitats or areas with marked or little to no seasonality of rainfall. We know what families and genera of plants are likely to replace an existing arboreal vegetation when it is manipulated or removed by humans. Species-specific identification of plants is not required to reveal these kinds of climatic and vegetational trends.

Armed with this new abundance of modern ecological data, large, modern pollen and phytolith collections, and using principles of uniformitarianism, we can identify major vegetational associations of many types and, through them, follow the course and causes of climatic and vegetational change through time. We discuss long paleoecological sequences from a large number of lowland regions, including Colombia, Panama, Belize, Brazil, and Guatemala.

The final issue we address here is the strength of the interdisciplinary approach in studying past human adaptations. In the New World, no longer is "paleoecology" the province of geologists, botanists, and researchers from the other natural sciences mainly interested in how the sequence of climatic and vegetational changes and other events affect the nonhuman biota over time. To do "archeology" now is more than to analyze the stones, tools, pots, and plant remains from sites where people lived. Archeological data sets are more likely to tell us about the diversity of plant species manipulated, how they were manipulated and changed morphologically, and how they were processed into food. The importance of paleoecological reconstruction lies in its relevance to the historical landscape. It determines the context of human activity and provides information on relationships between the environment and the evolution of subsistence strategies as well as on the organization of labor and demographic trends. None of the data logically aligned with each discipline are easily extractable from the others, and all are essential.

The proliferation of new techniques and approaches relevant to tropical archeology and the integration of their data sets represent a very major advance for all those prehistorians seriously concerned with the human–tropical forest relationship.

CHAPTER 2

The Neotropical Ecosystem in the Present and the Past

The land is one great wild, untidy, luxuriant hothouse, made by nature for herself.

Charles Darwin (1845)

INTRODUCTION

In this chapter, we dicuss the ecological contexts of foraging and farming in the New World lowland tropics. We first summarize and define the salient characteristics of the American tropical biome, especially as they relate to human settlement and resource exploitation. Then, we use paleoecological data to reconstruct tropical habitats when humans began to exploit their resources during the final stages of the Pleistocene. Lastly, we examine the ecological circumstances that immediately preceded the onset of food production in the Neotropics, and from these we identify some major selective pressures that were acting on human subsistence decisions on the eve of food production.

THE AMERICAN TROPICAL BIOME: SOME BASIC FEATURES

As one moves toward the equator from the temperate regions, one eventually enters the zone where the sun takes a path close to the zenith and will be directly overhead sometime during its annual "march" back and forth across the earth's surface. In the New World, this occurs between two parallels of latitude stretching across Mexico and Chile, each 23°27′ from the equator, called, respectively, the tropic of Cancer and the tropic of Capricorn (Fig. 2.1). Thus, the tropical lands can be defined as the only parts of the earth surface where the sun is directly overhead during any time of the year.

It is true that this astronomical delimitation is somewhat arbitrary and can have little meteorological meaning in those areas of the tropics that have decidedly untropical climates (see below). Air masses originating in the tropics also penetrate the temperate zones and for a short time cause the weather there to mimic the region from where they came (Hastenrath, 1985). Nonetheless, it is only within the Cancer and Capricorn boundaries that a number of physical factors interact to make the climate both warm and moist throughout the year over a great land area.

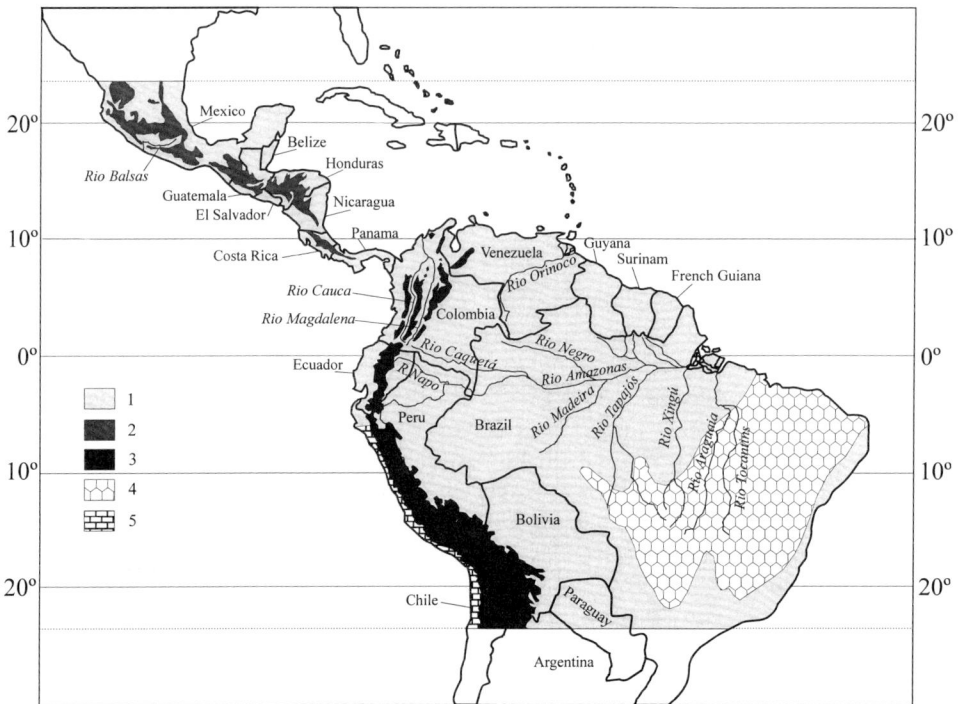

FIGURE 2.1   The major environmental zones of the American tropics. Sources used: Haffer (1969), Janzen (1986), Markgraf (1993), Prance (1987), Sarmiento (1975), Sarmiento and Monasterio (1975), and Wagner (1964). Mountain zones delimit areas above 1500 m. 1, Humid and mostly forested lowlands (areas north and south of 20° latitude may support a different climate and vegetation); 2, Central American mountain zones; 3, Andean mountain chain; 4, cerrado and caatinga; 5, coastal desert.

The high solar radiation constantly received in the tropical lands plays the greatest role in elevating temperatures. Also, because the solar radiation hitting equatorial areas has traveled through less atmosphere than it would at higher latitudes, the intensity of the radiation and the warming is greater. Further, because the changing tilt of the earth's rotational axis during the year is felt less nearer the equator, days and nights are about of equal length throughout the year (Forsyth and Miyata, 1984). One of the most unusual features of tropical life to the student who has been raised at higher latitudes is to have summer-like days that end between 6:00 and 7:00 PM the year through, with little warning that nightfall is coming by way of a long dusk. As a result, there is no protracted period with days of short length and little warmth from the sun, as happens during the temperate zone winter, or with long day length and extreme heat buildup. Thus, seasonal temperature fluctuations are negligible.

The high solar radiation and warm temperatures characteristic of the tropical zone also lead to an abundant amount of rainfall. Because warm air is less dense and has greater energy than cold air, it rises, and as it rises it expands outward. Then, when it cools off in the upper atmosphere, it loses its capacity to hold moisture, which then falls to the ground as precipitation. This steady rising of equatorial air accounts, in large part, for the high precipitation received by many areas of the tropics. Also, the air rising from the equator, having cooled and spread outward, suddenly plunges back to earth at about 30° N and S latitude. Because falling or subsiding air becomes warmer and capable of holding more water, the results are drier climates and the occurrence of many deserts at this latitude (Jackson, 1977; MacArthur, 1972).

In order for the earth's surface to be covered everywhere with an atmosphere, the rising equatorial air must be replaced, so surface winds from the temperate zone rush toward the equator and what we know as the trade winds are created (MacArthur, 1972). These winds have been the friends of mariners since the earliest seafaring days because they blow in a strong and predictable manner between latitudes 10 and 30° north and south of the equator. However, the trades do not blow in a direction indicated by a simple analysis of their point of emanation.

This is due to the Coriolis force, which is the deflective effect of the earth's rotation on any object in motion, including air. Because the earth rotates from west to east, and because it spins faster at the equator than at higher latitudes, air moving from the north to the equator will deflect right and come from the northeast. Air moving to the equator from the south will deflect left. Thus, the northeast and the southeast trade winds are created. Because the air arriving into the tropics also pushes ocean water, the major equatorial ocean currents also result (MacArthur, 1972).

The strength and directionality of the trade winds are important because they also have a considerable effect on tropical precipitation. Because they obey the dictum of the Coriolis force, they follow a long trajectory over tropical waters where they pick up a considerable amount of moisture evaporated from the warm ocean surface. When they finally hit equatorial land they rise and release their water. This effect is most sustained on the eastern side of the Neotropical landmass.

Thus, the basic patterns are simple and fairly easily explained. The warmth of the tropical lands comes from being placed on the center of a sphere and having a sun overhead, and the moisture is a combination of rising equatorial air and air rushing from elsewhere to replace it. There are variations on these patterns, however, that create significant differences in annual precipitation among some tropical areas, which are discussed later in the chapter.

Some important physical characteristics of the lowland tropics resulting from the phenomena previously described include mean monthly temperatures that do not drop below 18°C, eliminating the possibility of killing frosts, and precipitation values that usually exceed 1200 mm each year. Temperature variability from month to month is slight; in fact, diurnal (daily) temperature changes are greater than those occurring during the year. Seasonality does occur in many tropical areas, but it is marked by increases and decreases of precipitation and not of temperature (Schwerdtfeger, 1976; Walsh, 1996).

It is the year-round high temperature and humidity that allow the diverse and luxuriant plant growth we consider to have been manipulated and genetically altered at an early date by humans. It should be remembered that most of the higher plants themselves and plant diversity on the earth evolved under tropical conditions, which existed over large areas of the planet during the geologic epochs that came before the Quaternary period (the past 2.5 million years of earth's history) (Richards, 1996). Trees, especially, benefit from a tropical climate. One of the most striking biogeographical patterns on the earth is the strong, positive correlation between tree species diversity and annual rainfall (Gentry, 1988a). Consequently, much of the area that we consider is, or was before the onset of intensive human disturbance, covered by tall (20 m or more), high-diversity, multicanopied forest.

For example, a hectare (2.5 acres) of forest in Panama contains an average of 176 species of trees and shrubs greater than 1 cm diameter at breast height (DBH) (Foster and Hubbell, 1990). Upper Amazonian forest in Peru contains 300 tree species greater than 10 cm DBH in a 1-ha plot (Gentry, 1988b). In contrast, a 1-ha tract of land in an eastern North American forest will typically support between 15 and 25 tree species (Gentry, 1988a).

Favorable climatic conditions encourage experimentations by the flora in such things as tree architecture as well. One sees features in the tropics, such as buttresses around the bottoms of tree trunks and aerial roots, that are not present or rarely found in the temperate zone (Richards, 1996). Animals are also numerous in the tropical ecosystem, but there is little suitable plant food near the forest floor for grazing and browsing animals like deer and other larger game. The mammals, therefore, tend to be smaller creatures, many of whom live and feed in the trees (Eisenberg, 1989). The tree-living and terrestrial mammals alike have developed close feeding relationships with the numerous fruits and invertebrate fauna of the forest. Often living in small and dispersed family units, the fauna are not so obvious nor easily exploited as in zones outside of the tropical forest.

Substantial areas that fall within tropical latitudes do not manifest the typical climatic characteristics previously described; therefore, they lack certain defining floristic traits. In particular, high altitudes offset latitude and lead to cool temperatures. At elevations starting at about 1500 m the lowest mean monthly temperature falls below 17°C, preventing growth of many palms and other typically tropical flora. Here, in the "montane forest," (Richards, 1996) a vegetation zone that we do not consider in this book, there are many trees more characteristic of the temperate zone, such as oaks, elms, and *Magnolia*. These high-elevation areas are typified by the huge Andean region of South America and they also subsume much of Mesoamerica, which is dominated by the central and southern highland Mexican plateaus.

There are also areas of the lowland tropics where annual precipitation does not reach 1200 mm per annum and, consequently, a tall, high-diversity forest cannot flourish. Some of these areas are found along the Pacific coast of South America and northeastern South America, where subsiding or falling air masses plus unusual features of land–sea temperature and circulation patterns inhibit rainfall and create desert and savanna-like vegetation (Sarmiento, 1975; Schwerdtfeger, 1976). In central and eastern Brazil there are very large areas of savanna and xeric woodland vegetation called the cerrado and the caatinga, respectively (Eiten, 1972; Sampaio, 1995).

Based on this environmental variation, the American tropics can be divided into five broad natural zones: (1) the warm, low-elevation (0–1200 m), humid, and mostly forested lands, called here the greater American tropical lowlands, which stretch from the southern half of Mexico to central Brazil; (2) the cool highlands of Mexico, Guatemala, Honduras, and Nicaragua, whose arboreal associations, characterized by oak and conifers, are largely North American in affinity; (3) the Andean mountain chain, which is generally forested at elevations to 2300 m but that then yields to paramo vegetation at higher zones and whose floristic affinities reach north into the higher elevations of Panama and Costa Rica; (4) the cerrados and caatinga of Brazil and (5) the arid Peruvian coastal desert; (Fig. 2.1). As mentioned in Chapter 1, we do not focus on areas 2, 3, and 4, but we do consider their modern and past plant associations as they become relevant to discussions of archeological data from the lowlands.

It must be stressed that the major zones are areas of high environmental and ecological diversity. As mentioned previously, the greater tropical lowlands include regions along the Pacific coast of Ecuador and Peru too dry to support much vegetation. Major savanna areas can be found within high rainfall zones of southern Venezuela due to poor soil drainage (Prance, 1987). Especially in South America, the correlation between vegetation type and climate tends to be less distinct than in Central America owing to variation in edaphic and soil conditions. Also, the South American continent is very broad and much less broken up into land–water

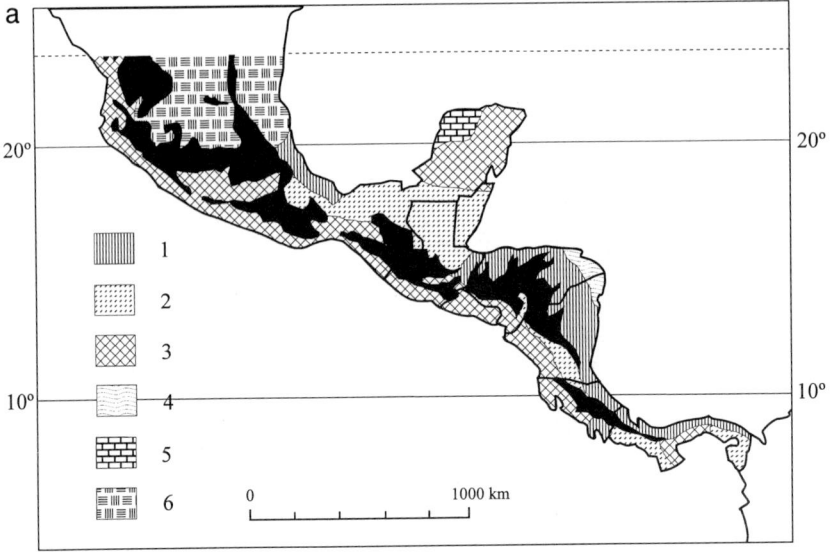

FIGURE 2.2   (a) The major types of forest and other vegetation types in the lowland American tropics of Middle and Central America. Sources used: Beard (1944), FAO (1971), Gentry (1995), Janzen (1986), Markgraf (1993), Murphy and Lugo (1995), and Wagner (1964). 1, Tropical evergreen forest; 2, tropical semi-evergreen forest; 3, tropical deciduous forest; 4, pine woodland and savanna; 5, low scrub/grass/desert; 6, mostly cactus scrub and desert. (b) The major types of forest and other vegetation types of South America. Sources used: Beard (1944), Cochrane and Jones (1981), Gentry (1995), Haffer (1969, 1974), Heppner (1991), Markgraf (1993), Prance (1987), Sarmiento (1975), Sarmiento and Monasterio (1975), and Huber (1995). 1, Tropical evergreen forest; 2, tropical semi-evergreen forest; 3, tropical deciduous forest; 4, mixtures of TEF, TSEF and TDF—TSEF and TDF grow over substantial areas of the southern Guianas and south of the Orinoco River; 5, mainly semi-evergreen forest and drier types of evergreen forest—floristic variability can be high in this zone, as indicated in the text; 6, savanna; 7, thorn scrub; 8, caatinga; 9, cerrado; 10, desert.

strips that help create the distinct differences in precipitation and vegetation seen on the Caribbean and Pacific sides of Central America (Prance, 1987).

The Amazon Basin possesses a diverse array of vegetation types, including dense, evergreen forest, vine and bamboo forest, and savanna, all of which may occur in high rainfall areas. (Balée, 1989; Prance, 1987). Throughout the book we define the limits of the Amazon as the entire drainage basin, an area of approximately 6 million k^2 that is approximately equivalent to the continental United States (Moran, 1993) Figures 2.2a and 2.2b provide a more detailed view of the major vegetation types that would probably be supported by the modern climate in the absence of human interference.

Despite the great habitat variation, it is clear that, by far, the greatest part of the Neotropical landmass was covered by forest when the earliest systems of food

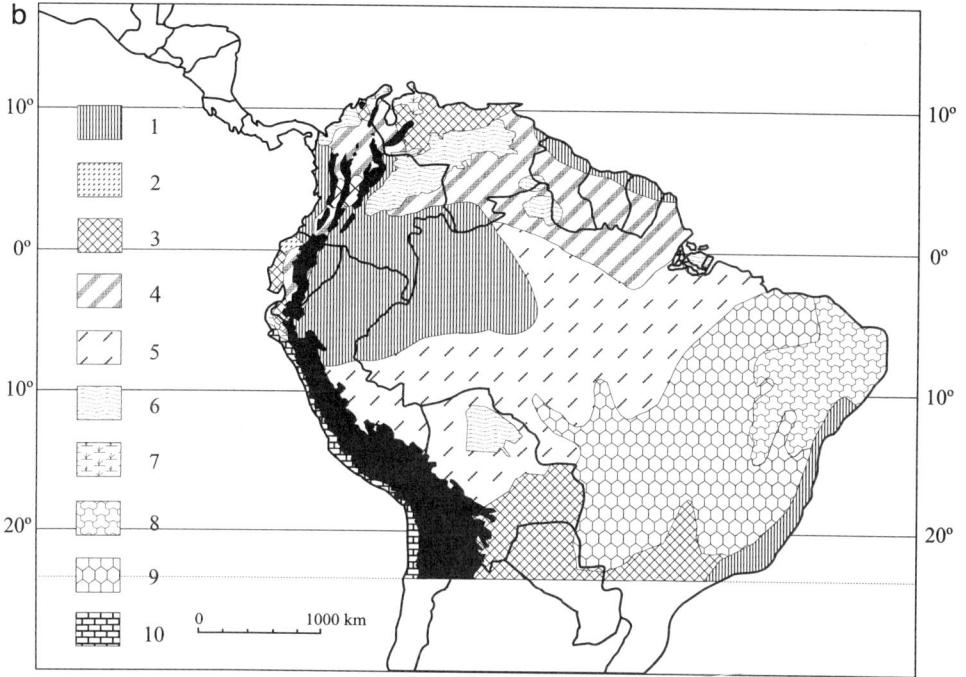

FIGURE 2.2   (*Continued*)

production were developed. We argue later in this chapter that it was within the most optimal zones of the drier forests that the transition from foraging to farming occurred. We now discuss these forests and begin to explore differences in their physical characteristics and biota that may affect human exploitation.

AMERICAN LOWLAND TROPICAL FOREST

General Considerations of the Major Forest Types

The American tropical landmass holds about half the global total of tropical forest, 4×10^6 km^2 in area, and one-sixth of the total broadleaf forest of the world (Whitmore, 1990). Older and/or more general accounts of tropical vegetation (e.g., Rumney, 1968) leave one with the impression that the Neotropical forest is a mostly ever-wet formation, or rain forest. However, before the destructive effects of human agriculture (discussed in Chapters 4 and 5), the Neotropics were characterized by different types of forest containing significant variation in overall design, plant species number, and floral composition, but that all maintained high tree number and richness under a closed canopy (Bullock *et al.*, 1995; Gentry, 1990, 1995; Leigh *et al.*, 1982; Richards, 1996).

Recent studies have shown that the distributions of these forests, their floristic compositions, and their plant and animal productivity can be fairly well predicted by a few associated physical features, mainly variations in the amount and distribution of rain received every year and soil fertility (e.g., Foster, 1990; Glanz, 1990). Thus, considerable floristic and faunal similarities can be found among forests of different regions that have similar physical traits (Foster, 1990; Foster and Hubbell, 1990; Gentry, 1990). For example, the floras and mammalian faunas of forests in Panama and Peru are "strikingly similar" (Terborgh and Wright, 1994, p. 1,829) despite their geographic separation. Both forests are characterized by similar rainfall regimes with marked annual periodicity, and they grow on fertile soils. There may also be considerable heterogeneity in design and species composition across small distance scales of a regional forest, although this feature appears to be less important for human exploition of a forest than are the overall differences among the major forest types, which we now describe.

Lowland tropical areas experience significant differences in the total amount of rain they receive every year. This is largely due to the fact that some experience an extended "dry season," 3–7 months long, during which <10% of the annual precipitation occurs and most of the days are characterized by full sunshine. The result is forests having different degrees of species diversity and canopy closure called here (from the driest and least diverse to the wettest and most diverse) deciduous, semi-evergreen, and evergreen types (Figs. 2.2a and 2.2b; Bullock *et al.*, 1995; Leigh *et al.*, 1982; Richards, 1996). We argue that these types of forests had dramatically different histories of associations with humans, and that the deciduous and semi-evergreen formations, particularly the former, were largely the home to early human settlement, the wild ancestors of many important crop plants, and the origin of plant husbandry.

Seasonal differences in rainfall have probably been closely observed, if not understood, since humans moved into low latitudes because they are critical to human use of the forest. They can be explained by shifts in a zone of equatorial low pressure called the Intertropical Convergence Zone (ITCZ). This is where the trade winds of the northern and summer hemisphere converge on their return to the equator. It was the hated "doldrums" described by the ancient mariners. The ITCZ consists of a huge trough of low pressure and unstable, rapidly rising air where, as discussed previously, much of the tropical precipitation is generated (Hastenrath, 1991; Reading *et al.*, 1995). The ITCZ follows the sun and thus shifts seasonally, being in a more northerly location from May to November when it reaches its maximum incursion into the Northern Hemisphere at 10° N, placing it over the western terminus of the Isthmus of Panama, and in a southerly location from December to April when it is centered near the equator in South America (Plates 2.1 and 2.2) (Schwerdtfeger, 1976).

Shifts in the location of the ITCZ are the major influences on the seasonality of tropical rainfall. The Northern Hemisphere winter and spring from January to May, when the ITCZ is positioned south of the equator, is the northern tropical dry season and the southern "wet" season. Other factors that influence precipitation have to do with proximity to mountain ranges and moving air masses and whether one is a windward or leeward location relative to these factors (Trewartha, 1961).

For example, the Caribbean side of Central America remains in a windward location from the easterly trade winds during the Northern Hemisphere winter when the ITCZ is actually furthest south. Because air is forced upward and cooled upon hitting the coast and nearby mountain ranges of the Continental Divide, it gives up its moisture, leading to substantial rain during the dry season (Trewartha, 1961; Walsh, 1996). Caribbean Central America is also subject to atmospheric disturbances from North American polar air masses, which heat up when they cross the Gulf of Mexico and deposit moisture when they hit the mainland. However, usually these air masses have already lost much of their moisture by the time they cross the divide and reach the Pacific slopes, leaving the dry season truly dry there (Trewartha, 1961).

Factors such as these make the eastern side of the tropical landmass generally much wetter and less seasonal than the western side, especially in Central America. A similar situation exists in relation to much of the Andean mountain chain. Moist, rain-bearing winds come from the east and rise to deposit their rain on the eastern slopes and inter-Andean basins.

Of the three major types of forest, tropical evergreen forest (TEF) is the only true "rain forest" because it receives more and, most important, more evenly distributed precipitation (generally at least 100 mm every month) every year (Plate 2.3). Consequently, plant growth never slows and the trees are always in full leaf (Hammell, 1990; Richards, 1996). Annual precipitation in TEF is usually greater than 3000 mm per year, although extensive areas in the central Amazon that receive substantially less rain support an evergreen forest (Richards, 1996; Whitmore and Prance, 1987). TEF typically occurs on the Caribbean side of Central America (Fig. 2.2a). It covers large areas in parts of South America, including the Venezuelan Amazon and northern parts of the Guianas (we use this term to refer to the countries of Guyana, Surinam, and French Guiana). TEF accounts for at least two-thirds of the huge Amazonian forest, including that on the eastern slopes of the Andes, and also occurs along the coastal strip of southeastern Brazil (Beard, 1944; Prance, 1987; Richards, 1996) (Fig. 2.2b). Possibly the rainiest climate and the wettest forests in the world occur along a strip of northwestern Colombia, where the lower elevation of the Andes and a seasonal weakening of the Pacific high pressure system allow rain-bearing winds to extend to the coast and into northern Ecuador (Richards, 1996) (Fig. 2.2b).

In contrast to rain (wet) forest, tropical semi-evergreen (also called "moist") and deciduous (also called dry) forest (TSEF and TDF, respectively) are adapted to areas that receive a marked dry season of between 3 and 7 months duration when little rain occurs. For this reason they are called "seasonal tropical forests" (Bullock *et al.*, 1995; Leigh *et al.*, 1982) (Plates 2.4 and 2.5). They generally receive, respectively, between approximately 2000 and 2800 and 1200 and 2000 mm of rain in a single year, although as little as about 800 mm of precipitation annually seems enough to create a closed-canopy deciduous forest in a few areas, such as coastal Jalisco province, Mexico (Lott, *et al.*, 1987). Conversely, the dessicating effects of the trade winds that blow strongly from north to south across Central America during the dry season create a highly deciduous forest in areas of the Pacific watershed that receive more than 2000 mm of annual rainfall (Janzen, 1986). The word deciduous applied to tropical forest has the same meaning as when it describes temperate trees, but it is a response to precipitation and not temperature changes. During that part of the year when rain and vegetative growth stop, the trees lose their leaves.

An important point not sufficiently appreciated by students of American agriculture is that seasonal tropical forest covers a very significant part of the Neotropical landscape. It has been estimated that more than 50% of the Central American forest and possibly 25% of the South American forest is, or would have been before human interference, deciduous and semi-evergreen forest (R. Foster, personal communication, 1996; Murphy and Lugo, 1986).

It is true that even most evergreen forests are to a certain extent seasonal with respect to rainfall in that they receive less precipitation when the ITCZ is farther away than when it is nearer. However, evergreen forests will still have a minimum mean monthly rainfall of more than 60 mm (Wright and van Schaik, 1994). When we refer to "seasonal forests" we indicate that rainfall there is highly punctuated. Another important and insufficiently understood fact is that in areas supporting deciduous and semi-evergreen forests the dry season drought is severe and, for several months, absolute. For example, in central Pacific Panama no rain at all will fall for several consecutive months between the end of December and the beginning of the wet season in late April.

Generally, the length of the dry season increases with latitude, so that in central Pacific Panama the dry season is 5 months long, whereas in coastal Pacific Mexico it is 6 or 7 months long. Highly seasonal tropical forests dominate the Pacific side of the Central American landmass where precipitation is typically lower than on the Caribbean watershed (FAO, 1971; Murphy and Lugo, 1995; Wagner, 1964; Fig. 2.2a). They also occupy substantial areas of Ecuador west of the Andes, Colombia, Venezuela, and the southern portions of the Guianas, and can be found in Bolivia (Gentry, 1995; Fig. 2.2b). They are found to a limited extent in the Amazon Basin, where they occur near the mouths of its southern tributaries such as the Tapajós and Xingú.

The Amazon Basin holds more than 75% of all of the lowland forest present in South America. Most of it appears to be evergreen and semi-evergreen forest (Fig. 2.2b). The western side of the basin in the shadow of the Andean mountain chain and the easternmost areas generally have the wettest forests (Salati and Vose, 1984; Prance, 1987). Also in the Amazon Basin can be found extensive areas of swamp forest ("*varzea* forest" and "igapo"), which occur on the margins of water courses that annually flood their banks (Prance, 1987).

Although the Amazonian forest is often used as the case study in discussions of interactions between prehistoric humans and the Neotropical biome, it is actually a very different entity from the Neotropical forest as a whole. Of primary importance in this regard is the soil character, which affects the productivity and nutrient value of the plants for primary consumers, including humans, and agricultural exploitation. We digress here to summarize some salient characteristics of the Amazonian system, especially as they compare to other Neotropical areas.

The Amazonian vs Other Neotropical Forests

We follow the time-held tradition of dividing the Amazonian ecosystem into two broad categories: the terra firme zone, which is the interfluve zone removed from the major watercourses that makes up 98% of the basin, and the *varzea*, the floodplain of the Amazon River and its major white-water tributaries, which make up approximately 2% of the area of the basin (e.g., Sioli, 1984; Meggers, 1971; 1987; Moran, 1993). Although this division may obscure some variability within each zone (Balée, 1989; Moran, 1993), it is still a useful dichotomy by which to judge the overall suitability of the zones for human exploitation. Much of the Amazon terra firme forest is on a deep, sandy, and infertile sediment derived from continental crusts, called the Brazilian and Guiana shields, that formed during pre-Cambrian geologic epochs. Consequently, the sediment has been subject to weathering and erosion for hundreds of millions of years (Putzer, 1984; Sombroek, 1984). The soils are relics from these ancient eras and are still on the landscapes only because no glaciers were ever present to scour them away (Van Wambeke, 1992). Classified mainly as "oxisols," they exhibit high acidity and are extremely poor in essential, dissolved minerals for crop plants such as nitrogen and phosphorous. Soils approximately west of Manaus not affected by the two ancient geological shields, called "ultisols," are nevertheless similarly of low fertility (Sanchez, 1981).

Still poorer soils called spodosols and entisols are found along the headwaters of the Rio Negro, the headwaters of the Tapajós River in Brazil, and in scattered areas around the Rio Negro basin in Venezuela and Brazil. These soils are derived from very deep or coarse sandy materials and are famous for their low fertility. Sanchez (1981) has pointed out, however, that the spodosols of the upper Rio Negro have acquired an undeserved notoriety because their areal extent is actually fairly small. The coarse nature of the soils accounts, in part, for the coffee color of the Rio Negro because water passing through them carries suspended organic matter that is let out into the river system.

The net result for the plants living on all these Amazonian soils is that they have inherently low nutrient value per unit of bulk and low mineral content. Consequently, fewer terrestrial vertebrate fauna can be supported, and even slash-and-burn systems that utilize the nutrients in the fallen vegetation to support crop plants must shift locations very frequently and emphasize crops such as manioc that are suited to low soil nutrients. Moran (1993) reminds us that there is variability in soil quality in the Amazonian terra firme region and that there are areas of more fertile soil belonging to the alfisol and vertisol groups. However, these more productive soils are extremely patchy in distribution and of limited extent, occurring as "spots" surrounded by oxisols and ultisols (Sanchez, 1981).

The situation in the terra firme region is contrasted by the soils of the *varzea* forests, which are probably some of the best in the world for agriculture because they are rejuvenated every year by sediment rich in nutrients originating in rivers of the volcanic Andes. We return again to the conclusion reached by many others that the Amazon Basin at once offers among the poorest and the richest environments for human exploitation. However, the best habitats comprise less than 2% of the total land area of the basin and, as will be shown, occupy mostly the middle and lower stretches of the Amazon River.

Forest soils in other parts of South America and in Central America are generally far richer in mineral nutrients than soils in Amazonia, excluding the floodplains and the eastern foothills of the Andes (Sanchez, 1981; Gentry and Terborgh, 1990). Central America contains substantial areas with young, little-weathered volcanic soils well supplied with nutrients, called inceptisols. Inceptisol regions are densely populated and heavily used for agriculture today. Soils of the Isthmus of Panama are the youngest in all the Neotropics, having emerged from the sea only 4 or 3 million years ago (Coates, in press). They generally are of high fertility, especially on the Pacific side, where forests are much drier and less weathered (FAO, 1971).

Rich, inceptisol soils are also found in the Cauca and Magdalena valleys in Colombia. Other large areas of fertile soils in lowland South America are found in the Guayas Basin and the outer, humid Pacific coastal plain of Ecuador and areas of Brazil near Altimira, Ouro Preto, and Rio Branco and the cacao growing region of coastal Bahia in Brazil. In these regions nonvolcanic but high base status soils called alfisols and vertisols are found (Sanchez, 1981).

Important Differences between Rain Forest and Seasonal Forest That May Affect Human Exploitation

Regarding the types of Neotropical forest outside of the Amazon Basin, we now consider in more detail the considerable differences among them in plant species

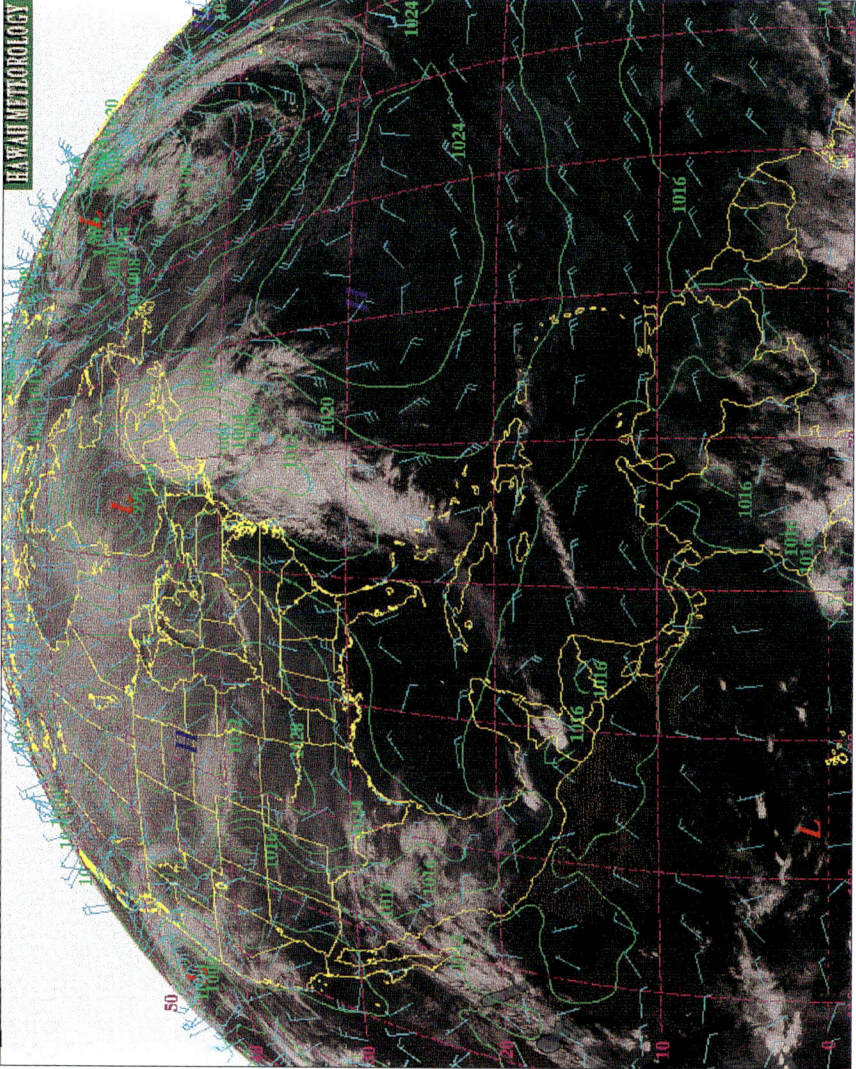

PLATE 2.1   The location of the ITCZ (the band of clouds over the equator) during the southern hemisphere rainy season, on March 16, 1997. Central America and northern South America are at the height of the dry season at this time.

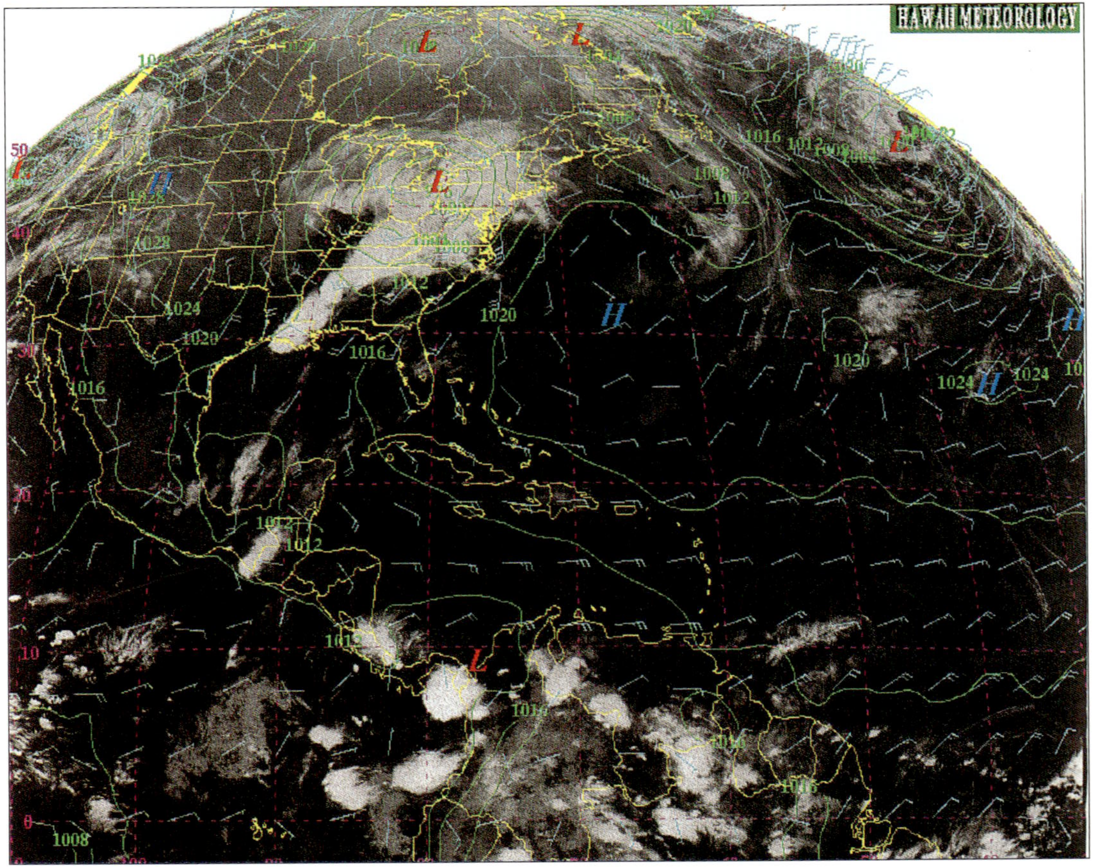

PLATE 2.2 The location of the ITCZ (the masses of clouds over Panama and northern South America) as it "moves" with the sun into the northern hemisphere at the beginning of its rainy season, on May 3, 1997.

PLATE 2.3   Tropical evergreen forest from Fort Sherman, Panama.

PLATE 2.4   Tropical semievergreen forest from Barro Colorado Island, Panama. The buttressed tree on the right is *Brosimum alicastrum* (the famous Mesoamerican ramón), which grows naturally in Panama. There is a considerable leaf drop on the forest floor from the dry season precipitation reduction.

PLATE 2.5   Tropical deciduous forest from the Azuero Peninsula, central Panama. Many of the trees have lost their leaves. Early tropical forest populations probably lived and farmed in forest like this, but little of it remains for study.

PLATE 2.6   Fruits of peach palm.

PLATE 2.7   A mother and daughter tapir from Barro Colorado Island, Panama. They regularly came to the forest clearing in the early evening for food.

PLATE 3.2   Manioc tubers from the Panamanian market.

PLATE 3.6   *Cucurbita ecuadorensis* in its habitat in southwestern Ecuador (Manabí province).

PLATE 3.8   A small sample of the fruits grown in the Neotropics today. 1, *Cucurbita moschata*; 2, *Bactris gasipaes* (peach palm); 3, *Persea americana* (avocado); 4, *Mammea americana* (mamey); 5, *Ananas comosus* (pineapple); and 6, *Carica papaya* (papaya).

PLATE 4.1   Coring Lake La Yeguada. The pine forest in the background is a result of replanting with the introduced tree *Pinus caribaea*.

PLATE 4.9   The now-deforested foothill zone (500–1000 m a.s.l.) of central Pacific Panama during the rainy season.

PLATE 4.10   The now-deforested foothill zone of central Pacific Panama during the dry season, showing the change in plant cover resulting from the prolonged absence of rain.

PLATE 5.3   The current degraded landscape on the coastal plain near the Aguadulce shelter. In the foreground is *Curatella americana*. Such environments severely disturbed by humans appear to have existed in this area of Panama during the past 4000 years.

diversity, plant type, faunal availability, and overall structure that may influence human adaptation. For the most part, ecology is now considered to be more important than geography in determining these differences, and the principal ecological drivers appear to be how the vegetation and fauna respond to precipitation and soil fertility (Gentry, 1990).

TEF holds a greater richness, by approximately 20–40%, of tree and other plant species than do seasonal forests, whereas the variety, if not the biomass, of insects and mammals may be about the same (Murphy and Lugo, 1986; Coley and Barone, 1996). As mentioned previously, there is a strong positive correlation between rainfall and species diversity in tropical forests, which explains why TEF holds more species than do seasonal forests (Gentry, 1988a). One explanation that may account for this phenomenon is that greater precipitation leads to higher rates of seedling damage by insects and other pests that require year-round high humidity and, thus, competitive exclusion occurs less, and plant speciation takes place to a greater degree in wetter forests (Wright, 1992). Seasonal and rain forests often exhibit major differences in species composition. One study comparing wet and dry forest in Costa Rica found that only 11 tree species out of a sampled total of 298 were shared by both habitats (Frankie et al., 1974).

During the dry season many trees in seasonal forest lose their leaves, allowing more sun to penetrate into the understory, the soil dries, and the environment takes on a decidedly untropical, and decidedly more comfortable look and feel (although some pests such as ticks are much more active during the dry season). Significantly, in seasonal forests, the flora manifest rhythms of flowering, fruiting, and growth of underground plant organs designed to be in tune with the cyclical patterns of rainfall and solar irradiance (available light) (Leigh et al., 1982; van Schaik et al., 1993; Wright and van Schaik, 1994). These annual "bursts" in forest resources would have represented the optimal time for exploitation by foragers and would have encouraged the development of scheduling and food storage, necessary prerequisites for societies that cultivate foodstuffs. As will be discussed later in the chapter, tuberous plants rich in edible starch for human consumption appear to be much more common in seasonal forest because the underground tuber is an organ adapted, in part, for energy storage during the long dry season.

Other factors of critical importance relating to seasonal forests have to do with their suitability for agriculture. Soils of seasonal forests appear to be far more fertile and suitable for plant cropping. A principal reason is that, all things being equal and soils not being inherently poor due to a sterile substrate as in the Amazon, seasonal forest soils are far less leached than rainforest soils (FAO, 1971; Murphy and Lugo, 1986). Long and marked dry seasons make burning of vegetation for field preparation a much easier task than in a perennially humid environment, injecting more nutrients into the soil. Agricultural weeds and pests are also less aggressive in less humid environments (FAO, 1971; Murphy and Lugo, 1986; Coley and Barone, 1996).

Other factors important for agriculture are that the canopy height of seasonal, particularly deciduous forest averages approximately only 50% of that of rain forest, trees rarely reach the considerable stoutness common in wetter areas, and there will be only one or two canopy strata compared with the three or more typically present in wetter forest (Beard, 1944; Murphy and Lugo, 1995). Smaller statured, more simple structured forests are likely to be easier to clear by simple slash-and-burn technology. They were also probably easier to penetrate by early humans using simple foraging economies because the forest is more open and would have been amenable to burning.

It is difficult to estimate the amount of land area that would support a tropical deciduous forest today because so much of it has already been cut and is in pasture or farmland. Gentry (1995, p. 146) comments that "there have been remarkably few attempts to focus on the distinctive floristic composition of dry forests as a whole or on how different dry forest plant communities differ from each other," despite the fact that they have been of greater use to humans. Janzen (cited in Holden, 1986, p. 809) has noted, "In any tropical country, look where its major crop and pasturelands are. That's the former dry forest."

That population densities today in the lowland tropics are higher and agriculture is more developed and sustained in areas of former drier forest is no accident and may well reflect an historical tendency for settlement preference there. It is noteworthy in this regard that the first Europeans to see the Isthmus of Panama described high population densities and extensive areas of deforested and farmed land on the dry Pacific but described these to a much less extent on the wetter Atlantic slopes (Sauer, 1966). Population was also much higher in the dry forest area of western, coastal Ecuador than in the humid, eastern lowlands. The population was so high, in fact, that agricultural production was increased by constructing artificial planting platforms—raised fields—in large areas of coastal swamps (Parsons, 1969).

Therefore, among the major forest types defined here, there appear to be significant and consistent differences in soil fertility, forest structure, species diversity, and degree of species "clumping" that seem capable of influencing the character of human exploitation. In the following section we discuss in more detail the characteristics of particular flora and fauna available for human exploitation.

AMERICAN TROPICAL FOREST: POTENTIAL FOR HUMAN EXPLOITATION IN THE PRESENT AND THE PAST

The past 15 years have seen a considerable shift in scientific thought about the character of tropical forests. This new information has also challenged conventional wisdom about the human–tropical forest association in many ways. Only 20 years ago the dogma held that tropical ecosystems were unchanging entities, unaffected by even the comings and goings of the ice ages, which during the past 2 million years were associated with profound climatic and vegetational changes everywhere else on the globe. This climatic constancy alone was thought sufficient to explain the great biological diversity of the tropical system. The favorable climate and diversity of plant resources, in turn, provided a benign environment in which humans could live.

Both propositions made sense. Environmental stability with high rainfall and humidity allowed luxuriant plant growth and the evolution of a remarkable number of species. High species diversity and plant productivity translated into an almost unlimited number of edible products that human populations could exploit the whole year through. However, when biologists and human ecologists began to collect data intensively and systematically make observations about the plants and animals in the forest, and when paleoecologists began to retrieve long-term records from areas now covered by dense tropical forest, opinions concerning environmental stasis and the quality and abundance of the food supply shifted dramatically.

The following is now believed:

1. The availability of high-quality plant and animal resources for human consumption, especially carbohydrates, is generally limited in tropical forest.

2. Marked seasonal rhythms in tropical forests create substantial variability in plant and animal abundance during the course of a year and result in peaks in resource production interspersed with lean periods.

3. There have been major natural and human disturbances of the American tropical forest during the late Pleistocene and Holocene periods (ca. 22,000 years ago to the present). The natural perturbations drastically changed wild resource density and distribution and probably necessitated significant adjustments in human subsistence after the close of the Pleistocene between 11,000 and 10,000 years ago.

Wild Resource Availability

The relationship between resource density and distribution and the size and permanence of human settlement has been of long-standing interest in anthropology. Numerous studies have shown that where wild edible foodstuffs are numerous, concentrated, and produced in predictable supplies from year to year, human populations tend to be larger and more stable (e.g., Hayden, 1981).

In the lowland tropics, the resource–population argument has traditionally been framed in terms of protein availability and has been largely inspired by the long-standing and fervent arguments concerning the ability of the Amazon Basin to support large, sedentary human populations (e.g., Carneiro, 1961; 1970a; Gross, 1975; Meggers, 1954, 1971; Beckerman, 1979; Roosevelt, 1980; 1991). However, in evaluating the overall suitability of the tropical forest to support human populations, we follow the dictum of Earle (1980, p. 3) that "the primary objective of all procurement strategies under investigation is their caloric yield inasmuch as energy is a most basic requirement of any population."

Stated quite simply, people need starch to stay alive, and a minimum requirement per person per day seems to be in the order of 400–600 g. Minimal protein requirements are approximately 30 g per person per day. We note the comments of Bailey and Headland (1991, p. 275) that, generally, "it is likely that humans eating all but the most specialized diets uncharacteristic of hunter–gatherers ingest adequate levels of protein if they are getting adequate levels of calories." We also recognize that low but "adequate" protein intake is associated with lower reproductive success in some animals, that higher protein levels are generally correlated with higher body weights and reproductive success in humans, and that lipids (fats) and fatty acids present in high quantities in animal tissue are essential to good human health, especially for pregnant and lactating women and young children, even if all protein needs are met with plant foods (Hill *et al.,* 1987).

Several studies have pointed to the demographic significance of having a considerable and stable source of calories. Milton (1984) presents evidence showing that fisher–farmers in Amazonia existed at much higher population densities than neighboring hunter–gatherers primarily because they had an ample supply of carbohydrates even though they experienced seasonal shortages of protein. The hunter–gatherers were limited mainly by severe seasonal and annual shortages of carbohydrates from wild resources. Hart and Hart (1986) documented a similar situation in the rain forest of Zaire, where Mbuti pygmies, although having ample meat resources year-round, experienced severe seasonal shortages of starch-dense foods from wild plants.

It may be the case that neither protein nor carbohydrates as such but the relative proportions of calories derived from plant and animal foods are important in determining overall population size. The demographic implications of a subsistence strategy that derives calories mainly from animal foods have received little attention from anthropologists, but they appear to be extremely important. Milton (1984) notes that, in general, populations oriented toward securing food from the first trophic level (plants) are numerically more abundant than populations relying primarily on animal resources. A primary reason for this relationship is the high metabolic costs of extracting calories from protein, which becomes necessary if meat is not high in fat (i.e., fish). Milton states,

> Typically, the higher the trophic level the fewer the organisms . . . a population meeting much of its energetic as well as protein needs from animal food will require considerable more meat each day to sustain itself than will a population meeting energy needs from plant foods and using animal foods only to meet protein requirements. Given the same amount of available protein the second population will be able to exist at a higher density, all else being equal . . . Indeed, catabolizing protein for energy is a wasteful process because of the costs involved in degradation of the individual carbon skeletons of the amino acids." (pp. 19-20)

Our evaluation of the tropical ecosystem also assumes that natural selection has shaped the relationships between humans and their resources, as it has those of nonhuman predators. Reiterating the discussion in Chapter 1, we rely on models drawn from the field of evolutionary ecology, particularly those that predict the range and importance of food items in a diet and why they should change, which are grouped under the heading "optimal foraging theory" (Hill *et al.*, 1987). We believe that efficient foraging strategies conferred a major selective advantage on human populations (Smith and Winterhalder, 1992; Hawkes *et al.*, 1982) and that, on average, those populations with the most efficient strategies will have had the most reproductive success over human history, have spread at the expense of those with less efficient strategies, and will by and large have left the archeological records that we study.

Wild Plants of the Tropical Forest

In the tropical forest there are probably thousands of "edible" species. Certainly, hundreds of species of mammals, birds, and insects derive a living from the panoply of animal and plant life that is available further down the food chain. Numerous investigators have been impressed by the sheer number of plants that appear to offer potential as a food source for humans or are actually eaten from time to time by native tropical people (e.g., Prance, 1986; Harris, 1987; Clement, 1993). This distinction between proven and potential food sources is important in this argument because although many surveys of edible plants contain species that may offer potential for sustainable development, they have not been demonstrated to be important human foods either today or in the past. Much of the anthropological literature, drawing from these impressive plant lists, has assumed that life in the tropical forest should be easy for foragers (e.g., Lee and Devore, 1968; MacNeish, 1991; Roosevelt, 1989; but see Perlman, 1980).[1]

We suggest that a checklist of edible species available in any ecosystem is a poor predictor of food abundance and quality and of the basic economy and demography of the people occupying that biome. This is the case because energetic considerations play major roles in food choices, often with the result that a narrow range of foods provides a surprisingly large proportion of calories in the diet. This has been shown to be true of tropical forest people (e.g., Alvard, 1993; Gragson,

[1] Paleoecological studies, including those of Piperno and colleagues (Piperno, 1985c, 1994; Piperno *et al.*, 1991b), indicate that in many areas what had been thought to be extant, "pristine" forest was heavily disturbed, and even removed, by indigenous slash-and-burn agriculture before European contact, especially highly seasonal forests on fertile soils (see Chapters 4 and 5). We are aware, therefore, that plant and animal species representation might be altered from that first encountered by native Americans. However, these considerations should neither influence the substance nor alter the conclusions of this discussion, which are based on the plant and animal species diversity, biomass, and nutrient content that are characteristic of the tropical forest as well as on the fertility of its soils. Furthermore, the forests used as case studies are old-growth vegetation, and some of them were apparently little disturbed by prehistoric humans (Piperno, 1990a; Piperno and Becker, 1996). They form acceptable, if imperfect, analogs for the assessment of their natural resources.

1993; Hill and Hawkes, 1983). To find, pursue, and process additional foods would present unacceptably high costs to human foragers by seriously constraining time available for other essential (fitness-enhancing) activities and also by risking overexposure to potential negative consequences of the food quest, such as accidents and encounters with dangerous prey.

Also because of energetic considerations, "high-quality" foods capable of supporting fairly large and at least semisedentary human populations typically come in large packages of calories (large animals) or occur in numerous and dense aggregations (plants such as wild wheat and barley). Such foods can be taken with relatively less effort expended in searching, pursuing, and, in the case of large animals, processing foodstuffs. Following from these arguments, there are two major characteristics of wild plant foods in the tropical forest that have important implications for human subsistence: Very few are starch rich and most are too broadly dispersed in space to provide a good energetic return.

That a unit area of land can support an astonishingly high number of species carries with it the rider that relatively few members of that species can be present as adults in the plant community. For example, in semi-evergreen forest on Barro Colorado Island, Panama, only 33 species of 186 inventoried in the flora had an average density of one or more trees per hectare (Hubbell and Foster, 1983). In the deciduous forest of Guanacaste, Costa Rica, the most highly clumped species can be expected to have no more than 12 conspecifics (members of the same species) occurring within 100 m of an adult of that species (Hubbell, 1979). In evergreen rain forest in Amazonian Peru, which may be the most diverse forest studied to date (although a recently begun census in an Ecuadorian evergreen forest may indicate an even greater number of species per hectare; R. Condit, personal communication, 1996), there were only twice as many individuals as species in a 1-ha patch (Gentry, 1988b). A total of 283 species were recorded per hectare.

It is illuminating to compare these tree density and clumping figures with Flannery's (1986b) estimates of at least several hundred individuals per hectare for the mesquite (*Prosopis* sp.), wild legumes, cacti, and grasses that fed early Holocene populations in the arid, highland Oaxaca Valley of Mexico. The return rates per unit time spent in foraging in a tropical forest must almost certainly be far lower than those in arid zones such as those previously described (later in this chapter we provide estimates of return rates for various vegetational associations).

A hallmark of the tropical forest is that it bears a high number of fruits and seeds, which have been estimated to support more than 80% of the nonhuman mammalian biomass in some areas (Terborgh, 1990). It is clear from various surveys that tropical fruits vary greatly in their nutritional value for humans (e.g., Clement, 1993; also see Chapter 3 for a discussion of tree fruits). Some, such as the Brazil nut (*Bertholletia excelsa*) and peach palm (*Bactris gasipaes*), native to the nonriverine forests of Amazonia, yield nuts (edible seeds) or starchy fruits that contain considerable quantities of carbohydrates, fats, and protein and, thus, are more cereal than fruit-like in their nutritional content. However, these seem to be exceptions to the rule.

Many other trees yield small, fleshy fruits that are designed to attract mammalian seed dispersal agents (Howe, 1982). These fruits are typically high in soluble sugars and certain vitamins and minerals. They constitute a prime food source to forest animals, but they do not contain the type of starchy energy required for a human staple food (Clement, 1993; Hladik *et al.,* 1993).

Hladik *et al.* (1993, p. 130) also remind us that because of the spatial diversity of species and other reasons, "Fruit . . . is a patchy, ephemeral resource." They note that accurate estimates of annual fruit production are difficult to make because many well-placed fruit traps are required to collect fruit before it hits the ground and is quickly eaten by the forest animals. They cite studies suggesting that annual fruitfall is much less than the number of individual mature trees in the forest would lead an investigator to believe, even considering the loss from arboreal frugivores (tree-living, fruit-eating animals) before the fruit hits the ground.

Despite the overriding tendency for high species diversity per unit area of tropical forest, there are cases of monodominant or otherwise dense aggregations of fruit trees occurring in patchy areas of an otherwise speciose forest (e.g., Clement, 1993). Palms, particularly, may form dense populations on poorly drained soils (Kahn, 1993). Some of these populations were surely utilized and/or extensively managed by past human populations (Balée, 1988). However, their nutritional content is often such that they cannot provide a staple food source.

Another factor to be considered is the relationship between fruit production and soil fertility that may cause marked variation in total fruit fall. In the Neotropics, measurements of annual fruit fall have been made in forests on fertile soil; namely, on Barro Colorado Island, Panama (Smythe, 1970), and in Manu National Park, southeastern Peru (Terborgh, 1986). Because fruit production is probably related to nutrient availability of the soil, and mammalian biomass seems to be correlated with fruit abundance (Gentry and Terborgh, 1990), it is likely that most of the Amazonia forest, growing on extensive areas of poor soils, produces substantially less fruit and supports fewer mammmals than areas surveyed to date.

Palms are considered crucial to the economies of tropical forest people. Clearly, palms provided in the past, as they do today, many necessities of daily life, including thatch, fibers, and implements such as spears and bows, as well as food (e.g., Balick, 1984, 1986, 1988; Boom, 1987). An important point in this discussion is that they constitute a staple food source much more commonly in the Old World than in the Neotropics.

The Asian tropics support the growth of several types of palms that yield a concentrated supply of edible starch in their trunks, commonly called "sago starch," that is heavily exploited by people (Ellen, 1988; Hladik *et al.,* 1993; Ruddle *et al.,* 1978). As with tuberous plants, the palm starch provides a mechanism for storing resources needed for reproduction, in this case terminal flowering, which occurs every 5–10 years. In contrast, in the Neotropics palm fruits are primarily consumed, quite often in fermented drinks, providing essential fat, protein, and vitamin-rich sources (e.g., Clement, 1993) but little in the way of bulk or as a source of a staple carbohydrate.

There are examples of New World palms with edible starch. One of them, *Mauritia flexuosa,* forms dense aggregations around the edges of swamps and on other poorly drained soils in South America but is currently unexploited by Amazonian Indians (Kahn, 1993). Cases of reported sago exploitation are few. The Warao Indians of the Orinoco Delta in Venezuela consume starch from the stems of *Manicaria saccifera* (Wilbert, 1976). Palm starch makes the Aché optimal diet list and is less expensive to procure than are capuchin monkeys (Hill *et al.,* 1987). Interestingly, Yuquí foragers of Bolivia also consume palm starch when little other food is available (Stearman, 1991).

The practice of extracting palm starch, so important in the Old World and apparently an energetically efficient food procurement practice (Ruddle *et al.,* 1978; Ellen, 1988; Ulijaszek and Poraituk, 1993), is limited in extent today in the Neotropics. Whether these differences in consumption represent historical factors, availability of palm vs other starch sources, or biological differences of the palms themselves, and whether they can be extended into the past, is unclear.

One palm species that can be considered a major food item throughout most of the Neotropics is *Bactris gasipaes,* called "pehibaye" or "peach palm" (Patiño, 1958; Prance, 1984; Clement, 1988) (Plate 2.6). This plant, which is probably native to the Amazon Basin (Chapter 3), is the only convincing example of a New World palm that was domesticated in prehistory (Clement, 1988).

It has nutritionally balanced fruits that are rich in starch and also have considerable quantities of fat, protein, and some vitamins (Brücher, 1989). The peach palm was important in prehistoric diets throughout South America and lower Central America by the time of the Conquest but apparently had not spread north of Costa Rica (e.g., Clement and Mora Urpi, 1987; Patiño, 1958; Prance, 1984). Ethnohistoric references point to large plantations of peach palm when the Spanish arrived. In the Sixaola Valley of Costa Rica early in the 16th century, the Spanish cut stands of peach palms said to number 30,000 to erect a fortress and feed their slave labor (Prance, 1984).

Other important palm fruits in the Neotropics include coyol (*Acrocomia mexicana*) and *Elaeis oleifera* (American oil palm) in Central America and the genera *Astrocaryum, Euterpe, Jessenia, Orbigyna,* and *Mauritia,* as well as the oil palm, in the Amazon (Brücher, 1989; Clement, 1993). Many fruits are especially valued for cooking oil and for making "palm wine" and juices and, therefore, they supply a good source of fat, which otherwise might be hard to obtain in the forest. Palm "hearts," commonly called "palmitos," are also eaten, but these are low in protein and apparently only marginally nutritious (Brücher, 1989). Exploitation of the heart, which actually is a bundle of very young leaves at the base of the stem, is usually possible only after the tree has aged considerably—10–15 years—and results in the death of the tree. Thus, the harvest is not sustainable.

If, as it appears, energy-rich fruit or seed sources are uncommon, what about other options from plants? Significantly, in terms of human exploitation of the flora, wild plants with large underground storage organs (rhizomes, tubers, and corms) that might serve as good energy sources, such as *Dioscorea, Calathea, Ipomoea,* and *Xanthosoma,* appear to grow at low densities in the mature forest (Croat, 1978; Gentry, 1990; Milton, 1984; Headland, 1987; Hart and Hart, 1986; Hladik and Dounias, 1993; D. Piperno, personal observation). However, tubers seem to be more common in highly seasonal forest, anthropogenically disturbed forest, on the margins of water courses and lakes, and at the forest–savanna ecotones (a point that will be discussed further later in the chaper). In addition to being scarce and spatially dispersed, tubers are often difficult to locate and, when located, may reside deep in the ground and contain high amounts of defensive, toxic chemicals (Hladik and Dounias, 1993).

Estimates of wild tuber density, especially in the Neotropics, can be criticized as being largely impressionistic because they are seldom based on quantitative studies. However, Bahuchet *et al.* (1991) and Hladik and Dounias (1993) measured wild yam density and standing biomass in African tropical forest of various types. The standing biomass of edible tubers was between 1 and 3 kg per hectare, which is much lower than the biomass in an African savanna (Vincent, 1985) (Table 2.1) and, probably, in the Venezuelan llanos, where they form a major food for foragers (Hurtado and Hill, 1987). Yams were completely absent in one forest plot they studied. Hart and Hart (1986) similarly found that yams were absent in a closed forest in eastern Zaire.

Given that New World studies would benefit from the same kinds of quantitative estimations of tuber abundance, what do the Old World figures mean? Interestingly, Hladik *et al.* (1993) regard yields of the forest wild yam to be much lower than those for sago palms, considered to be one of the best energy sources in the Old World forest. The implication is that, generally, wild tuber abundance is not

TABLE 2.1   Wild Tuber Density in Various African Habitats

Habitat	Stems/ha	Standing biomass/ha (kg)
Semi-evergreen forest (Central African Republic)[a]	45	1.0
Continental evergreen forest (Gabon, Africa)[a]	22	<1.0
Coastal evergreen forest (Cameroon, Africa)[b]	40	<1.0
Semi-evergreen forest (Cameroon, Africa)[a]	53	3.0
Forest edge (Gabon, Africa)[a]	175	Not computed, but highest density of stems ha^{-1} recorded
Fallows within forest (Cameroon, Africa)[a]	160	12.5
Savanna and savanna bushland (Northern Tanzania, Africa)[b]	1,850	57,141
	445	4,535

[a] From Hladik and Dounias (1993).
[b] From Vincent (1985).

sufficient to support significant human populations and that large foraging ranges would be required to exploit the tuberous resources. We add, however, that we agree with Bahuchet et al. (1991) and Hladik et al. (1993) that these figures do not support the notion that foragers cannot live independently in tropical forest without a cultivated food supply. We discuss this point in detail later in the chapter.

It is also important to note that although generally low wild tuber density may vary significantly among different types of forest. For example, Hladik and Dounias (1993) found that yams and other tuberous plants were much more common in seasonal than in evergreen African forest (Table 2.1). This is a result of two factors: (i) Tubers are storage organs that enable plants to survive the dry season in the seasonal tropics and to reproduce quickly when the rains begin (e.g., Sauer, 1950; Hawkes, 1989), and (ii) tuberous plants need sunlight to regenerate and grow, and the floor of seasonal forest typically receives more light.

The role that seasonal differences in the availability of sunlight plays in the biology of forest tubers has received far less attention than that of rainfall seasonality. However, Bahuchet et al. (1991) and Hladik and Dounias (1993) observed that yams adapted to the African evergreen forest do exist, and that they appeared to be designed for rapid growth and reproduction on the normally shaded forest floor following the admission of sunlight after a tree had fallen. The production of these storage organs probalby had little to do with rainfall periodicity. It follows that some tubers may successfully reproduce in any forest where gaps are continually opened and sunlight is allowed to penetrate to the forest floor, as in the situations in which clearings were made by humans for living areas or for horticultural plots.

Hladik and Dounias (1993) also considered that their finding of more abundant wild tubers in seasonal forest was attributable to the fact that the semi-evergreen forest they studied has a long history of known human settlement and associated environmental disturbance that conceivably increased yam availability. It is clear from these various studies that yams and other tubers are most abundant where more light penetrates the canopy, as in the seasonal and disturbed forest.

Another important contrast between tubers in evergreen and seasonal forest discussed by Headland (1987) and Hladik and Dounias (1993) has to do with growth habit. Unlike seasonal forests and other, more open habitats, evergreen forests typically support the growth of perennial forms of yams, meaning that they do not complete a growth and reproductive cycle every year, as do the annual forms. Selection for perennial tubers appears to relate to the characteristics of sunlight availability on the rain forest floor just discussed.

In rain forests, most plants that contain underground storage organs seem to be vines. Generally, vines, including yams, must grow into the canopy to reproduce successfully, and they need sunlight at some point during the juvenile stage to accomplish this (Bahuchet et al., 1991). In an aseasonal environment where "waiting periods" for sunlight might be long and unpredictable, perennial forms would be best equipped to take instant advantage of light gaps randomly created by tree falls or other disturbances and mature quickly (Bahuchet et al., 1991).

Of interest for human subsistence is that, as perennial tubers simply must persist over relatively long periods of time before they are called on to play their part in the plant's reproductive cycle and are not required to survive long dry seasons, it does not make sense for them to build up reserves at any time during a year. Consequently, they tend to be smaller, harder, and more difficult to digest than annual varieties, although Hladik and Dounias (1993) discovered a few varieties of large and nutritionally significant perennial yams.

Other differences exist between tubers of different habitats. Those of more open ground tend to be chemically defended and require costly processing, whereas those of rain forest are usually mechanically defended by way of deeper burial or surface spines (Bahuchet *et al.,* 1991; Hladik *et al.,* 1993), potentially making them extremely laborious to dig up. The implication is that tuber exploitation, generally, is energetically costly (a point we discuss later in the chapter).

In addition to the high toxic content of some wild tubers, a high proportion of other plants in the tropical forest contain "secondary" or defensive chemical compounds (Janzen, 1969; Freeland and Janzen, 1974; Coley *et al.,* 1985), which would either render them inedible or lead to increased processing time to achieve edible foodstuffs.

To summarize, tropical forest plant resources are generally poor in starch, broadly dispersed in space, and may often require considerable processing before being eaten. Exploitation by human foragers would seemingly require small and highly mobile groups. Can the available animal fauna improve on this picture? We leave, for a time, our discussion of wild plant harvesting in the tropical forest and turn to the animal prey that would be available.

Wild Animals of the Tropical Forest

As with the flora, impressions of food abundance drawn from surveys of species lists can be deceiving. For example, in forests at Manu National Park, Peru, Terborgh *et al.* (1984) reported nearly 100 species of mammals, including 9 species of monkeys, and more than 500 species of birds. However, assessing the behavior, body mass, and demographic characteristics of the terrestrial vertebrates on such a list yields a more pessimistic outlook for human exploitation than species number alone would indicate.

As noted previously, in most of the American tropical zone there are few large terrestrial game animals. The limited condition of light on the floor of a tropical forest means that there is little herbaceous plant growth on which ground-dwelling herbivores can feed and, consequently, terrestrial herbivory is a minor way of life. Tropical "herbivores" are mainly insects, such as beetles and caterpillars who feed in the canopy, preferentially on young leaves of plants because they have a higher nutritional value than older leaves (Coley and Barone, 1996).

Many of the terrestrial mammalian fauna must derive much of their diet from the numerous fruits and seeds of the forest. Many tropical trees selfishly cooperate by making fleshy fruits designed to attract dispersers, and numerous animals eat the fruits and either disperse or destroy the seeds (Howe, 1982). Today, the largest of these animals and the only ungulates (hoofed animals) native to lowland tropical America are the tapir, two species of peccary (a hog-like creature), and three species of deer, one type of which (the "brocket deer") is rather small (Figs. 2.3a, and 2.3b; Plate 2.7). Tapirs, peccaries, and brocket deer depend heavily on fruits (frugivory) and seeds, and will eat leaves (folivory), browse, and in the case of collared peccaries, roots, especially in times of fruit shortages (Eisenberg, 1989).

Other characteristic ground-dwelling animals of the tropical forest include capybaras, agoutis, and pacas [large rodents, the latter two of which also live primarily on fallen fruit (Smythe *et al.,* 1982)], and coatis (a carnivorous racoon-like animal), armadillos, and rabbits (Figs. 2.3a, and 2.3b) (Eisenberg, 1989). Except for white-lipped peccaries, which may travel in groups of more than 100 individuals, these animals tend to forage alone or in small family units and to be shy. The brocket deer and paca are nocturnal.

In contrast to the situation for terrestrial herbivores, as the density of the forest canopy increases, the habitat's suitability for arboreal leaf and fruit eaters also increases (Glanz, 1990). Thus, monkeys and birds may be present in substantial numbers. The Amazon Basin is thought to have the greatest number of primate species in the New World, although their densities in the central portion of the terra firme forest north of Manaus appear to be the lowest of any Neotropical forest surveyed (Malcolm 1990) (below). In more fertile forests studied, it has been estimated that monkeys along with other arboreal species make up over half the mammalian biomass (Eisenberg and Thorington 1973; Glanz 1982).

However, an arboreal life style places serious constraints on body size because animals that are too large can't travel about in the canopy and reach food on smaller branches (Coley and Barone 1996). Hence, although the number of arboreal primates in tropical forests is impressive, when compared to other habitats such as savannas the overall mammalian biomass remains very low.

Monkeys are amongst the most commonly hunted type of animal today in the lowland Neotropical zone and were certainly important food items in some areas in prehistoric time (e.g., Redford 1993. Energetic studies have indicated, however, that some species may yield fewer calories per unit effort than ground dwelling mammals, including smaller agoutis and pacas, quite possibly because their arboreal habit necessitates a longer search and pursuit effort (Hill *et al.* 1987; Alvard 1993). Hill and Hawkes (1983) commented that pursuit lasted quite some time after capuchin monkeys were located by Aché hunters of Paraguay because the monkeys were skilled at hiding in tree branches.

For the Aché, monkeys were more costly to take than plant resources like palm starch and certain tree fruits, which typically rank below animals (Hill *et al.* 1987). Killing monkeys requires skilled use of bows and arrows or, preferably, blow guns, while agoutis, pacas, and armadillos are easily dispatched with clubs, digging sticks and other hand-held weapons, or even by hand (Hill and Hawkes, 1983; Alvard, 1993). Few of the other species of mammals and birds in the Neotropical forest are major prey species, probably because their small size leads to nonfavorable return rates (e.g., Hill *et al.*, 1987; Alvard, 1993). A variety of ethnographic data confirm that the most commonly captured game animals in tropical forest are those that have the largest body size (Redford, 1993). These are not conditions conducive to group hunts and the capture of many pounds of game fairly quickly. Rather, in comparison to more open habitats with herds of large and medium-sized game, the forests would seem to require larger foraging ranges and greater investments in search and pursuit.

This brings us to the question of overall game availability and its potential to support humans in a tropical forest. Unfortunately, many discussions of this issue have revolved around the Amazon Basin terra firme region, where the Neotropical mammalian biomass appears to be at its lowest because of poor soil fertility and low annual fruit production (Malcolm, 1990; Gentry and Terborgh, 1990). Also, these discussions have been largely concerned with the question of protein acquisition per se by fairly large, sedentary populations rather than the overall abundance and quality of the food supply, especially as they relate to foragers (Gross, 1975; Beckerman, 1979; Meggers, 1984).

It is a fairly easy matter to accumulate diversity data on tropical fauna by surveying long enough in any forest, but biomass and density data more relevant for assessing the human exploitation question are more difficult to calculate because of the cryptic and/or nocturnal habits of species, uncertain effects of recent habitat degradation and hunting pressure, and elimination of natural predators such as large cats from some of the forests studied (e.g., Glanz, 1982, 1990). Fortunately, recent studies (see Wright *et al.*, 1994) have used standardized and sophisticated methods of observation and recording and have estimated animal densities in forests of higher soil fertility outside of the Amazonian terra firme, that appear to be more characteristic of the tropical zone as a whole.

Based on these density data, we provide estimates for animal biomass per unit area and human carrying capacity in Neotropical forest from Central America and from outside the interior of the Amazon Basin in South America. Although they are necessarily initial approximations, we believe they lay the basis for some interesting and counterintuitive conclusions about prehistoric life in the forest.

The data for animal densities and biomasses in Table 2.2 are for five lowland forests: Sierra de Chame, Guatemala; Barro Colorado Island, Panama; Guatopo/ Masaguaral, Venezuela (younger forests on the edge of llanos/woodland); Cocha Cashu, Peru; and north of Manaus, central Brazilian Amazon. Included are arboreal and terrestrial mammalian taxa (monkeys, agoutis, tapirs, peccaries, etc.) commonly taken by Neotropical hunters today. Our calculations of biomass per unit area used body mass and species density data provided in Robinson and Redford (1989), Glanz (1990), and Wright *et al.* (1994).

FIGURE 2.3   Some of the major mammalian game animals hunted by Neotropical peoples in the present and during the past 10,000 years. (a) a, agouti (*Dasyprocta punctata*); b, nine-banded armadillo (*Dasypus novemcinctus*); c, tapir (*Tapirus bairdii*); d, coati (*Nasua nasua*); e, squirrel (*Sciurus granatensis*); f, howler monkey (*Alouatta palliata*).

FIGURE 2.3   (*Continued*) (b) a, paca (*Agouti paca*); b, capybara (*Hydrochaeris hydrochaeris*); c, white-lipped peccary (*Tayassu pecari*); d, brocket deer (*Mazama americana*); e, collared peccary (*Tayassu tajacu*); f, white-faced monkey (*Cebus capucinus*). The capybara, the world's largest rodent, feeds on herbaceous vegetation in marshes and on the edges of lakes. All animals are drawn after Eisenberg (1989).

TABLE 2.2 Density and Biomass of Game Animals in Tropical Forests

Game animal	Density (no./km²)					Biomass (kg/km²)				
	SC	G/M	BCI	CC	M	SC	G/M	BCI	CC	M
Large Rodents										
Agouti (*Dasyprocta* spp.)	30	81.5/80	59	5.2	<1	108	293/288	212	19	3
Paca (*Agouti paca*)	30	106/12	8.2	3.5	<1	246	869/98	67	29	6
Squirrel (*Sciurus* spp.)	100	32.5/50	7.9	25	—	30	10/15	2	8	—
Ungulates										
Tapir (*Tapirus* spp.)	—	0.6/—	0.5	—	—	—	90/—	75	—	—
Peccary (*Tayassu tajacu, T. pecari*)	10	3/10	11	6	—	300	90/300	330	180	—
Brocket deer (*Mazama* spp.)	20	7.5/—	6.5	2.6	<1	520	195/—	169	68	26
Primates										
Howler monkey (*Alouatta* spp.)	5	18/150	82	27	10.5	33	119/990	541	178	69
White-faced monkey (*Cebus* spp.)	—	31/35	15.6	68	2	—	90/102	45	197	6
Carnivores										
Coati (*Nasua nasua*)	15	—	56	2	—	68	—	252	9	—
Lagomorphs										
Rabbit (*Sylvilagos* spp.)	10	4.3/10	5.7	—	—	10	4/10	6	—	—
Edentates										
Nine-banded armadillo (*Dasypus* spp.)	77	5/11.3	27	<1	1.8	270	18/40	95	3.5	6
Total						1585	1778/1843	1794	692	116
People	1.5	1.6/1.7	1.6	0.8	0.1					

Note. Estimated maximum supportable human population in people/km². Abbreviations used: SC, Sierra de Chame, Guatemala; BCI, Barro Colorado Island, Panama; GM, Guatopo/Masagual, Venezuela (a younger forest on the edge of llanos/woodland); CC, Cocha Cashu, Peru; and M, north of Manaus, central Brazilian Amazon.

All forests except the terra firme example north of Manaus are in regions with marked seasonality of precipitation and grow on fairly fertile to fertile soils (the Manaus forest receives relatively low rainfall—ca. 2200 mm per annum—but it is distributed on a more even basis and the forest is evergreen). None has an annual precipitation greater than 2600 mm and none is currently under active human interference, although poaching in the recent past may be expected to have lowered mammalian biomass in some of them.

The biomass projections for all the forests err on the low side because density figures were not available for several monkey species, a few terrestrial mammals, and larger birds, such as curassows, guans, and tinamous, that are commonly hunted today (Redford, 1993). Also, aquatic resources, such as fish, large rodents (capybara), and reptiles (caiman), that even in small rivers would be exploited to some extent and would supply considerable inputs into the diet were not included. The latter factor might particularly affect the biomass calculated for the forest at Cocha Cashu, Peru, which is in a meander belt of the Rio Manu.

It can be seen that all the forests outside of the terra firme except for that at Cocha Cashu, Peru, have very similar biomass figures. This is another indication that there are broad similarities among the resources of different forests with similar physical environments and rainfall. The lower terrestrial biomass at Cocha Cashu than at other sites not in the terra firme might be explained by the fact that seasonal flooding limits terrestrial mammals, such as armadillos, pacas, and peccaries, that have small home ranges and make burrows to live in (Wright *et al.,* 1994). The terra firme forest study site north of Manaus, Brazil, has the lowest edible biomass of any Neotropical forest studied, being an order of magnitude lower than most of the others. The study site data are remarkably similar to data obtained earlier by E. B. Ross and discussed by Meggers (1984, p. 630), and it is likely that the site characterizes large regions of the interfluve zone throughout the Amazon Basin.

Biomass data obtained for the Neotropical forest, ranging between 116 and 1843 kg of edible meat/km^2, can be compared with mammalian ungulate biomass data for East African tropical savannas and parklands, which vary between 5000 and 20,000 kg/km^2 (Butzer, 1971). The comparison is striking. The impression of limited availability of large and medium-sized mammals in a tropical forest is a valid one. More than 80% of all the edible biomass in the forest comes from animals no larger, and often smaller than, average-sized dogs.

Assuming (i) 25% of the calculated biomass is available to human hunters after accounting for predation by other carnivores, the proportion of edible meat weight, and other factors; (ii) an average annual caloric requirement of 750 g per person per day [we use a higher figure than usual based on the Hill *et al.* (1984) observation that foraging in tropical forest is a particularly strenuous activity and their calculations of Aché energy expenditure]; and that (iii) meat serves as a substantial caloric base, a "maximum supportable population" (MSP) (Jochim, 1976) of between 0.8 and 1.7 people/km^2 is reached for seasonal forests outside of the Amazonian terra firme region (Table 2.2). The MSP calculated for the Amazonian forest north of Manaus is 0.1 person/km^2.

We assume that an actual supportable population would be lower, mainly because of natural variations in the availability of animals, including fish, and plants during the year (discussed later) that probably act as significant constraints on human population density and set the carrying capacity (e.g., the censusing was done during the dry season when animals are more abundant and active and tend to weigh more). Conversely, plants and invertebrate products of the forest (tubers, honey, and grubs) would also inject calories into the diet on a regular basis and supplement game.

When adjusted for the leanest resource periods of the year, the MSP figures calculated for most tropical forests are unlikely to fall below population densities calculated for hunters and gatherers, such as the Bushmen and Hadza (.01–0.16/km^2), who are occupying "marginal" habitats today in Africa and Asia, but who are maintaining their subsistence needs without problem (Hayden, 1981). We reach the conclusion that there appears to be enough terrestrial meat to sustain foragers in seasonal tropical forest on fertile soils away from major water courses where large supplies of fish and other aquatic fauna are not available. The data listed in Table 2.2 do not indicate a rich resource base and affluent foraging. They strongly suggest, however, that small and mobile groups of foragers can satisfy many of their nutritional needs and survive independently in seasonal tropical forests by focussing on meat and deriving supplemental inputs from plants and invertebrate products (a point which we discuss later in this chapter).

This conclusion is not entirely surprising. Empirical data demonstrate that the foraging Aché of Paraguay derive more than 50% of their dietary calories from terrestrial game (Hawkes et al., 1982). Yuquí foragers of the Bolivian forest also consume high amounts of meat (Stearman, 1991). Aché hunters achieve a very respectable return rate of 910 calories/person/hour from hunting (Hill and Hawkes, 1983). Similar, and even higher, return rates have been reported for the Yanomamo (Hill and Hawkes, 1983).

Other Neotropical indigenous groups studied generally have good return rates of game using simple technologies and appear to be meeting their daily nutritional requirements (e.g., Hill and Hawkes, 1983; Hames and Vickers, 1982; Milton, 1984; Yost and Kelley, 1983). We add that all these hunters work long hours most days procuring game and, hence, are hardly "affluent" in the Sahlins (1972) sense of the word. Also, Neotropical foragers shift locations almost daily to ensure good hunting returns.

Game depletion in relatively short periods of time, particularly of the important larger sized fauna, has also been demonstrated in areas occupied by horticultural peoples in the Amazon who maintain permanent settlements for several years at a time (Hames, 1980). To the extent that the same would be true of other Neotropical forests is unclear, but it is likely that game depletion would generally occur near settlements that did not shift locations frequently.

Given the differences in plant productivity and agricultural potential that seem to be characteristic of seasonal and aseasonal forests, it would be useful to have similar comparisons of faunal density among these forest types. None is currently available, but censuses are due to start in aseasonal forests of Panama early next year (J. Wright, personal communication, 1996). It might be expected that seasonal forests have a higher biomass of mammalian fauna because, as they generally have less leached and more fertile soils than aseasonal forest, fruit production by the flora may be considerably higher. Also, the trees of seasonal forest are generally faster growing and, thus, shorter lived than those of evergreen forest so they need to invest less in toxic and other compounds meant to deter animals that might feed on them (Coley et al., 1985; Coley and Barone, 1996). Lower levels of plant defenses in fruits and leaves might also contribute to higher animal biomass.

In the following sections, we discuss other aspects of the tropical biome affecting human exploitation. Clearly, hunting in the Neotropical forest must be considered on the basis of resources ocurring outside the Amazon Basin and when evaluating the needs of small and mobile groups who occupy the deciduous and semi-evergreen forests. Thus considered, protein and caloric resources of the forest are less limiting for certain types of human occupation than has often been supposed.

Seasonality as a Factor in Tropical Human Subsistence

A factor that may be highly significant in assessing the resources of the tropical forest is the seasonal distribution of plant and animal foods. Licbig's Law of the Minimum suggests that the period of lowest resource availability in a year will have an important effect on human adaptation.

It was once thought that because of high year-round temperature, biomass production of the tropical biota experienced little seasonality and that plants leafed, flowered, and fruited throughout the year in no particular rhythm.

Recently, a group of ecologists began to find that a pronounced seasonality of rainfall and solar radiation caused variation in the reproductive and leafing patterns of the plants, called phenology or phenological activity by ecologists (Leigh et al., 1982; van Schaik et al., 1993; Wright and van Schaik, 1994). This variation in the production of fruits, nuts, leaves, and underground plant parts was parallel to that experienced in temperate zones, but it was paced by the comings and goings of the rains and by changes in the availability of sunlight, not the rise and fall of temperature.

Because many important game animals are frugivorous, the seasonal rhythms of plant production have an important effect on animal availability. There are well-documented "seasons of scarcity" in tropical forest when wild plant sources, including fruits, nuts, and young leaves, are at a minimum and, in response, mammals are fewer, leaner, and more dispersed (e.g., Leigh et al., 1982; Yost and Kelley, 1983). Leaner animals posess less body fat and offer fewer calories to foragers.

In semideciduous forests of Panama there are two peaks of fruiting, occurring in September–October and March–June. The late wet and early dry seasons (November–February) seem to be the leanest times of the year for plant productivity and the time of most stress for animal populations (Smythe, 1970; Foster, 1982). Wet season rains, in and of themselves, may also be expected to diminish hunting yields because animals are less active and more difficult to hear and track. In Costa Rican deciduous forest, there is a single peak of fruiting occurring during March–June and especially at the end of the dry season in April (Frankie *et al.*, 1974; Opler *et al.*, 1980).

Although data are few as to how seasonal fluctuations in resources might affect human foragers, the lean periods in the forest have been shown to be times of lower food availability for some human populations. Parts of the dry season appear to be the most difficult time for Yuquí foragers of Bolivia (Stearman, 1991). Some animals and plants exploited during other times of the year are scarce and alternative foods must be sought. Fish become a more important resource for the Yuquí at this time, especially as the fish become concentrated in shallow ponds and oxbow lakes and are easily caught.

Several months of the late wet/early dry season have been cited as a time of wild plant food diminishment for foragers in the forests of northwestern Amazonia (Milton, 1984) as well as in the tropical savannas of Venezuela (Hurtado and Hill, 1987). Reduced phenological activity of trees is implicated in all cases as the primary factor causing resource reduction.

All the examples previously discussed come from people occupying wet forest with a short and relatively wet dry season—types of forest for which terrestrial animal density figures are currently unavailable. Also, all have agricultural foodstuffs to fall back on, if needed, so the cultural response to the seasonal bursts of wild plant foods is compromised. It can be postulated that among prehistoric foragers with no recourse to a cultivated plant supply, responses to the periodicity of wild resource production in the tropical forest may have included settlement movement, scheduling of resource acquisition, storage, and use of "starvation foods" that ordinarily would not be consumed (Stearman, 1991).

Studying the foraging Aché of Paraguay, Hill *et al.* (1984) found little seasonal variation in the amount of calories consumed. Differences in consumption from season to season were minor and largely qualitative. This is potentially important because the Aché occupy a much drier type of forest than do the other foraging groups studied who experienced significant seasonal fluctuations of resources. The implication might be that resources important to humans in drier tropical forest, especially game animals, occur in higher densities than in wet forests and effectively are available in higher number throughout the year. Also important, lower rainfall and fewer days with heavy precipitation may have contributed to substantially more days with high hunting success.

It should also be noted that although rainfall in the Aché region is typical of that of highly seasonal, mostly deciduous tropical forests (between 1200 and 2100 mm per annum), the periodicity of annual rainfall may be less pronounced than that of many seasonal forests discussed here, probably because the region borders the subtropics (although the flora and fauna are clearly lowland tropical)

(Hill *et al.*, 1984). It is unclear whether fruit and leaf production there has the seasonal bursts and valleys characteristic of forests of lower latitudes that may cause fluctuations in terrestrial game availability. This factor may also, in part, account for more consistent resource availability throughout the year.

It may never be possible to study modern human foraging in a tropical deciduous forest because virtually none of it is left uncut. However, terrestrial faunal censuses to be started soon in wet, aseasonal forests of Panama will at least give us some idea of the overall biomass differences between seasonal and aseasonal forest, which we expect may be significant.

Summary

There appears to be a growing consensus, which the authors join, that the tropical forest is not an easy environment in which foragers can make a living. However, there appear to be significant differences in the types and densities of useful plants and possibly animals among the major tropical forest types. In general, forests experiencing marked seasonality appear to offer the most potential to human foragers. In all the forests, wild protein and carbohydrate availability is limited and the resources are too spatially dispersed to permit large and stable human populations.

The implication is that most foragers must exist at low population densities and be sufficiently mobile so as not to deplete the resource base. Therefore, "complex" hunters and gatherers (e.g., Fritz, 1994) could not have emerged in this biome during the Holocene except, perhaps, in such few places as along the Amazon River and its productive *varzea* zone. It also follows that some antecedents and corrolaries of food production development proposed for other areas of the world, such as population pressure and declining yields to labor, may not have been applicable in the tropical forest. However, we firmly conclude that the tropical forest was capable of supporting foragers before the emergence of food production, *contra* the suppositions of Bailey *et al.* (1989) and others who argue that the biome can provide a full-time living only for people with some access to a cultivated food supply. This view is discussed at length later in the chapter.

Bettinger (1991, pp. 99–100) reminds us that measures of foraging efficiency provide a useful way to compare hunting and gathering across markedly different ecozones and, if used as indicators of "affluence," they suggest that the tropical forest Aché are better off than hunters and gatherers occupying savanna environments. Furthermore, the Aché eat substantially more meat than do the Kung Bushmen or the Alyawara of Australia. Cuiva foragers of the Venezuelan llanos (Hurtado and Hill, 1987) and, quite possibly, the Yuquí (Stearman, 1991) also derive most of their calories from meat, not plants. This leads us to wonder if this counterintuitive subsistence pattern at low latitudes was typical of tropical forest foragers of the past (Ranere and Cooke, 1991). It is qualified by the fact that few Neotropical foragers have been studied. Nevertheless, it makes good sense given that wild game is available, if not abundant, that high-quality vegetable carbohydrates are limited, and that game is typically less expensive than plants to procure.

Also, Hill *et al.* (1984) make the important point that many South American (and, we assume, Central American) forest game animals have a much higher caloric value per gram of live weight than do African or North American ungulates that account for most of the game taken in those parts of the world. In other words, peccaries, rodents, monkeys, and armadillos are fatter than North American and African game animals and, pound for pound, represent a superior source of animal carbohydrates. The implications are twofold; terrestrial game becomes a better energy substitute for plant starch than commonly assumed, and higher meat intake levels are possible before adverse effects from eating too much meat protein occur (e.g., Speth and Spielmann, 1983; Speth, 1987). It follows that tropical meat would support larger numbers of people than "low-fat" varieties.

Still, we would not regard the Aché as affluent foragers because they live in small groups and shift their living sites almost daily. A similar situation has been documented for the Yuquí foragers of Bolivia, who occupy a relatively productive forest zone on the western edge of the Amazon Basin (Stearman, 1991). This leaves the impression that any significant increase in population size, decrease in mobility, or decline in resources would, before too long, overtax the resource base and necessitate some adjustments.

A commitment to the exploitation and manipulation of plants requires some kind of concentration of appropriate resources and familiarity with them, itself a function of how much time is spent in any particular area from year to year (Moran, 1983). Given the previous discussion concerning resource quality and density in the tropical forest, our thesis that food production emerged independently there may require that certain zones of the forest were more productive than the forest at large and permitted more protracted periods of settlement during a year. We now proceed to discuss those areas likely to have been foci of early Holocene semisedentary and sedentary settlement, where resources were more abundant and where successful experimentation with the plant world could probably more easily succeed: aquatic ecozones and forest recovering from disturbance.

MORE FAVORABLE TROPICAL HABITATS: LAKES, RIVERS, AND COASTS AND THE REGENERATING FOREST

Aquatic Ecozones

Many investigators have looked to aquatic ecozones as places where high productivity of wild tropical resources, especially animals, allowed a sedentary existence and development of cultural complexity (e.g., Lathrap, 1970, 1977a; Meggers, 1984;

Moran, 1993; Roosevelt, 1989; Sauer, 1952). Many of these researchers focus on the Amazonian *varzea,* the floodplain of Amazonian white-water rivers, where aquatic fauna important for human consumption are abundant and soils are rich for agriculture. Similar environments can be found in the Orinoco Basin (Roosevelt, 1980). The extremely favorable ecological circumstances of the *varzea* led Lathrap (1970) to propose that agriculture, dense settlement, and ceramic technology had primacy along the middle Amazon, and populations carrying these traits then diffused into other regions of the tropics.

The *varzea* is typified by the presence of large aquatic mammals, which provide not only a stable source of protein but also essential lipids and calories from fats. They include manatees, capybaras, the world's largest living rodent, and giant otters. Large reptiles include turtles and caimans (Best, 1984). Turtle eggs were, and still are, a delicacy along the rivers of Amazonia. Fish, of course, are also abundant, at least seasonally. The biomass and densities of aquatic mammals and reptiles are many times higher than the game situation in the forest (Best, 1984). For example, a single manatee can supply up to 90 kg of meat (Linares, 1976), which is about equal to that of the largest terrestrial animal (the tapir). During low water levels fish may be easily trapped in remnant *varzea* lakes and other standing pools of water, and mammals aggregate in easily taken groups near remnant stands of water. That these animals once played major roles in human subsistence along some major water courses of Amazonia is very clear. Beckerman (1979) and Best (1984) supply excellent discussions of this issue.

It is likely, however, that the *varzea* has not always been the very rich resource zone that we know it to be today. As Junk (1984) explains, the Amazonian river still has not filled the valley created during the low sea-level stand of the last glacial period. This means that sedimentation and flooding that create the highly productive conditions of the *varzea* are mainly occurring today only on the middle to lower stretches of the Amazon River, from approximately Manaus eastward. The clear-water rivers flowing from the Guyana and Brazilian shields are also still filling their valleys and carry low sediment loads. As we discuss in Chapter 4, a *varzea* environment was probably not born until sometime during the late-early to middle Holocene, at which point river levels had risen sufficiently to bear sediment and flood banks.

Where the land meets the sea is another ecozone where resources occur in favorable quantities. Yesner (1987) explains that coastal resources are neither second-rate resources nor the "Garden of Eden." There is marked variability in the costs of acquiring coastal resources that is dependent on the class of resource and the particular species being exploited. For example, sea mammals probably have more favorable caloric return rates than fish, which will often yield higher returns than shellfish. Estuaries are particularly important because they are rich in fish, shellfish, crabs, small "terrestrial" animals, and shore birds, and they permit simple, energetically efficient land-based technologies (Cooke and Ranere, 1997) (for a prehistoric example, see the discussion of central Pacific Panama in Chapter

4). Yesner (1987) notes that a real commitment to marine resources appears to postdate the Pleistocene around the world. This suggests that the terrestrial Pleistocene resources, discussed in detail below, were more efficient to exploit than were the coastal zones.

Rivers and the edges of lakes and swamps also offer more favorable conditions of settlement and resource supply than those of the interior forest. They may hold significant supplies of native fish as well as capybaras, turtles, iguanas, shore birds, and other high-quality resources (e.g., Hurtado and Hill, 1987). Many species of palms form dense aggregations on swampy soils and around the edges of shallow or seasonal water bodies, whereas they are much more dispersed in the dryland forest. Peccaries, tapirs, pacas, and other frugivorous mammals will congregate around these areas in order to feast on the copious palm fruits that are available.

After the large game was lost as a consequence to environmental change and/or human predation 10,000 years ago and before sea level stabilized and use of coastal resources became common or necessary 7000 years ago, the edges of lakes and water courses were prime places of early forest settlement. Also, because the margins of lakes and swamps were small and circumscribed, they must have quickly filled with people soon after they were first discovered by humans.

It should be noted that fishing and aquatic game hunting in many of the major Amazonian rivers and tributaries are probably substantially more productive than they are in places such as Panama and other areas outside of the Amazon Basin, where water courses and their floodplains are significantly smaller. In Panama, the fish fauna are depauperate because of the recent age of the isthmus (R. Cooke, personal communication, 1996). In these regions, where, as discussed previously, terrestrial game biomass is generally much higher than that in the Amazon Basin, forest animals may have assumed a greater importance in diets than they did in Amazonia.

The Regenerating Forest

Another important area where wild resource character is of far greater use to humans is secondary or disturbed vegetation. First of all, we must comment on the growing sentiment and body of literature in anthropology suggesting that much of the "climax" tropical forest is in an early successional state by virtue of its own internal dynamics and, therefore, offers more food to humans than has been supposed in the absence of human interference (see Bailey and Headland, 1991; Colinvaux and Bush, 1991; Stearman, 1991; Hoopes, 1995).

This argument is derived from Connell's (1978) "intermediate disturbance" hypothesis for the existence of high species diversity in the tropics. This hypothesis posits that the continuous creation of small gaps by fallen trees, blowdowns, and other natural disturbances in the forest promotes high species richness by preventing competitive interactions among species from lasting long enough for competitive exclusion and, hence, extinction to occur. The hypothesis has been supported by various studies of coral reefs and it may hold explanatory power for the existence of high species diversity in the tropics. However, it confounds time and spatial scales relevant for evaluating the natural cycles of forest structure vis-à-vis human exploitation and mistakenly creates the impression that no active human alteration of a forest is needed to increase the amount of successional growth to levels useful to human foragers.

Tree falls and blowdowns are certainly fairly regular occurrences [although the authors have never observed "terrain riddled with openings created by tree falls" (Hoopes, 1995, p. 194)], but in one semi-evergreen forest studied, only one gap over 150 m^2 was formed per hectare every 5 years (Brokaw, 1982). The mean time between the formation of successive gaps in any one area has been estimated to be approximately 100 years in several types of Neotropical forest (Brokaw, 1982). Furthermore, if no further disturbance takes place after the tree fall, plant succession is rapid and results in the elimination of the first gap invaders within a few years. These conditions are not likely to result in considerable enhancement of resources for people.

Blowdowns and landslides as a result of intense storms may create larger successional spaces of perhaps a few hectares in area, but these are much less common than tree-fall gaps. In one forest studied, they occurred perhaps once every 5 or 20 years, respectively, and the heavy damage was extremely localized (Foster, 1982).

The tropical forest is being increasingly described as a mosaic of patches in different successional stages, and at face value this is an accurate depiction. However, what this actually means is that a 50-m^2 area of 100-year-old-trees with a canopy sufficiently closed to allow little light penetration may lie side by side with a 200-m^2 area of old-growth forest 300 years old. It does not indicate, as increasingly thought by anthropologists, that considerable areas of a forest undisturbed by humans are receiving full sunlight at the forest floor and attracting early successional growth.

Having noted these factors, we can discuss the ecology of plants most useful in the human diet and their response to persistent human disturbance, which we hold to be the critical ecological factor making the tropical forest more attractive to human foragers. Many studies have noted how useful wild plants, such as tubers and palms, are more common in growth regenerating from human disturbance than in primary forest (e.g., Bye, 1981; Messer, 1978; Hart and Hart, 1986; Headland, 1987). As discussed previously, such plants are adapted to reproducing quickly once sunlight becomes available, and disturbed places in the forest receive more such light. In semi-evergreen forest on Barro Colorado Island, Panama, even fairly casual human disturbance, so long as it is persistent, such as the maintenance of a small trail through mature forest, results in increases in the densities of important tubers and plants used for utilitarian purposes (D. Piperno, personal observation).

Hladik and Dounias (1993) recorded the highest standing biomass of wild yams in fallows (regenerating agricultural plots) in African tropical forest (Table 2.1).

Many other herbaceous and woody plants valued for their leaves, fruits, and other parts that can be hard to find in little disturbed contexts are common in regrowth vegetation and they form important dietary supplements and utilitarian items for indigenous groups.

The consequences of secondary vegetation for animal availability appear to be great as well. The seedlings and young saplings of even small tree-fall gaps are attractive browse for terrestrial herbivores such as tapirs and peccaries (Hartshorn, 1990). Secondary forest and abandoned gardens attract substantially more game animals, concentrating them around human settlements and effectively increasing their numbers over those found in undisturbed situations. Agoutis, paca, deer, peccaries, and many species of birds head the long list of fauna that would normally shy away from human beings but feed on garden and regrowth vegetation and are "hunted" there (e.g., Holmberg, 1969; Ross, 1978; Hames, 1980; Balée and Gély, 1989). This tradition is ancient. Linares (1976) has documented a pattern of "garden hunting" among prehistoric groups of Caribbean western Panama.

Recent studies provide an explanation for the relationship between resource quality and disturbed growth and indicate a sound ecological basis for the human preference to feed on regenerating vegetation. Successional plants have less tough leaves and tubers and higher protein content, as well as lower concentrations of digestion-inhibiting fiber, proteinase-inhibiting secondary compounds, and other antiherbivory substances, than do mature forest plants (Milton, 1984; Coley, 1983; Coley *et al.*, 1985).

The major underlying factor for these differences appears to be the nature of resource availability in tropical habitats (Coley *et al.*, 1985). When resources (mainly light) are in low supply (primary situations) slow growers are favored over fast growers (successional plants), but the former need to invest heavily in antiherbivory compounds simply because they tend to be longer lived. When humans entered the tropical forest and fired and cleared the vegetation, they unconsciously increased the reproductive fitness of many wild plants and animals most beneficial in their diets and set the stage for control of the reproduction of these plants through cultivation and domestication.

MORE OPTIMISTIC AND PESSIMISTIC VIEWS OF THE TROPICAL FOREST HABITAT

Several views have been offered about food abundance and quality in the tropical forest that we believe are too extreme. Beckerman's (1979) very positive outlook on resources was a reply to Gross (1975), who had questioned the general availability of protein in the Amazon Basin because of low animal numbers. Beckerman did not address the overall viability of the tropical system to support human life, focussing instead on vegetable foods as alternative sources of protein and not as food staples. Interestingly, most of the plants he lists as providing relatively good sources of protein are energy poor.

Colinvaux and Bush (1991) offer one of the most optimistic assessments of the tropical forest environment, calling it a "prime habitat" for hunters and gatherers. These two ecologists contribute a wealth of useful information on biomass, annual primary productivity, and other aspects of tropical forest dynamics, but for all the reasons discussed here, these data are most useful for assessing exploitation by nonhuman primates and other mammals and not for evaluations of human food quality. Although we agree with Colinvaux and Bush that human foragers certainly can subsist in the tropical forest, we would not go so far as to say that it constitutes an optimal habitat for non-food-producing people.

More "marginal" environments are, of course, occupied today by many groups of hunters and gatherers, who cannot be said to have an abundant and stable resource base but who, under the proper conditions of population density and movement, appear capable of long-term occupation of these areas. Nonetheless, several investigators have recently questioned whether humans could have survived in the tropical forest biome independent of agriculture (Hart and Hart, 1986; Headland, 1987; Bailey *et al.*, 1989). They note that most "hunters and gatherers" living in tropical forest today either derive a significant proportion of their calories through reciprocal exchanges of energy-rich food with nearby agriculturalists or they cultivate small garden plots (see also Milton, 1984). Supporters of this view, which we call the "foraging exclusion hypothesis," argue that symbiotic relation-ships between foragers and farmers have characterized the tropical forest since prehistoric time and allowed people who were not committed farmers to live there.

This most provocative question is obviously relevant here. If the foraging exclusionists are correct, there has been no independent evolution of food produc-tion in the tropical forest because foraging never existed independently. All prehis-toric manifestations of agriculture, no matter how early, must therefore be secondary developments. We think we have already provided an answer to this argument by providing data figures on animal biomass in seasonal Neotropical forests that, when added to the plant and other calories available, appear more than sufficient to support low-density foraging. We take this opportunity to address other features of the foraging exclusion hypothesis that we find problematic.

First of all, from the previous discussion it is clear that one must be very careful to specify the type of "tropical forest" used to estimate food availability. It is estimated that less than 50% of the Neotropical forest is aseasonal or true rain forest. Nonseasonal forests may offer the poorest plant food choices for humans, particularly in terms of tuber availability, and forests on fertile soils will differ dramatically in fruit, nut, and faunal availability from forests on infertile soils.

Some advocates of foraging exclusion carefully state that their conclusions are by and large drawn from, and relevant to, the situation in primary (undisturbed) evergreen forest (Headland, 1987), although Headland also questions whether the Paraguayan Aché, who live on the drier tropical/subtropical forest boundary, were ever "pure" foragers before contact with outsiders. Others (Bailey *et al.*, 1989) seem to refer to the entire spectrum of forest types with no regard for their potential differences to support humans.

All the exclusionist advocates assume that prehistoric foragers were passive actors in their landscapes and did not, through firing and other methods, alter the densities of useful plants and animals. However, paleoecological data discussed at length in Chapter 4 show that humans were firing and making small-scale clearings in tropical forest shortly after 11,000 years ago. The creation of secondary forest through even small-scale human interference has been demonstrated to increase useful plant and animal production in all forest types.

Finally, exclusionists assume that the present day distribution of forests on the landscape is relevant to that of the past. We show later in this chapter that during the late Pleistocene and early Holocene the climate and vegetation were far different from those of the present. Although open landscapes were much more common then than now, forests were still widespread during the Pleistocene and humans could not have easily penetrated southern Central America and entered South America without living in tropical forest some of the time. In fact, more than one-fourth of the Paleoindian sites identified with the archeological record appear to have been located in areas reconstructed on the basis of paleoecological evidence as some type of forest (see Chapter 4). In short, the modern vegetation and its resources offer few clues to the lifeways of late Pleistocene and early Holocene foragers who, as will be demonstrated later, entered a "tropical" zone that few of us would recognize today and immediately began to modify it to suit their own purposes.

Hladik and Dounias (1993) quite logically drew no definite conclusions on this matter from their measurements of wild yam densities, but they note that because tubers and other nutritionally rich foods are found in most forests, and that foragers typically make good use of what they are given and exist at low densities on the landscape anyway, the entire self-sufficiency question might be spurious and not really answerable through analyses of modern resources and modern populations. Hladik *et al.* (1993, p. 128) quite perceptively add that "How well rain forests can sustain human beings depends on how well they are managed."

We wholeheartedly agree on all counts. The resolution of this issue should be sought in the archeological and paleobotanical records. With the increasing availability of these kinds of records from the humid, lowland tropics, researchers will be able to study past processes using empirical information left by past peoples and vegetation.

DRIER TROPICAL HABITATS

Thorn Woodland and Other Drier Vegetation Types

Some areas of the Neotropics possess climates with well-drained soils where rainfall amounts are between 500 and 1000 mm per annum. Here, types of more open woodlands usually occur. Where they are present, these dry woods are in contact with and usually grade into deciduous and moister forests and can be considered the end stage of the gradient from wet to moist to dry to semiarid adapted tropical woody vegetational formations. Areas holding these woodlands today include the Pacific slope of southwestern Mexico, northern Colombia, northern Venezuela, the Santa Elena Peninsula of southwestern Ecuador, northern Peru, and northeastern Brazil, where the vegetation is called "caatinga" (Sampaio, 1995) (Figs. 2.2a and 2.2b).

There is also a narrow zone of thorny scrub vegetation along the Pacific littoral of Panama, where the ocean salt dries the air and creates locally arid conditions. The western slopes of the central Andes in Peru between the coastal desert and high-elevation grasslands and deep, inter-Andean mountain valleys in rainshadows from Colombia to southern Peru also hold dry formations similar to the ones described here (Sarmiento, 1975).

Although trees and other woody plants may be conspicuous in these formations, they are distinguished structurally from tropical forest by the "absence of a continuous tree canopy at any height" (Sarmiento, 1975, p. 236). Trees are typically between 3 and 10 m tall, often belong to the legume and euphorb families (Leguminoseae and Euphorbiaceae), and commonly have thorns. Perennial grasses are usually common in the ground cover in addition to terrestrial bromeliads. The driest of these areas may have considerable growth of cacti and other arid-adapted scrubby plants. Prance (1987) considers that the arid floras of Central and South America were once connected because so many of the same genera, and in some cases species, of the depauperate floras of the regions are shared. As we discuss later, the obvious time for this relationship would have been during the Pleistocene.

In contrast to the tropical forest, thorn woodland and thorn scrub environments may offer a rich variety of wild resources for human consumption. Edible pods on legume trees that fruit en masse and require little processing, succulents such as maguey and cacti, and other plants are abundant. These plants support large populations of a variety of medium- and small-sized animals such as peccaries, deer, opposum, and rabbits. These environments also once supported large browsing and grazing herbivores (Mares, 1992).

Thorn and other drier woodlands, as well as adjacent areas of deciduous forest, are also home today to the closest wild relatives of some important crop plants, including a squash species, *Cucurbita argyrosperma,* maize, and manioc (Sauer, 1936; Rogers and Appan, 1973; see Chapter 3). However, it is unclear how much of the area described as "thorn scrub" vegetation in parts of the Balsas River Valley, where teosinte (maize's wild ancestor) was probably domesticated, would have been under a deciduous forest before being severely disrupted by humans (Miranda, 1947). Rainfall over most of the area is high enough to support a deciduous forest, and descriptions by Miranda, (1947) of arboreal associations growing on the edge of thorn scrub in currently more favorable habitats for trees are floristically the same as those in deciduous forest remnants in central Pacific Panama and Guanacaste province, Costa Rica (Hartshorn and Poveda, 1983; FAO, 1971).

Similarly, Sarmiento (1975) considers that a large part of the Brazilian caatinga may once have supported a deciduous forest before the higher arboreal component of the vegetation became impoverished by heavy human impact. In the caatinga today, there are numerous wild species of *Manihot,* which have led some researchers to conclude that manioc was originally taken under cultivation there. Northern South America (Venezuela, Guianas, and northern Brazil) is, perhaps, a more likely area of origin (see Chapter 3).

As discussed later, thorn and other dry woodlands were probably considerably more widespread during the late Pleistocene, when the climate was much cooler and drier. The moisture and vegetation gradients present today suggest that thorn woodland with cacti and scrub elements probably replaced the tropical deciduous forest that would have grown under the modern climate in major parts of the Pacific slope of western Mexico, including the Balsas River valley, Pacific-side Central America, southwestern Ecuador, certain areas of Colombia, Venezuela, and the Guianas, and the southern and eastern Amazon Basin.

We end our discussion of drier, wooded tropical vegetation types with the "cerrado" formation. Cerrado is a Brazilian word for the savanna and savanna forest vegetation that covers nearly one-fourth of the land area of Brazil (Neto *et al.,* 1994) (Fig. 2.2b). Lying directly below the southeastern limit of the humid Amazonian forest, it is a complex mixture of arboreal woodland with an open canopy and continuous grass undercover, open scrub grassland with scattered trees, and closed scrub. Small areas of forest with a closed tree canopy occur in the uplands where precipitation is higher and on patches of terrain where the soils are richer (Eiten, 1972). Precipitation over much of the cerrado area averages between 1100 and 1600 mm. Cerrado tree species have open crowns, "tortuous" trunks, and fewer branches for their size than do tropical forest trees. The plants of the understorey are xeromorphic and often include cacti.

Although the cerrado area is large, we do not consider it in any detail here. Other than wild plants related to manioc, no other potential ancestors of crop plants are known to grow in the region. The soils are acid and apparently suited to agriculture only in the gallery forests along river courses (Eiten, 1972). Prehistoric populations appear not to have evolved out of foraging subsistence modes until late in time (Schmitz, 1987). We suspect, although archeological data are not available at this time, that manioc was originally domesticated in the northern South American hearth (see Fig. 3.18). The importance of the cerrado to the subject at hand is that certain of its more open-land elements may also have expanded into the southern parts of the now-forested Amazon Basin during the late Pleistocene.

A Neotropical Desert: Coastal Peru

Although the coast of Peru falls within tropical latitudes, it contains one of the driest deserts in the world (Fig. 2.2). In most years no rain falls. The only green on an otherwise barren landscape is provided by the rivers that descend from the Andes to the east and cross the narrow coastal plain and by patches of vegetation, called lomas, supported by dense fog from the Pacific. The low abundance of wild plant and animal resources in the desert environment contrasts to the richness of marine life off the Peruvian shores. These factors led archeologist Michael Moseley to propose that the early appearance of complex societies and monumental architecture along the north and central coast was based not on agriculture but on marine resources (Mosely, 1975; 1992). The "maritime hypothesis" fueled a debate that continues today and that we discuss in Chapter 5.

Why it is so dry along the Peruvian coast, why there are so many fish and other marine resources, and what happens when warm water and moist air from the north—El Niño—intrude into the cool desert zone are well understood.

Along the Peruvian coast, winds are southerly or southeasterly all year and tend to blow parallel to the coast, in contrast to areas of southwestern Ecuador just north where the wind patterns shift seasonally (Martyn, 1992). This strong anticyclone (counterclockwise) circulation is reinforced by the height of the nearby Andes, which act as a 7000-km-long barrier between the Pacific and Atlantic air circulation systems, inhibiting exchange processes in the atmosphere (Schwabe, 1969). Moist air from the east and northeast (Amazon Basin), which produces convection and afternoon rain within the mountains, is blocked from reaching the coast by the height of the western cordillera; in essence, the entire coast of Peru is in the rainshadow of the Andes.

The prevailing southerly winds cannot bring rain to the Peruvian coast because of the permanent anticyclone circulation and the lee-side barrier (i.e., the Andes) and, very important, the suppression of updrafts caused by the temperature inversion above the cold offshore water. Water evaporated from the ocean surfaces is prevented from condensing as rain because it cannot rise. Heavy fog, known as garua, forms during the winter, however, and supports lomas vegetation.

The cold waters off the coast of Peru are known as the Peru or Humboldt current. This northward-moving current is part of the South Pacific gyre, a counterclockwise current system created by the prevailing winds and the rotation of the earth (the Coriolis force) (Levington, 1982). The Coriolis force also deflects water that is moved by wind air so that warmer surface waters near the coast deflect left and move away from shore, where they are replaced by the deep, cold-water current (upwelling).

The cold water is nutrient rich, full of nitrates and phosphates in the excreta of grazers from the surface, and it supports abundant phytoplankton that in turn, are eaten by zooplankton and fish. The dominant fish grazer in the Peruvian upwelling system is the anchoveta (*Engraulis ringens*), which is 10 times more abundant then any other grazer (Whitledge, 1978). The combination of dissolved minerals and nitrates from the excreta of zooplankton and anchovetas brought up during upwelling provides the nutrient base for one of the most productive fishing areas in the world.

Strength of upwelling varies along the coast, however. There are two areas of especially strong upwelling, 4–6°S latitude (the northern or Paita area) and 14–16°S (the southern or San Juan area) (Zuta *et al.*, 1978). Intensity of upwelling also varies throughout the year; it is strongest in May and September in the northern area and in June and August in the southern area. In both areas the cold, upwelling water occurs as tongues extending 70–130 miles from the coast or in patches 10–30 miles in diameter, with warmer water between.

Upwelling is stronger in the southern area because winds parallel a long, straight coast. In an El Niño year (see below and Chapter 5), upwelling persists in the southern area while it weakens in the others. These factors must be considered when discussing human exploitation of the Peruvian coastal habitat.

The Peru current is deflected westward just south of the equator, where it forms part of the south equatorial current (the equatorial countercurrent flows eastward on the equator). Coastal waters north of the Peru current are warmer. The warm offshore waters and the mangrove and estuarine environments support a very different marine fauna, considered part of the Panamic province (Reitz, 1994). The boundary between the warm-water and cold-water ecosystems shifts seasonally in concert with changes in wind circulation brought about by the migration of the sun. For southern Ecuador, this means warm water replaces the cold, bringing a rainy season during the January–April months.

When warm, nutrient-poor surface waters from the equatorial latitudes move south along the coasts of Ecuador and Peru, an El Niño event occurs. Because we believe that the modern El Niño probably dates no earlier than 7000–5000 B.P., we leave discussion of it for chapter 5.

With the exception of the northern extremity of the Peruvian desert, where wild relatives of cotton, jack bean, and a species of squash *Cucurbita ecuadorensis* now confined to southwest Ecuador, may have occurred during the early Holocene if the climate was substantially moister than that of the present day's (a subject discussed in Chapter 4), it is very unlikely that the zone was home to any of the wild crop plant ancestors pertinent to this book.

THE RETURN TO LABOR FROM FORAGING AND FOOD PRODUCTION IN NEOTROPICAL HABITATS

We have stated several times that we believe people in the Neotropics started to grow food because their returns to labor using methods of food procurement based exclusively on foraging declined sufficiently after the environmental changes that marked the close of the Pleistocene to have threatened their existing lifeways and demographic balance [we distinguish an altered life mode (changes in mobility and labor) and demographic balance (inability to support as many children as previously able) from severe food shortages leading to ill health or starvation]. We discuss the paleoecological evidence for this sequence of changes in the Neotropics in the next section.

If people abandoned an existence dedicated to foraging for at least part-time gardening because of the reasons we propose, then the horticulture they adapted must have resulted in higher returns per unit of labor time. Comparative data measuring the returns to labor of foraging vs farming are needed to assess the problem. It is also necessary to know the costs and benefits of exploiting various kinds of wild resources in order to model the patterns of resource change over time under environmental change that preceded food production. Here, we present the data currently available on this subject. Although more information is obviously needed, the data appear to support our contentions about the relative costs of foraging and farming in a tropical forest habitat.

Tables 2.3 and 2.4 present the returns to labor from collecting tubers and various wild resources of a tropical forest. The examples from tuber collecting in a forest are from the Old World, because the only data available for Neotropical exploitation come from the Venezuelan llanos. Also listed are tuber return rates from Australian and African desertic environments in which tubers typically occur in higher densities than in forest. Table 2.5 presents return rate data for some large and medium-sized animals from Australia and the Great Basin, North America. These provide some useful comparisons of hunting in different life zones and on animals of different sizes.

TABLE 2.3   Wild Tuber Return Rates

Wild tuber	Cal/hr/person	Remarks
Dioscorea luzonensis	484[a]	Digging and processing, not cooking; Batak, tropical
Dioscorea hispida	1739[a]	forest, Phillipines
	1125[b]	99% roots, no processing; Cuiva, Venezuelan llanos
Dioscorea	855[c]	No processing; Batek, Peninsular Malaysia
Dioscorea	2000–2500[d]	Gidjingali, coastal Australia
Dioscorea	1300[e]	Anbarra, coastal Australia; only includes collecting
Ipomoea	1700[f]	Australian desert
Vigna	1967[g]	Savanna and savanna bushland, northern Tazmania
Vigna	3240[g]	
Vigna	884[g]	
Ipomoea	3290–3759[h]	Australian desert

[a] From Eder (1978).
[b] From Hurtado and Hill (1987).
[c] From Endicott and Bellwood (1991).
[d] From Jones and Meehan (1989).
[e] From Jones (1980).
[f] From Cane (1989).
[g] From Vincent (1985).
[h] From O'Connell and Hawkes (1981).

TABLE 2.4   Returns in Calories per Hour of Various Neotropical Forest Resources after Encounter

Resource	Scientific name	Cal/hr/person
Honey	Unknown	22,411[a]
Honey	Apis melifera	20,609[b]
Deer	Mazama americana	15,398
Nine-banded armadillo	Dasypus novemcinctus	13,782; 2,662[c]
Fruit	Philodendron sellam (ripe)	10,078
White-lipped peccary	Tayassu pecari	8,755; 5,323[d]
Honey	Unknown	8,666
Coati	Nasua nasua	7,547
Fruit	Campomanesia zanthocarpa	6,417
Collared peccary	Tayassu tajacu	6,120
Paca	Cuniculus paca	4,705
Fruit	Ficus sp.	4,419
Fruit	Casimiroa sinesis	4,181
Fruit	Rheedia brasilense	3,245
Starch (palm)	Arecastrum romanzolfianum	3,219; 2,246[e]
Fruit	Chrysophyllum ganocarpum	2,884
Fruit	Annona sp.	2,835
Fruit	Philodendron sellam (unripe)	2,708
Fruit	Jacaratia sp.	2,549
Fiber and shoot (palm)	Arecastrum romanzolfianum	2,436
Growing shoot (palm)	Arecastrum	2,356; 1,584[f]
Nut (palm)	Acromia totai	2,243
Large palm larva	Calandra plamarum	2,133
Capuchin monkey	Cebus apella	1,370
Small palm larva	Rhynophorus palmarum	1,331

Note. All data and notes from Hill et al. (1987).

[a] Possible introduced species.

[b] Introduced species.

[c] First number is for animals encountered on the surface; second number is for animals dug up.

[d] First number includes time spent following tracks; second number includes only time after animal is heard or seen.

[e] Second number includes optional processing time.

[f] Men's return rate listed first.

The following general observations may be made. First, if the return rates from tuber collecting in the Old World tropical forest and Neotropical savanna can be extrapolated to the Neotropical forest (clearly, empirical data are needed on this issue), then tuber exploitation is a costly activity, ranking below hunting and other forms of plant collecting. For the Cuiva of the Venezuelan llanos, where tubers occur at greater density than in forest, men's return rates from hunting were still more than twice as high as women's rates from digging tubers (Hurtado and Hill, 1987).

That acquiring tubers is hard work is also attested to by remarks made by people under study who describe the heavy labor involved in finding, digging,

TABLE 2.5   Returns from Hunting Large Game and Other Game Animals of Non-Tropical Forest Environments

Animal	Cal/hr/person	Environment
Buffalo	200,000[a]	Coastal savannas of Australia
Wallaby	50,000[a]	Coastal savannas of Australia
Mule deer and bighorn sheep	17,971–31,450[b]	Great Basin desert
Antelope	15,725–31,450[b]	Great Basin desert
Cottontail rabbit	8,983–9,800[b]	Great Basin desert

[a] From Jones (1980).
[b] From Simms (1987).

and processing tubers (e.g., Endicott and Bellwood, 1991). Keegan (1986) uses a return rate of 5280 Cal/person/hr to estimate the cost of harvesting and cooking a hypothetical wild sweet manioc, but we believe this figure is much too high because it is based on harvesting domesticated manioc from a cultivated field and, therefore, may more accurately reflect planting returns. Tuber return rates are higher from plants of open, desertic habitats than from plants of forest habitats but are still not impressive.

Of the approximately 23 major resources in the Aché diet listed in Table 2.4, tubers would rank third or fourth from the bottom but would make the optimal diet list, assuming that return rates from tubers are generally the same as reported from modern-day hunters and gatherers in tropical forests, savanna, and coastal formations (between 484 and 2500 Cal/hr/person). In a hypothetical early Holocene forest, tubers were likely to have been regular components of the diet, although among the lowest ranked of the resource set. (Remember that resource rankings derived from the diet breadth model predict in what order resources will enter and leave the diet as conditions change but indicate little about the quantitative importance of any dietary item unless encounter rates are known.) However, in environments such as those of the late Pleistocene that were filled with resources ranked higher than the present day's, tubers may not have entered diets at all or they likely were very minor components of diets.

Given the discussion of tropical forest resources presented in this chapter and the Aché optimal diet list, tubers and palm starch are among the very few wild plant resources that combine a source of high-quality calories with acceptable energetic efficiencies, making both of them strong candidates for an early cultivation focus once food production is initiated. However, one might expect that given a choice between growing tubers and cultivating/managing palms for their starch, one would put one's money on tubers because palms would have a much longer turnaround time before harvest yields were seen (possibly explaining the inattention to palms as a source of starch today in the Neotropics). Furthermore, if some substantial part of the costs involving tubers is related to search time and processing, then planting varieties that do not need much processing might immediately and significantly raise their return rates, prompting an even greater emphasis on them.

In the tropical forest, fauna constitute many of the highest ranked food items, but a few fruits also make the "top ten" (although the quality of calories in them is unclear). Honey may have been a very attractive resource, although the return rates may be skewed because the technology used to obtain it is modern. As noted, previously, palm starch can be exploited at a fairly reasonable cost. These factors reinforce our contention that making a living purely from foraging is a viable subsistence strategy in the tropical forest. Game will be preferred because of its generally lower cost and protein content, but the plants and invertebrate products of the forest will contribute significantly with regard to energy.

The highest rates of returns typically come from larger animals. The northern Australian buffalo and wallaby yielded some of the highest return rates recorded thus far for extant resources. This is no surprise because large "packages" of food, especially meat that needs no processing, can be easily handled and turned into food. We will never derive a quantitative figure for the yield of calories/hour of such hunted and now extinct fauna as mastodons, horses, and giant ground sloths, but we can assume from studies of large animals still on the earth that they were perched at or near the top of the optimal resource set.

What about teosinte, maize's wild ancestor? Generally speaking, return rates from exploiting seeds are the lowest among all those reported to date, often being under 1000 Cal/person/hr (Cane, 1989; O'Connell and Hawkes, 1981; Simms, 1987). However, Russell (1988) found that the costs of exploiting large stands of robust and large-seeded Old World einkorn wheat, and presumably emmer wheat and barley as well, were considerably lower than those of exploiting many other wild grasses (2300–2744 Cal/person/hr). This perhaps accounts for their participation in the earliest systems of food production in the Near East and development into domesticated strains.

The only productivity data available for Balsas teosinte are those of Flannery and Ford (Flannery, 1973). Their teosinte harvest yielded 152.5 kg per hectare after deductions from the total yield to account for inedible roughage of the fruit cases. Data on the overall returns expected per person per hour on such a harvest are lacking, so it is not possible to evaluate the energetics of exploiting teosinte vis-à-vis other resources. Given that teosinte occurs in dense stands and that its abundance increases after human disturbance (Flannery, 1973), it seems possible that exploiting it during the early Holocene may have yielded higher returns than exploiting other grasses and plants of the forested Balsas watershed.

Flannery's Oaxaca data (discussed later), indicate that collecting teosinte is more costly than collecting the resources of thorny scrub vegetation, and there is little doubt that teosinte is dramatically more expensive than hunting large animals. Teosinte may be another important crop plant ancestor that was not or was only infrequently incorporated into the diets of late-glacial hunters and gatherers but that entered the diets of early Holocene foragers.

How do the costs of foraging compare to those of farming? Table 2.6 provides examples of comparisons among total energy returns from tropical foraging and

TABLE 2.6   Return Rates of Alternative Subsistence Strategies in the Neotropics

Subsistence strategy	Cal/hr/person[a]	Protein capture (g/hr)[a]
Machiguenga gardens (food production)	3842	45
Machiguenga forest (hunting and gathering)	116	7.3
Machiguenga fishing	214	38
		Cal/kg[b]
Machiguenga gardens		80
Machiguenga fishing		740
Machiguenga hunting and gathering		1150

Aché mean foraging returns[c]	
	Cal/hr/person
Men	1253
Women	1087

Cuiva foraging returns[d]		
	Cal/hr/person	Remarks
Men	3001	Mostly hunted game; handling time not included. Processing not included
Women	1125	99% Roots

[a] All numbers from Keegan (1986); based on data provided by Johnson and Behrens (1982) and Johnson (1983).
[b] Energy expended by the household/year, obtained by dividing total energy input by food input. From Johnson and Behrens (1982).
[c] From Hill et al. (1987).
[d] From Hurtado and Hill (1987).

farming that are currently available. It can be seen that for the Machiguenga of southern Peru, both the calorie and protein returns from horticulture far exceed those from either hunting and gathering in the forest or fishing. We note that much of the garden protein is accounted for by the maize that the Machiguenga plant, and we reach the surprising conclusion that maize alone is a more efficient source of protein than the hunted game in the diet [although maize cannot provide the essential amino acids and lipids that meat does (Hill et al., 1987)].

Hames (1990) states that Yanomamö horticulture is five or six times more efficient than hunting and adds that similar trends probably hold for many other Amazonian groups. Although available data are not quantified in net caloric or protein returns, the Aché also have higher subsistence returns per unit effort when they farm than when they forage (Hawkes et al., 1987). Again, differences could be substantial. Aché men spent 6.5 hr/day acquiring and processing food in the forest and only 3 hr/day when farming. Women spent 3.75 hr/day foraging and 2.7 hr/day farming. It is also significant that Aché and Cuiva foraging return rates were lower than returns from Machiguenga farming because these foraging returns are among the highest reported for hunters and gatherers.

Although data are still few, they suggest that returns for tropical horticulture are likely to be greater than returns from hunting and gathering in the forest. This clearly is not in accordance with the long-held notion that the shift to food production carried with it declining returns to labor and was initiated only with great reluctance as a response to a deteriorating food base or real food shortages (Sahlins, 1972; Cohen, 1977a; Harris, 1977a; Hayden, 1995).

We propose that people were "pulled" and not "pushed" (Stark, 1986) into food production because during the early Holocene the costs of foraging became too high relative to costs of previous foraging strategies. In the Neotropics, the post-Pleistocene food base was still capable of providing good nutrition to small and mobile groups of foragers, but people had to work harder than before to produce an equivalent amount of food. Because less food often means fewer grandchildren (Hawkes, 1987), and because the incorporation into the diet of low-ranked and especially low-density foods may also lower forager population densities (Winterhalder and Goland, 1993), these significant declines in return rates would eventually lead to a demographic decline.

Furthermore, there was no guarantee that working harder would raise the overall return of food to acceptable levels again, particularly where resources were not abundant to begin with. Combining foraging theory and population biology (prey response to exploitation) models, Winterhalder (1993; et al., 1988) found that working long hours quickly led to declining prey availability and lower human fitness and population size for foragers. Apparently, resource intensification of this type is an option largely open only to populations that are cultivating because, although their per capita return to labor declines, the total yield increases and more people can be supported. This is often not possible under a foraging existence because the total yield of wild resources will be quickly depleted, leading to population reduction. Winterhalder's studies elegantly explained why hunter and gatherer work effort is often "modest" and why they have limited material trappings without resorting to illusions that they have limited "wants" (Sahlins, 1972).

The particular ecology of the habitat that foragers live in will also influence how many hours they can profitably spend acquiring food. For example, in a tropical forest the "safe area" for small children may be no larger than the camp, and women's time allocation to food procurement may be constrained by the quality of child care they can provide when foraging (Hurtado et al., 1985; Hawkes, 1987). Longer hours in the food chase may also expose foragers to fitness-decreasing factors such as predation and hunting accidents.

The obvious alternative to bail a population out of this predicament is food production. Planting even small gardens would at once lead to more favorable rates of returns and would significantly increase the total yield of food. Decreased mobility would become possible. Moran (1983) reminds us that high mobility is associated with negative factors in the tropics, and studies of the foraging Ache (Hawkes *et al.,* 1987) make clear the difficult work demands that women face in moving camps every day and caring for children. Increased food yields from house gardens would allow populations to grow substantially, and the processes leading to the fixation of food-producing behavior and intensification of this behavior would advance.

This now brings us to the question of the costs of foraging before the onset of the Holocene which, if we are correct, must have been relatively low compared to foraging costs in Holocene forests. There are very few data relevant to the costs of exploiting tropical low woodland/thorny scrub/savanna habitats. These habitats, or something like them, were the kinds occupied by some late Pleistocene occupants of tropical America who were to engage in early food production. These habitats were replaced by forests at the beginning of the Holocene. In order to provide some information on this question we rely on (i) Flannery's (1986b) density and yield data for the resources in the thorn woodland and scrub zones of the Oaxaca Valley, which probably make an acceptable analog for some of the late Pleistocene environments in question; (ii) return rates for Neotropical foragers exploiting savanna habitats today; and (iii) return rates published for wild fauna occupying the desertic habitats of the Great Basin (Simms, 1987) that are known to have been present and hunted in tropical America during the late Pleistocene (Table 2.5).

Flannery's (1986b) estimates of densities for white-tailed deer and rabbits in Oaxaca and other similar (and degraded) environments today (12 and 320/km², respectively) are much higher than deer and rabbit densities reported from the forests in Table 2.2. The return rates for mule deer and bighorn sheep, similar in size to white-tailed deer, and cottontail rabbits in the Great Basin are, respectively, between 18,000 and 31,000 and 9000 Cal/hr. The return rate for deer and sheep is far higher than any scored for tropical forest fauna, whereas the return rate for rabbits would place them near the top of the optimal diet list for tropical forest resources (Table 2.4).

Similarly, the foraging return rate in the Venezuelan llanos of predominantly hunted products, 3000 Cal/person/hr, is one of the highest reported for foragers anywhere (Hurtado and Hill, 1987). This is probably due largely to the fact that large and medium-sized game items tend to be found clustered at high density around the margins of rivers and ponds.

Wild plant resources of at least certain types of thorny woodland and scrub are also far more productive than those of forest because they are clumped, are good sources of starch and protein because many taxa are legumes, and often do not require extensive processing. We repeat Flannery's (1986b) estimates that several hundred individuals per hectare of mesquite, other legumes, and cacti may be found in the dry vegetation facies of the Oaxaca Valley. It is likely that foraging return rates in such zones would be much higher than those obtained from foraging in forests.

To study this question more rigorously will clearly require input–output analyses of varied types of woody, dry growth as well as additional studies of wild and cultivated tropical forest resources. At the very least, it appears that significant shifts in cost/benefit ratios occur as a tropical forager moves across habitat and food procurement boundaries. How tropical foragers of the past were faced with such shifts and when, why, and to what extent they occurred is the subject of the next section.

THE LATE PLEISTOCENE AND EARLY HOLOCENE NEOTROPICAL WORLD

General Considerations

After a lengthy period when discussions of the natural environment in relation to early food production were viewed with skepticism (Braidwood, 1951; Flannery, 1986a) or, worse, as irrelevant "environmental determinism" (Wagner, 1977), serious considerations of the relationship between environmental change and the beginnings of food production are becoming more common in the anthropological literature (e.g., Henry, 1989; McCorriston and Hole, 1991; Moore and Hillman, 1992; Bar-Yosef and Belfer-Cohen, 1992; Wright, 1993). We believe that they are vital to the question of agricultural origins for a number of reasons.

The environment in large part determines the type, quality, and abundance of wild food resources and the distribution of plants that potentially can be brought under cultivation and domesticated. Dramatic oscillations of climate and vegetation may result in changes in resource density and distribution and, as is clear from the preceding discussion, necessitate a series of new options for humans with regard to which and how many plants and animals are available to them and how to procure them.

During the past 2.5 million years, in the geological epoch known as the Pleistocene, the earth's climate and land surface have been dominated by the ice ages, those cold times when vast ice sheets moved north and south from the poles. The last great advance of glaciers is considered to have started about 120,000 years ago and to have reached its maximum extent 18,000 radiocarbon years ago, a time called the "Last Glacial Maximum" (LGM). The glaciers then started their final retreat about 14,000 years ago, with the glacial climatic effects ending by 11,000 or 10,000 years ago in all areas of the world studied.

The timing and vegetational shifts associated with the major cycles of the last ice age are crucial in assessing early resource use and resource adjustments in the lowland Neotropics. If one accepts a pre-Clovis (pre-11,000 B.P.) entry of people into tropical America, it probably occurred sometime during the full glacial conditions between 20,000 and 11,000 years ago, and a period of only several thousand years elapsed before rapid environmental change ensued. If the primacy of the

Clovis horizon is accepted, entry was still, but just barely, at the full glacial climes and a period of only a few hundred years elapsed before the severe climatic changes leading to the termination of the Pleistocene. Either way, the first humans did not enjoy their newfound flora and fauna for very long before adjustments to profoundly changing conditions were required.

For a long time it was thought that the tropical lands were immune to the effects of the ice ages. After all, because glacial advance from the poles was limited to temperate regions, such as Europe, North America, and China, environmental responses at low latitudes must have been minimal, being limited to the localized effects of expansion of montane glaciers in the Andean region and lowered tree lines. The argument of environmental stasis went hand in hand with the condition of high species diversity in the tropics, the theory being that unchanging climate over long geologic time resulted in a lack of physical stress on the biota, led to low rates of extinction, and allowed the continuous accumulation of species (e.g., Fischer, 1960; Slobodkin and Sanders, 1969).

During the 1960s a few pioneering investigators demonstrated that the tropical highland regions of America and Africa had experienced dramatic climatic and vegetational change apparently synchronous with Northern Hemispheric glaciations (Van der Hammen and Gonzalez, 1960; Martin, 1964; Livingstone, 1975). Temperatures were reduced on the order of 9°C, resulting in large-scale downslope movement of vegetation, and precipitation was lower as well. These studies forced a revision of the environmental stasis hypothesis, at least for the highlands, and spurred other investigators to study the environmental histories of the lowland tropical regions.

Although data points are still relatively few compared to those for the temperate zone and some other regions, there is now considerable evidence from paleoecological research that the whole of the Neotropical world was undergoing dramatic climatic and vegetational change during the last stages of the Pleistocene and on the eve of food production between 12,000 and 10,000 years ago.

The Ice Ages in the Neotropics

The Neotropical world during the final 12,000 years of the last glacial cycle (between 22,000 and 10,000 B.P.) was a much cooler and drier place than it is today. Lowland temperatures were depressed by approximately 5–7°C, and precipitation was generally reduced by at least 25–40%. These changes resulted, depending on the area and elevation considered, in (i) an 800- to 1200-m downslope expansion of forest generally limited today to cool and high mountainous areas above 1500 m; (ii) partial replacement and reduction of lowland evergreen rain forest by montane and/or drier lowland forest elements; and (iii) replacement of some of the seasonal tropical forest by types of open vegetation similar to, but not exactly like, today's thorn woodland, thorn scrub, and savanna growth (e.g., Binford *et al.,* 1981; Absy *et al.,* 1991; Markgraf, 1993; Leyden, 1984; Leyden *et al.,* 1994; Piperno *et al.,* 1991a; Schubert, 1988; Bush and Colinvaux, 1990; Bush *et al.,* 1992; Van der Hammen and Absy, 1994; Colinvaux *et al.,* 1996a,b; Thompson *et al.,* 1995).

Figures 2.4a and 2.4b present the reconstructed vegetation of the American lowland tropics between 20,000 and 10,500 years ago based on paleoecological sequences referenced previously and discussed throughout this section. The number of data points available for evaluating the late-glacial vegetation is still small but growing, and it now includes direct evidence from the interior of the Amazon Basin. Because the main determinants of vegetation today are temperature and rainfall, and because the glacial climatic conditions discussed previously and later in the chapter appear to have been pantropical, we believe that the vegetation map accurately presents the broad outlines of the glacial environment. Regions particularly in need of more study are discussed below.

There was initially considerable skepticism about the magnitude of glacial period environmental change in the lowland tropics that was evidenced from the first studies. However, the conditions creating the contrasts between glacial and interglacial periods are much better understood than they were 20 years ago, and they clearly predict the dramatically altered conditions seen in tropical paleoecological records.

Variations in the earth's orbital geometry, called the "Milankovitch" cycles after their discoverer, result in changes in distance between the sun and earth and in the tilt and orientation of the earth's spin axis. These are considered to be major drivers of glacial cycles because they alter the amount and seasonal distribution of solar radiation received by the earth, which translates into varying degrees of heat reception by the earth's atmosphere and oscillations of surface temperatures through time (Imbrie and Imbrie, 1979).

Other related and powerful forces at the earth's surface involved in triggering glacial cycles have recently come to light. They involve massive reorganizations between the atmosphere and the ocean whereby the oceans, the source of most of the preindustrial CO_2 on the planet, absorb CO_2 from the atmosphere instead of sustaining atmospheric levels as they currently do. In order for this to happen, something called the North Atlantic Conveyor Belt, which today drives a powerful current deep in the ocean and brings CO_2 from dead organisms to the surface where it is released into the air, probably stopped functioning during glacial cycles (Broecker and Denton, 1990).

This amounted to a reduction in the greenhouse capacity, meaning far lower levels of atmospheric CO_2, air capable of holding much less heat, and lower sea surface and land temperatures—in short, ice ages (Broecker and Denton, 1990; Guilderson *et al.* 1994). It is not yet understood whose hand is on the oceanic conveyor belt and CO_2 switches, but it is probably related in some way to changes in atmospheric circulation brought on by the Milankovitch changes in seasonal radiation.

Although these changes in the deep water circulation of the oceans do not explain all the climatic differences becoming apparent between the glacial and interglacial earth, it does seem that interactions between the oceans and the atmosphere may hold the key to the onset and termination of the ice ages. Because these interactions were global in extent, it comes as no surprise that the tropical land areas should have also participated in the phenonomena. Much light has also been shed in recent years on exactly how and why the tropical environmental conditions were so profoundly different from the modern day's. The milestone CLIMAP (Climate, Long-Range Investigation, Mapping, and Prediction) project numerically modeled the general circulation pattern of the glacial atmosphere. Using the remains of small marine organisms, called foraminifera, retrieved from ocean mud, and assuming that their current climatic tolerances were the same as they were in the past, CLIMAP predicted very little lowering of ocean surface temperatures at low latitudes (CLIMAP project members, 1976, 1981). For many years this view colored interpretations of the magnitude of change over land, because today tropical land surface and sea surface temperatures (SST) are in equilibrium, and it is highly unlikely that different conditions existed in the past.

Hence, as more data began to emerge from the lowland tropics indicating cooling much more significant than that over the oceans, a problematic contradiction became apparent between terrestrial and oceanic data sets (e.g., Rind and Peteet, 1985; Colinvaux, 1987). Fortunately, recent studies of fossil corals, which apparently are more sensitive to temperature changes than are the smaller organisms used by CLIMAP, are revising the CLIMAP interpretation and indicating a 4 or 5°C lowering of tropical SSTs (Guilderson et al., 1994; Emiliani, 1992; Emiliani and Erickson, 1991). These results are more in line with reconstructions based on terrestrial paleoecological data and the contradiction has apparently been resolved. Profound changes in the ocean deep-water circulation that brought less CO_2 to the surface resulted in much cooler seawater and land surfaces at low latitudes.

Tropical sea surface temperatures also play major roles in the amount of rain received by adjacent land areas. Today, a substantial part of the yearly precipitation that falls over tropical land surfaces comes from water that was initially evaporated over warm oceans and then transported to land by prevailing winds. Significant reduction in tropical SSTs would lead to a decrease in available moisture from the oceans to feed land surfaces as precipitation, and it may have been a primary cause of the changes in tropical rainfall patterns. Other investigators have also proposed that the ITCZ was located in a more southerly position than it is today, which would particularly lower Northern Hemisphere precipitation (Hodell et al., 1991) but would not explain why many parts of northern South America were dry. Markgraf (1993) believes that the subtropical high-pressure systems, which are zones of descending dry air, were weakened but perhaps shifted closer to the equator.

Estimates of temperature depression in the tropical lands based on paleoecological data from terrestrial sites generally range between 5 and 7°C between 20,000 and 11,000 years ago (e.g., Behling, 1996; Bush and Colinvaux, 1990; Bush et al.,

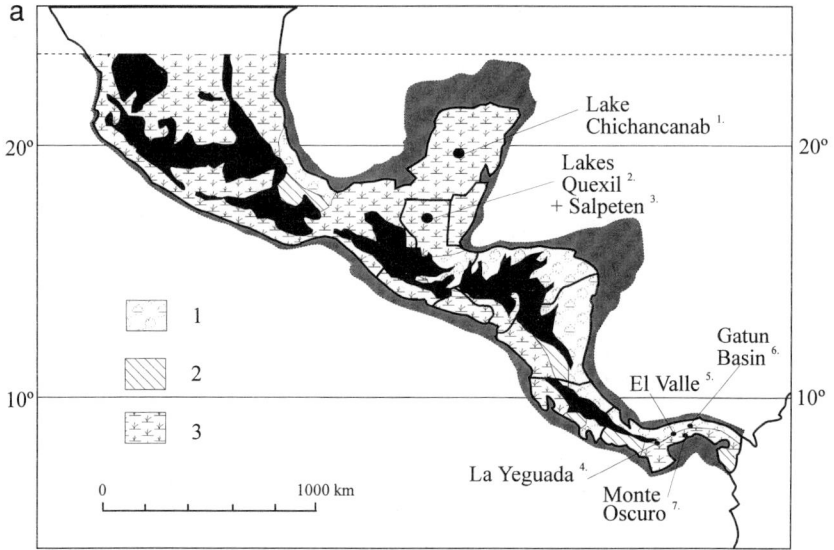

FIGURE 2.4   (a) Reconstructed vegetation of lowland tropical middle and Central America between 20,000 and ca. 10,500 B.P. with location of paleoecological sites. 1, Hodell *et al.* (1991); 2, Leyden *et al.* (1993, 1994); 3, Leyden (1987); 4, Piperno *et al.* (1990) and Bush *et al.* (1992); 5, Bush and Colinvaux (1990) and Piperno *et al.* (1991a); 6, Bartlett and Barghoorn (1973) and Piperno *et al.* (1992); 7, Piperno (1995b). (b) Reconstructed vegetation of lowland tropical South America between 20,000 and ca. 10,500 B.P. with location of paleoecological sites. 8, Binford *et al.* (1981) and Leyden (1985); 9, Colinvaux *et al.* (1996a); 10, Van der Hammen (1974); 11, Wijmstra and Van der Hammen (1966); 12, Van der Hammen *et al.* (1991); 13, Liu and Colinvaux (1985); 14, Bush *et al.* (1990); 15, Haberle (1997) and Piperno (1997b); 16, Behling (1996); 17, Absy *et al.* (1991); 18, Vincentini (1993); 19, De Oliveira (1992); 20, Ledru (1993); 21, Van der Hammen and Absy (1994). (1) Largely unbroken moist forest, often with a mixture of current high-elevation and lowland forest elements. In some areas, montane forest elements (e.g., *Podocarpus, Quercus, Alnus,* and *Ilex*) are conspicuous. Annual precipitation is lower than it is today but sufficient precipitation exists to support a forest. (2) Forest containing drier elements than characteristic today. High-elevation forest elements occur, especially in moister areas of the zone. Areas near the 2000-mm precipitation isohyet and areas with sandy soils may contain savanna woodland. The vegetation may be patchy. (3) Mostly undifferentiated thorn woodland, low scrub, and wooded savanna vegetation. Some regions (e.g., Guatemala) have temperate elements (e.g., *Juniperus*). Areas receiving greater than 2000 mm of rainfall today may still support a drier forest, as in No. 2. River- and streamside locations support a forest. (4) Probably substantially drier vegetational formations than No. 5, with fewer trees and more open-land cerrado and caatinga taxa. Paleoecological data are lacking for the zone. (5) Fairly open and humid forest containing many current high-elevation taxa (e.g., *Ilex, Podocarpus, Rapanea,* and *Symplocus*) combined with elements of the modern semi-evergreen forest and cerrado. Northward shifts in the southern polar fronts and other factors ameliorate the general precipitation reduction experienced elsewhere. The modern, seasonal forest–cerrado vegetational formations of the region are not present until approximately 10,000 B.P. (6) Desert/cactus scrub.

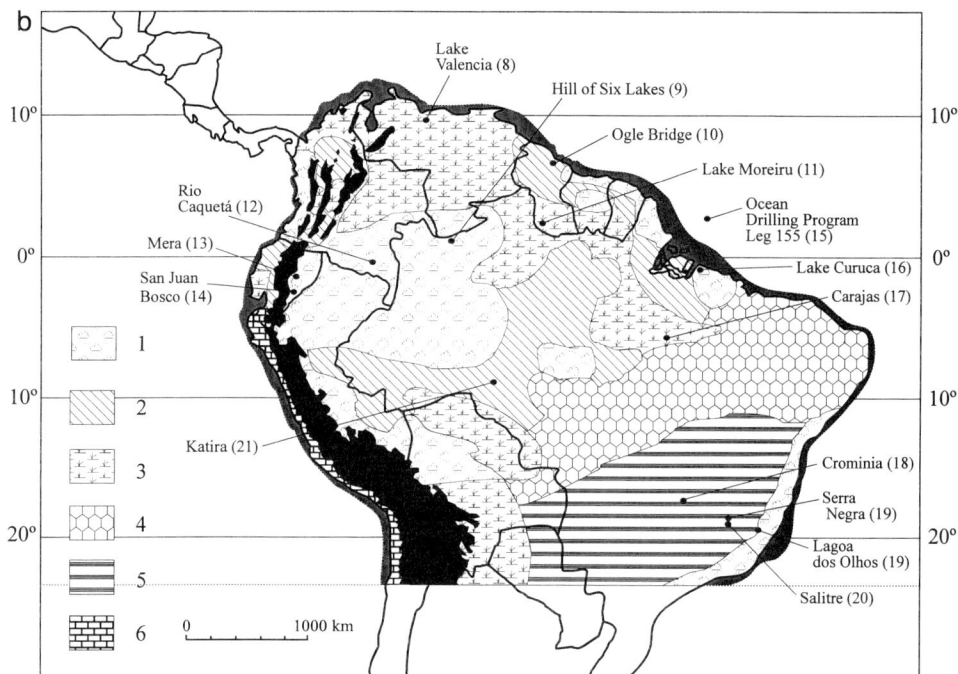

FIGURE 2.4 (*Continued*)

1990, 1992; Liu and Colinvaux, 1985; Leyden *et al.,* 1994; Markgraf, 1993; Van der Hammen and Absy, 1994). A major consequence of reduced temperatures was that the kinds of forest found in mountains at elevations between 1500 and 2300 m, called montane formations, moved downward approximately 1000 m and partially replaced some of the lowland forests in areas with annual rainfall greater than 2000–2500 mm. The result was what have been called "disharmonious" biotic communities—mixtures of plants and animals that today cannot be found together. The term is catchy but misleading because such associations were probably together for the 80% of the time during the past 2.5 million years when glaciers were advancing and because the modern vegetation can be considered a short-term, interglacial aberration.

Although moisture patterns may have been subject to more regional differentiation, with somewhat less dry climates prevailing nearer subtropical latitudes, especially in the Southern Hemisphere, a substantial reduction in precipitation is evidenced at many highland and lowland tropical sites from Mexico to Brazil (Bradbury, 1989; Binford *et al.,* 1981; Bush and Colinvaux, 1990; Leyden, 1984, 1985; De Oliveira, 1992; Ledru, 1993; Ledru *et al.,* 1996; Leyden *et al.,* 1993, 1994; Markgraf, 1993; Piperno, 1995b; Piperno *et al.,* 1990, 1991a; Van der Hammen, 1974; Van der Hammen and Absy, 1994; Watts and Bradbury, 1982; Wijmstra and Van der Hammen, 1966).

Like that for temperature reduction, evidence on the matter is strong and concordant and comes by way of demonstrated vegetational change at the onset of the Holocene from dry plant associations to mesic ones and/or variation in a lake's pattern of sedimentation indicative of times when the sediments were under much less standing water. The major consequence of precipitation reduction was to turn some areas of seasonal forest receiving less than approximately 2000–2500 mm of rain per year into open thorn woodland, thorn scrub, and savanna woodland vegetation.

Estimates of the amount of precipitation reduction have ranged from 25 to 50%. Leyden (1984), considering the nature of the vegetation change recorded in Venezuela and Guatemala, believes that reduction may have been on the order of approximately 50%. We believe that, on the basis of the presence of late-glacial thorn woodlands and savannas on the Panamanian Pacific coastal plain, precipitation was reduced by at least 35% in lower Central America, whereas Van der Hammen and Absy (1994) use an estimate of between 25 and 40% for Amazonia, which we discuss further later in the chapter.

Colinvaux (1993) believes that precipitation reduction was more modest, being on the order of no more than 10–20%. However, his estimate is based on data derived from the CLIMAP modeling of sea surface temperatures that have been shown to be too low by at least 50%, and the data fail to account for the drastic vegetational changes observed in many lowland areas. A reduction of sea surface temperature on the order of 5°C, which now appears likely, would almost certainly result in a larger precipitation decrease than that predicted by CLIMAP.

In the lowlands of Guatemala, Haiti, northern Venezuela, Guyana, Pacific-side Panama, and eastern and southern Brazil, where the modern potential vegetation is deciduous or semi-evergreen forest and the annual rainfall does not exceed 2000–2500 m, the drier climate resulted in an apparent replacement of much of the forest by types of open vegetation dominated by low woodland and scrub, cacti, and grasses, whose resulting associations may not have present-day analogs (e.g., Absy *et al.*, 1991; Binford *et al.*, 1981; Hodell *et al.*, 1991; Leyden *et al.*, 1993; Leyden, 1984; Van der Hammen, 1974; Van der Hammen and Absy, 1994; Piperno, 1995b). It follows that other areas with similar, highly seasonal climates and vegetation today, such as the Balsas River Valley of Mexico, southwestern Ecuador, and northwestern Costa Rica, also had more open types of vegetation, although confirmation with empirical data is needed (Figs 2.4a and 2.4b).

At several sites, sediment hiatuses during the late Pleistocene are recorded; these are situations in which sediment buildup did not occur at the bottoms of lakes. This almost certainly occurred because the climate was too dry to allow the lake to hold water and associated aquatic life, of which ancient lake mud is partially derived, and the absence of water permitted erosion of any sediment to occur. Such was the case in Pacific Panama (Piperno, 1995b), eastern Amazonia (Absy *et al.*, 1991), and southern Brazil (De Oliveira, 1992; Ledru, 1993).

The Pleistocene vegetation of the Amazon Basin has been the subject of considerable debate and speculation. We must remember that the basin is an area approximately the size of the continental United States, that it may have been subjected to regional patterns of change, and that paleoecological data are still few and scattered. Nevertheless, data are beginning to emerge that strongly suggest that the forests of the basin were changed considerably by the glacial climate but in ways unlike those suggested by advocates of "refugial theory" (Prance, 1982; Whitmore and Prance, 1987).

Discussion of the Pleistocene in Amazonia is now centered on whether cooling (Colinvaux, 1987; 1993) or cooling with drying (Van der Hammen and Absy, 1994) were the forcers of vegetational change. The bulk of the evidence suggests that the latter scenario is correct, and that precipitation may have been reduced by approximately 30–35%, somewhat lower than values estimated for other regions (Van de Hammen and Absy, 1994; Haberle, 1997; Piperno, 1997b). Such a reduction would have converted some of the seasonal forest currently receiving less than 2000–2500 mm of rain annually into more open terrain and low, scrubby forest, especially because the soils in question are sandy and might yield their trees to somewhat higher levels of precipitation. However, this scenario still leaves an extensive and largely unfragmented block of forest in the central and western parts of the basin (Fig. 2.4b).

Colinvaux's team (Colinvaux et al., 1996a,b) has demonstrated that before approximately 11,000–10,000 years ago the vegetation of areas receiving between 3 and 5 m of annual precipitation in the northern and western parts of the Amazon Basin was forest whose associations comprised many trees now largely restricted to mountain forests at elevations above 1100 m, such as Podocarpus, Ilex, and Hedyosmum. Many lowland forest taxa still found near the sites today apparently persisted in these forests during glacial time. Podocarpus pollen has also been found in late Pleistocene levels from Lake Curuca near the mouth of the Amazon and was recently described by Behling (1996). Thus, cooling appears to have been the principal forcer of vegetational change in these areas of the basin.

It is important to note that sites investigated by Colinvaux's team (Colinvaux et al., 1996a,b) that were shown to persist in forest throughout glacial times are in some of the wettest regions of the Amazon today. They would not be expected to become open terrain even under a 50% reduction of rainfall, which most investigators agree is an improbably high estimate for the Amazon.

Few of the paleobotanical information derived from Amazonia can be reconciled with refugial theory, an elegant and influential hypothesis that sought to explain the distribution of the Pleistocene forest in Amazonia and other areas but that was built largely on modern patterns of endemism primarily among some species of passerine birds, lizards, and Heliconius butterflies before actual paleoecological data began to be retrieved (e.g., Haffer, 1969; 1974; Prance, 1982). Basically, refugial theory holds that during the Pleistocene wetter, elevated areas of tropical America where rainfall was maintained by orographic mechanisms were the habitats for rain forest species, which were ousted from lowland areas because of arid conditions (e.g., Prance, 1982; Whitmore and Prance, 1987). The nonforested lowland zones were postulated to have been savanna. Whitmore and Prance modified this reconstruction, hypothesizing that vegetation of these areas was probably some kind of transitional, very dry forest.

Refugial theory has been used to model cultural movement in South America during the Pleistocene and Holocene (Meggers, 1982, 1987), and it has been invoked as support for the foraging exclusionary hypothesis because the areal extent of forests was supposedly too limited in glacial time to offer its resources to humans. For example, Bailey *et al.* (1989) predict that Turrialba, a Costa Rican Paleoindian site now underneath evergreen forest at an elevation of 900 m above sea level (a.s.l.), was in an open environment at the time of occupation. This reconstruction is an aberration of even refugial theory because the upland location of Turrialba and the region's high rainfall would have made it a likely location for a refugium. The influence of refugial theory has extended far beyond Amazonia. Vaughn *et al.* (1985, p. 83) stated the formerly commonly held assumption that "geologic data prove tropical Pleistocene climates to have been too arid, during glacial ages, to permit the existence of rain forests."

The generation of empirical data from paleoecology has resulted in a much clearer understanding of the distribution and composition of forests during the late Pleistocene. Refugial theory does not appear to work because cooling was a principal forcer of vegetational change, and the hilltops that captured orographic rainfall (the refugia) supported novel tree associations in which some lowland trees could not compete (Colinvaux, 1993; Colinvaux *et al.*, 1996a,b; Piperno, 1997b). In some areas, including the Amazon Basin, many lowland taxa were apparently capable of surviving even the coldest periods and coexisted with certain highland elements to form forests comprised of novel species associations. This latter finding makes sense if one remembers that lowland forest taxa have spent 80% of the past 2 million years under colder and drier conditions.

Also, the Amazonian Pleistocene climate, although drier, cannot be called arid. Reduction and fragmentation of the terra firme forest on the scale proposed by refugial theory and its prediction of extreme glacial aridity in the Amazonian lowlands have not been supported by terrestrial paleobotanical data (e.g., Colinvaux *et al.*, 1996a,b). An examination of pollen and phytoliths in deep sea cores from near the mouth of the Amazon River, in which the "catchment area" consists of all the watersheds of the basin, showed no large increase in grass frequencies during the LGM consistent with a large-scale reduction of the terra firme forest in the central part of the basin, as required by refugial theory (Haberle, 1997; Piperno, 1997b).

It should also be remembered that the basin is an area of continental proportions and that many parts of it are far from the sea. Unlike other neotropical regions, the Amazonian forest internally recycles more than 50% of its own precipitation today (Salati and Vose, 1984) and may be said to partially control its own climate.

It is possible that factors such as these, in addition to an equatorward shift of mid-latitude cloud cover (Guilderson *et al.*, 1994) and a northern shift of the southern polar front (De Oliveira, 1992), mitigated the moisture loss from lower sea surface temperatures in the central and western portions of the Amazon Basin.

In conclusion, it appears that the Amazon Basin held the greatest continuous stretch of forest during the Pleistocene, as it does today. Open types of vegetation may have been found primarily in the broad, drier corridor in the east that runs northwest to southeast, crossing the Amazon River between Obidos-Santarem and the mouth of the river Xingú, and in other more seasonal parts of Amazonia, especially those fringing the present-day cerrados of central Brazil (Fig. 2.4b). The stretch of drier and open land between Obidos-Santarem and the Xingú mouth would have been a convenient route for Paleoindians to enter the interior of the Amazon region because it connected to the open vegetation of Venezuela and Colombia to the north and northwest. That an early archeological site is located near this route (Roosevelt *et al.*, 1996; see chapter 4) may not be a coincidence. We repeat, however, that Amazonia is so huge that much more data are needed to enable this and other questions concerning environments and early human occupation to be addressed with confidence.

We have been hesitant to use the word savanna in describing all the open types of late-glacial vegetation. Palynological and phytolith data indicate that in some areas these were not simply tracts of grasses and sedges dotted by low trees and shrubs as are the savannas of today (Sarmiento and Monasterio, 1975) but rather unusual combinations of tropical thorn-scrub, temperate shrub, and herbacious plants not seen today (e.g., Piperno *et al.*, 1991a; Hodell *et al.*, 1991; Leyden *et al.*, 1993).

In some areas, the closest modern representatives of these formations may be the thorn woodland and scrub associations seen along the very dry Pacific littoral of lower Central America and in dry zones of northern South America. These associations contain many useful plants such as mesquite, acacia, and prickly pear and other cacti. Sarmiento and Monasterio (1975) and Prance (1987) have commented that the presence of many of the same genera of thorn–scrub vegetation in three now floristically separated areas (Panama, Colombia/Venezuela, and Brazil) argues for a formerly more widespread growth and connection of communities. Paleoecological data tend to be consonant with this assessment.

Thus, it seems that in some areas precipitation and temperature may have fallen beneath minimum levels required to support continuous grass cover that today are approximately 1000 mm of rain per annum and mean air temperatures of approximately 24°C (Nix, 1983; Cole, 1986). In other areas, vegetation with extensive grass ground cover may have existed where there was well-drained soil (Van der Hammen and Absy, 1994; Wijmstra and Van der Hammen, 1966).

In wetter areas of the lowland tropics where evergreen forest is the modern potential vegetation, it seems that temperature, not precipitation, was the primary climatic forcer of the late-glacial vegetation, as discussed for the Amazon. For example, sites from currently very wet areas of Pacific watershed Panama (annual precipitation 3 m or higher) located at elevations between 500 and 650 m show a water table reduction and changes in sediment chemistry between 20,000 and 10,700 B.P. attributable to precipitation decrease. However, the vegetation during this time was a mixed montane–lowland forest (Piperno *et al.*, 1991a; Bush and Colinvaux, 1990).

This scenario makes sense because today precipitation generally must fall below approximately 1500 mm per annum to preclude the growth of tropical forest and support open types of plant associations (Nix, 1983; Cole, 1986). In areas receiving more than 3 m of rain, even an extreme 50% reduction in precipitation would still have brought sufficient moisture to support a tropical forest, especially given the magnitude of temperature reduction.

Implications of Late Pleistocene Vegetation for Human Exploitation

The data in Figures 2.4a and 2.4b suggest that considerable areas of the Neotropical lowlands were covered by open, thorny woodland/scrub and/or grassy vegetation, and that considerable areas supported some kind of tropical forest despite the overall precipitation decrease. In the extensive open vegetational communities that occupied the late-glacial Central American and northern South American lowlands and a smaller proportion of the Amazonian landscape, plants such as maguey (*Agave*) and the cacti *Lamaireocereus* and *Opuntia* (prickly pear) probably occurred in association with stands of *Prosopis juliflora* (mesquite) and other legumes and useful taxa, as they do today. Such plants offerred a considerable high-quality, low-cost edible biomass, especially compared to the tropical forest flora, and were probably exploited by early human populations.

One can still find the thorny scrub plants in some number today along the dry Pacific littoral of Panama and similar areas of lowland South America, where they are surviving in their Holocene "refugia" in the expectation that this interglacial period will be another short-term event (which is not likely, given the rate at which fossil fuels are being injected into the atmosphere) and that conditions will soon become more favorable.

Regardless of their actual floristic compositions, it can be predicted that areas not covered by tropical forest were homes to big, herbivorous game animals roaming in some abundance. During the late Pleistocene, the Neotropical world supported more than 15 genera of large herbivores that disappeared from the landscape by 10,000 B.P. (Janzen and Martin, 1982; Webb, 1997). The largest animals that disappeared include mammoths (*Mammuthus*), mastodont-like gomphotheres (*Cuvieronius* and *Haplomastodon*), giant ground sloths (*Eremotherium, Megatherium,* and the Mylodontidae), and giant capybaras (*Neochoerus*).

Also present on the landscape before the advent of the modern climate were an array (more than 40 genera) of medium- and large-sized grazers, browsers, and omnivores such as horses, giant peccaries, and armadillo-like beasts. Today, only a single species, the tapir, survives from this assemblage. The El Jobo, Clovis, and "fishtail" points were obviously made to spear and kill such animals (Ranere and Cooke, 1991), and hunting in drier and more open areas during the late Pleistocene must have been a highly profitable pursuit.

On the other hand, one might expect that, in the forests, large game animals were far fewer and that human exploitation there must have been more oriented toward smaller game and plants. This certainly makes good ecological sense on the basis of modern-day distributions of the biota (e.g., Meltzer and Smith, 1987), but we offer the caveats that we are not well informed about the ecological requirements of some of the larger game that went extinct (e.g., giant peccaries), and that some of them may have been able to maintain viable breeding populations in forest habitats. Janzen and Wilson (1983) comment that even today in Costa Rica horses and cows appear to be able to maintain solid breeding populations in deep (deciduous) forest. Recent data from southern Amazonia indicate that some Pleistocene forest hunters would have had access to a now-extinct monkey nearly twice the size of any living today (Cartelle and Hartwig, 1996).

Related to this issue, it is often assumed that the late Pleistocene forests exhibited degrees of canopy closure as extreme as today's semi-evergreen and evergreen rain forest. However, we are uncertain at this time if this was indeed the case. Owen-Smith (1987) and Schule (1992) comment that much like the Old World elephant and rhinoceros today, the large American Pleistocene fauna may have disrupted the structure of the forest and prevented closed canopies by knocking down trees and generally tramping vegetation. A consequent increase in light availability at the forest floor would have provided more favorable habitats for grass and other edible herbaceous growth and perhaps attracted more species of large and medium-sized mammals.

Similarly, Janzen and Martin (1982) argue that certain characteristics of modern tropical trees indicate their past exploitation by large, fairly abundant mammals. For example, there are substantial crops of big-seeded and large, freshy fruits that are not eaten by contemporary mammals, have no effective dispersal mechanisms, and, thus, appear to be anachronisms (but see Howe, 1985). Many plants have obvious mechanical (spined trunks and thorns high in canopy trees) and chemical defense mechanisms but no obvious modern predators. If paleoecological data come to substantiate these suggestions, then hunting in some of the late-glacial tropical forests was more productive than it is in the forest today.

It should also be emphasized that in some regions, such as the Pacific watershed of Central America, the late-glacial environment was probably a patchy and hetero-geneous one in which different kinds of high-quality animal and plant resources could have been easily exploited from well-placed base camps (a point we explore further in Chapter 4) (Piperno *et al.,* 1991a; Piperno, 1995b).

In light of these considerations concerning the resources of modern tropical forest and thorn woodland habitats, we think that the late-glacial Neotropical landscape was far more productive per unit of labor effort in terms of both wild animals and plants than were Holocene environments. Changes in the flora and fauna consequent to climatic change at the close of the Pleistocene when lowland tropical forest advanced on the landscape would have necessitated some marked adjustments in foraging strategies, and these strategies almost certainly involved a higher investment in searching for and processing food items than those of Pleistocene antecedents.

Other Important Elements of the Late-Glacial Landscape

Other important but little understood factors that must eventually be considered in assessing the state of the late-glacial vegetation and its suitability for human exploitation include the seasonal distribution of temperature and rainfall and atmospheric levels of CO_2, which were reduced by almost one-third during glacial times (Shackelton *et al.,* 1983; Barnola *et al.,* 1987). Atmospheric CO_2 started to rise substantially between 15,000 and 12,000 B.P.

Sesonality is an important issue because it has the potential to influence the distribution of some crop plant ancestors. For example, in the Near East, the late-glacial rainfall was distributed fairly evenly throughout the year, although overall annual rainfall appears to have been reduced (Wright, 1993). The "Mediterranean climate," characterized by hot, dry summers and cooler, moister winters, is an interglacial climate that first made its effects felt only 11,500 years ago (Wright, 1993). The reasons for these phenomena are discussed below.

Based on these paleoecological data, Wright (1993) argues that wild wheat and barley, whose abundant, dense stands are adapted to the modern summer drought, did not become common on the landscape until after 11,500 years ago when the Mediterranean climate clicked on. This is also about the time when the wild wheat and barley-based Natufian culture was starting to extend its influence (Bar-Yosef and Belfer-Cohen, 1992). (In Wright's view availability of wild cereals is probably more important than varying energetic efficiencies associated with exploiting resources on a changing landscape in the transition to food production.)

Like wild Near Eastern cereals, teosinte and many tuberous plants rely on marked seasonality of rainfall to grow and mature. If the lowland tropical climate was more equable during the Pleistocene, with rainfall more evenly distributed throughout the year, the numbers of these plants on the landscape may have been similarly reduced.

How contrasts between the seasonal regimes of rainfall and temperature during the late glacial and early Holocene may have affected lowland tropical vegetation cannot be assessed via paleoecological data because the major indicator species are not well represented in existing pollen and phytolith records. (A series of cores soon to be taken from lakes in the Balsas Valley may inform this issue by tracking the frequencies of *Zea* pollen and phytoliths through time.) However, incorporation of Milankovitch orbital forcing factors into climatic models strongly predicts marked increases in seasonality during the early Holocene that would have affected the tropical zone (Kutzbach and Webb, 1993; Kutzbach *et al.*, 1993). They are discussed in detail in the section on the Pleistocene–Holocene transition.

How the glacial vegetation may have responded to lowered levels of CO_2 is not currently well understood. Generally, plants that are better able to compete in a lower CO_2 environment are plants that use the C_4 and CAM photosynthetic pathways because their leaf anatomy allows them to photosynthesize carbon more efficiently (Dippery *et al.*, 1995). Such plants include many lowland grasses, excluding bamboos, many sedges, and all the cacti.

On the other hand, plants (particularly annuals) that use the C_3 photosynthetic pathway, which include virtually all the tree and understorey growth of a tropical forest, including various wild tubers, may have been at a competitive disadvantage in the late-glacial world. C_3 annuals experience significantly lowered productivity when grown in a low CO_2 environment (Dippery *et al.*, 1995). Lowered CO_2 would have exacerbated drought stress and further promoted the expansion of C_4 and CAM open-land plants at the expense of some C_3 forest species. It may have led to more open forest canopies because of lowered light use efficiencies during photosynthesis (Sage, 1995).

Sage (1995) proposes that low Pleistocene CO_2 was, in itself, a limiting factor for the development of food production because significantly reduced C_3 plant productivity inhibited the development of effective cultivation systems.

It has become clear that the late-glacial world in lands now covered by warm and lush tropical forest was a far different one than originally envisaged. We are just beginning to come to grips with the implications of the various climatic factors for resource distribution and use during this time.

Toward the Modern Climate and Vegetation, and Food Production

Following the maximum advance of glaciers 18,000 radiocarbon years ago, the deep water ocean circulation underwent another reorganization and again became a major supplier of CO_2 to the atmosphere. A gradual overall warming of land and sea surfaces followed. With the retreat of the most powerful glacial climate forcers, the ice sheets and cold sea surface temperatures, changes in the seasonal and latitudinal distributions of solar radiation (insolation) received at the earth's surface—two of the Milankovitch factors—became more important in determining temperature and rainfall (Kutzbach and Ruddiman, 1993). However, in many regions, including the tropics, ice age circulation patterns apparently persisted until 12,000 to shortly after 11,000 years ago.

The transition to the climates and vegetation that characterize the current interglacial period was neither gradual nor simple. It was not marked by an even and steady vegetation succession but was characterized instead by fits and spurts and by rapid change and unstable climates. Climatic and vegetational conditions still deviated from those of today 2000 years after the final termination of the Pleistocene between 11,000 and 10,000 years ago.

In many of the paleoecological records that span the Pleistocene–Holocene transition, a marked change in climate and vegetation is evident by 11,000–10,500 years ago. There is some variability in the chronology and direction of change, called the "regionalization" of the environmental record by Markgraf (1993). As the extreme full-glacial forcing of the climate ended, the regional climatic phenomena evidently began to exert their own muscle in determining environmental conditions.

Temperatures appear to have reached near-modern levels in highland Colombia by 12,000 years ago. In the central highlands of Mexico, a marked precipitation and temperature increase occurred between 11,000 and 10,000 B.P. (Markgraf, 1993). Environmental reversals back to glacial-like conditions but lasting less than 1000 years have been registered between 11,000 and 10,000 years ago primarily in highland records from Colombia and Costa Rica (Islebe *et al.,* 1995; Van der Hammen and Hooghiemstra, 1995). They have recently been identified in lowland Guatemala (Leyden, 1995).

This phenomenon, called the "Younger Dryas" event, is also recorded in many sequences from the temperate zone. What caused the Younger Dryas is not well understood. In Central America, as elsewhere, it is thought to be associated with glacier meltwater events that discharged cold water into the ocean and changed the deep water circulation of the sea, although it is unlikely that the "conveyor belt" ceased to function altogether as it appears to have done during glacial times (Broecker, 1994; Goslar *et al.,* 1995; Leyden, 1995). In the Old World, the Younger Dryas had a very pronounced effect on climate and vegetation in the Jordan Valley and it has been implicated by several investigators as the precipitator of large-scale cereal cultivation (Henry, 1989; Moore and Hillman, 1992; Wright, 1994).

The Younger Dryas does not appear to have had much effect on the lowland Neotropical vegetation. A single record from the Peten, Guatemala, signals a brief and incomplete return to late-glacial conditions that reinitiate sometime shortly before and end at 10,300 B.P. (Leyden, 1995). Temperature dropped by approximately 2 or 3°C and the expansion of oak forest was reversed. Precipitation continued to increase, however, through the period of cooling. South of the Guatemalan lowlands, in Panama, Venezuela, Guyana, and Brazil, the Younger Dryas cannot be detected. The temperature appears to have risen in a single step, and quite steeply, between 10,800 and 10,500 years ago. Signals for a marked increase in precipitation are synchronous or slightly later, occurring by 10,800 to 10,200 B.P. in most records. Leyden (1995) believes that the apparent absence of the Younger Dryas in the lowlands south of Guatemala may, in part, be a result of more intense monsoonal activity from increasing solar insolation in these areas, which buffered them from the effects of the climatic reversal.

Wherever and whenever it occurred, the climatic snap initiating the Holocene appears to have been remarkably rapid, accounting for much of the temperature change in perhaps less than 100 years (Bush *et al.,* 1992). Paleoecological sequences simultaneously record abrupt and dramatic shifts of vegetation as the glacial types began to be lost from the landscape and plant associations more familiar to us today appeared or expanded.

When tropical forest reoccupied the open terrain that had expanded under the late–glacial climate is an important question because it is tied to the disappearance of the megafauna and open-land plants, and it necessitated full-time exploitation of forests by human groups. We believe that these factors are directly associated with tropical forest resource intensification, changing cost/benefit ratios, and the beginning of food-producing economies.

In the 36,000-year-old record from the Peten, Guatemala, which was a cool, arid, and largely treeless place during the late Pleistocene, elements of the current semi-evergreen forest make an appearance shortly after 10,300 B.P., and high tropical forest was probably established several hundred years or more later (Leyden *et al.,* 1993). At Monte Oscuro, Panama, pollen and phytolith records indicate that a tropical deciduous forest began to invade the area about the same time as it did in Guatemala (Piperno, 1995b). In eastern Amazonia, savanna elements experience a sharp decline and arboreal elements increase significantly at 10,460 B.P.(Absy *et al.,* 1991), whereas at Lake Valencia, Venezuela, the driest site investigated, the establishment of a high forest is not indicated until approximately 9,000 B.P., although arboreal representation becomes significant after 10,200 B.P.

It appears that in most areas options of exploiting open-land resources were beginning to disappear between 10,500 and 10,000 years ago. The diet breadth theory predicts that a decline in the abundance of high-ranked resources leads to increasing search costs and declining foraging efficiency and an increase of diet breadth as lower ranked resources are added to the diet. At this point, the wild ancestors of the various root and tuber crops and teosinte would have been increasingly incorporated into diets.

As the high-ranked items were increasingly lost from the landscape, the exploitation of tropical forest resources would diversify but overall foraging return rates would continue to decline. They would eventually fall equal to or below the return rates expected from the production of certain plants, at which point food-producing behavior was initiated. As noted previously, tubers with minimal processing costs and large-seeded grasses common on the landscape such as teosinte were likely to have been among the first plants to have been taken into cultivated plots.

In thinking about human subsistence during this period, we must emphasize that between 11,000 and 9000 years ago the lowland climate was more unstable and the vegetation was undergoing more changes than at any time before or since during the past 120,000 years. Resources that were advancing on the landscape were generally not those that people, given a broad range of both glacial and nonglacial resources to consider, would have chosen to exploit. This fact should help dispel any notions of "climatic amelioration" or "climatic improvement" that have often been used to characterize the conditions for the biota that occupied the landscape immediately following glacial time (e.g., Hayden, 1995).

At this juncture, we note that care must be taken in following the chronology of early Holocene forest development and associated subsistence strategies because radiocarbon dates and calendar dates diverge substantially during this period. The radiocarbon age range of 10,700–10,000 B.P., representing the major climatic snap terminating the Pleistocene and the subsequent expansion of forests, is equivalent to a much longer interval of 1500 years, 12,750–11,250 calendar years B.P., during which human populations negotiated a changing landscape (Bartlein *et al.,* 1995). The establishment of modern vegetational formations between 10,500 and 8600 B.P. is equivalent to the longer interval of 12,500–9700 calendar years B.P. What does not change much is the interval between the establishment of forest on the landscape and what we believe marks the onset of cultivation: 10,000 B.P.–~9500 B.P. vs 11,250–~10,750 calendar year B.P. The curve during the 10,500–9500 B.P. period is volatile and small differences in radiocarbon age will potentially slow or increase the rate of change by hundreds of calendar years.

The development of subsistence strategies and associated settlement and demographic factors during the early Holocene should be evaluated in the context of a much longer period of time than has been commonly viewed. These data also bear obvious implications for determining migration rates of trees during the early Holocene and for other important ecological questions.

What accounted for the early Holocene environmental chaos evident from the records? Implicated are fascinating variations in the earth's orbital geometry—the Milankovitch factors that, now fully liberated from the glacial-age climatic forcers, converged in an unusual way between 11,000 and 9000 B.P. to increase the seasonality of radiation and, thus, of temperature and precipitation. There is an orbital factor called perihelion, which is the time of year the earth is closest to the sun, and it changes because of variations in the earth's orbit around the sun.

Today, perihelion occurs in January, but between 11,000 and 9000 years ago it occurred from May to July, a time of heavy rain (Northern Hemisphere) or substantial dryness (southern hemisphere) in the tropical zone (Kutzbach and Webb, 1993). Also, the tilt of the earth's axis of rotation is constantly changing, resulting in variations in the latitudinal distribution of solar radiation. During the early Holocene, the axial tilt increased so as to point the Northern Hemisphere more directly in the path of the sun's rays and it received 8% more radiation in the summer and 8% less in the winter than it does today (Kutzbach and Webb, 1993).

The net effect of these phenomena was to increase greatly the seasonal contrasts in temperature and precipitation in the Northern Hemisphere tropics (the summers—the wet season—were warmer and winters were cooler), although effects may have been more pronounced in the Old World because of the greater land areas (Kutzback and Webb, 1993). In the Southern Hemisphere, the seasonal contrasts were less pronounced. An outcome in the Northern Hemisphere tropics was an overall annual precipitation increase because much of the rain that falls during the wet season in the seasonal tropics is a result of air warmed by solar radiation being forced upward (convection). Between 10,000 and 9000 years ago, climates were probably wetter than they are today.

Heightened seasonality of rainfall 10,000 years ago may have furthered the expansion of teosinte and various ancestors of the tuber crops during the early Holocene north of the equator. These plants had also probably found warmer overall temperatures more to their liking. Evidence for simultaneous changes in the Southern Hemisphere caused by orbital forcing comes from pollen sequences from two areas in the present-day cerrado region of southeastern Brazil near the southern latitudinal limit of the tropics. A reduction in winter cooling and decreased rainfall apparently occurred at approximately 10,000 B.P., when the modern seasonal forest–cerrado mosaic was established (De Oliveira, 1992).

It appears that the earth's biota, in effect, experienced a double-whammy shock between 12,000 and 9000 years ago. The massive reorganization of atmospheric and oceanic conditions that brought the end of the ice age warmed the climate, raised precipitation, and drastically altered the vegetation and fauna. Also, an unusual confluence of the major features of the earth's orbital geometry resulted in possibly the most extreme seasonality of climate ever recorded, shortly following the termination of the Pleistocene.

In conclusion, we have attempted to show that although people in the tropical lands never lived in the shadow of the great northern ice caps or saw continental glaciers advance and retreat, they were no less affected by the dramatic environmental oscillations associated with the end of the last ice age and the coming of the modern climate than were peoples in the temperate regions. We think that a consideration of the forces now known to create the ice ages should, in and of themselves, serve as proof that the low-latitude biota must have been markedly altered by glacial conditions. Finally, paleoecological data appear to indicate beyond doubt that tropical climate and vegetation were profoundly changed between 11,000 and 10,000 years ago.

CHAPTER 3

The Phytogeography of Neotropical Crops and Their Putative Wild Ancestors

INTRODUCTION

The goal of this chapter is to place the domestication of crops native to the lowland tropics "on the landscape" of the Neotropical world by reviewing what is known about the distribution of wild species considered ancestral to the major crops. In the cases in which allozyme and DNA data are available for crops and related wild species, we can talk with confidence about the nature of the populations that gave rise to the crops. With these insights, we use modern plant distribution data, accounts of contact-period crop distributions, and the climatic and vegetation reconstructions discussed in Chapter 2 to delineate areas of the lowlands where plant domestication was most likely to have occurred (i.e., Figs. 3.18 and 3.19). Archeological data presented in Chapters 4 and 5 demonstrate early Holocene dates for a number of crop domestications.

The reconstruction of crop phytogeography we present here is by necessity a broad-stroke treatment because the ancestors of a number of crops are still unknown, and distributions of known related species are often imprecise. Often more is known about the genetics of a crop and its related species than about their ecological requirements and current distributions. The increasing pace of deforestation makes it uncertain that our knowledge in the latter areas will improve greatly. Our treatment, then, remains a model for testing through the archeological and geological records.

There are a number of excellent compendia that include discussions of the origin, improvement, and cultivation of Neotropical crops (i.e., Harlan, 1992; Purseglove, 1968, 1972; Sauer, 1993; Simmonds, 1976; Smartt and Simmonds, 1995; Smith *et al.*, 1992; Brücher, 1989; Hancock, 1992). These overviews have been consulted, as has recent literature not covered in these sources. Our goal is not to repeat the details of what the interested reader can find elsewhere. Rather, we consider lowland crops within the context of the hypotheses developed in Chapter 1 and focus on those species whose domestication led to the development of tropical forest agriculture in all its variability. We use as our starting point the five crops immortalized on the Obelisk Tello.

THE GIFTS OF THE CAYMAN

At Chavín de Huantar, the best known ceremonial center of the Peruvian Chavin culture (early Horizon or middle Formative, 900–200 BC) archeologist Julio C. Tello discovered the important stone sculpture that now bears his name (Lumbraras, 1974; Burger, 1988). Lathrap (1973a) argues that on the Obelisk Tello are represented, among other things, foods brought to humanity by the dual god, the Cayman of the Water and Cayman of the Sky, through his intermediary, the jaquer (Fig. 3.1). The Water Cayman brings achira (edible canna, Queensland arrowroot, *Canna edulis* Kerr.), manioc (yuca, cassava, *Manihot esculenta* Crantz), and peanut (*Arachis hypogaea* L.); the Sky Cayman brings bottle gourd (*Lagenaria siceraria* [Molina] Standl.) and aji (chili peppers, *Capsicum* ssp.).

These were not the staple crops of Chavín de Huantar, located at 3100 m elevation in the Andes; a different suite of tubers and legumes, with some maize, fulfilled this function. Bottle gourd and aji can hardly be considered important crops in any subsistence system. Lathrap (1973a) argues that these plants are depicted on the Tello Obelisk because they were at the core of the subsistence system of the tropical lowland ancestors of the Chavin people. They are the base from which Chavin subsistence is derived because the lowlands is a source of Chavin iconography and culture.

Using this image as a starting point, we discuss crops related to those depicted on the Obelisk Tello in function, taxonomic relationship, or propagation practice. This gives us the major taxa of the lowland tropical agricultural system, a system viable throughout the lowlands and into the mid-elevation Andes and subtropical zones. We consider plants conspicuous by their absence on the obelisk in the following section.

Achira and Other Monocot Tubers

Achira (*Canna edulis* Ker.) is one of two root/tuber crops identified on the Obelisk Tello; the other is manioc (Fig. 3.2). As Lathrap (1973a) acknowledges, the

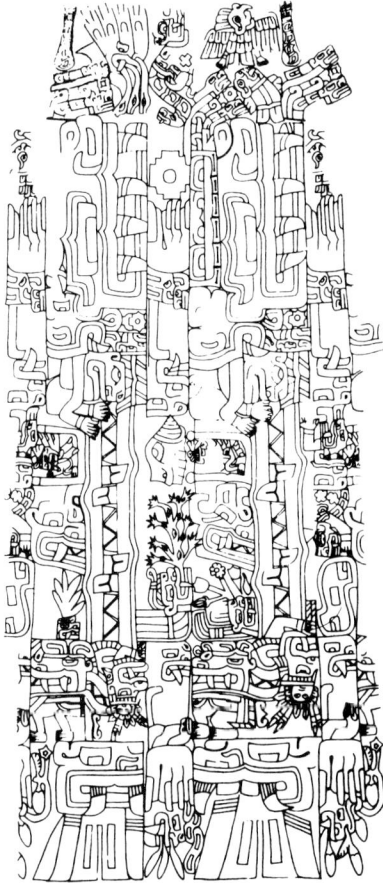

FIGURE 3.1   Rollout of the reliefs on the Obelisk Tello, depicting the dual cayman deities. Left, Cayman of the Water; Right, Cayman of the Sky. Reproduced with permission from Rowe (1967).

image identified as achira—a compact clump of broad, simple leaves—may be subject to other interpretations. For example, *Maranta arundinacea* L. (arrowroot) is quite similar. We take this image of a robust, broad-leafed herb to represent an important group of food plants domesticated in the seasonally dry, low to mid-elevation tropics: root/tuber/rhizome (i.e., subterranean storage organ-producing) foods from the monocot families Araceae (aroids; *Xanthosoma* spp.), Cannaceae (achira; *Canna* spp.), Dioscoreaceae (yams; *Dioscorea* spp.), and Marantaceae (arrow-roots; *Maranta* spp. and *Calathea* spp.) (Figs. 3.3a, 3.3b, 3.4a, and 3.4b).

The similarities among these types of plants, both in ecological requirements and in appearance, would not have been lost on native peoples. With one exception (*Calathea*), all are grown by replanting small pieces of the underground portion

FIGURE 3.2   Depiction of achira (*Canna edulis*) or a similar monocot tuber-producing plant (Cayman of the Water). Reproduced with permission from Lathrap (1973a).

that was utilized—a simple propagation technique. In fact, vegetative propagation can increase plant growth and production (Sauer, 1952; Salick, 1995). All the tubers discussed here are excellent starch producers requiring minimal processing. Indeed, arrowroot and achira are known for their easily digestible starch. It is likely that wild ancestors of these monocot root/tuber plants grew in lower biomasses (due to lower CO_2 levels and reduced seasonality, as discussed in Chapter 2) on the landscape of the late Pleistocene Neotropics. As has been pointed out by numerous authors (e.g., Sauer, 1952; Harris, 1969; Hawkes, 1989), the formation of underground storage organs is a response to a marked dry season (5 months or more). Thus, although root crops are grown today in well-drained soils throughout the wet tropics, this is not where we should look for initial domestication.

FIGURE 3.3   (a) *Calathea allouia* (leren). Modified from Noda *et al.* (1994), with permission from the Food and Agriculture Organization of the United Nations. (b) *Maranta arundinaceae* (arrowroot). Modified from Purseglove (1972), with permission from Blackwell Science Ltd. Arrows indicate the utilized parts of the plants.

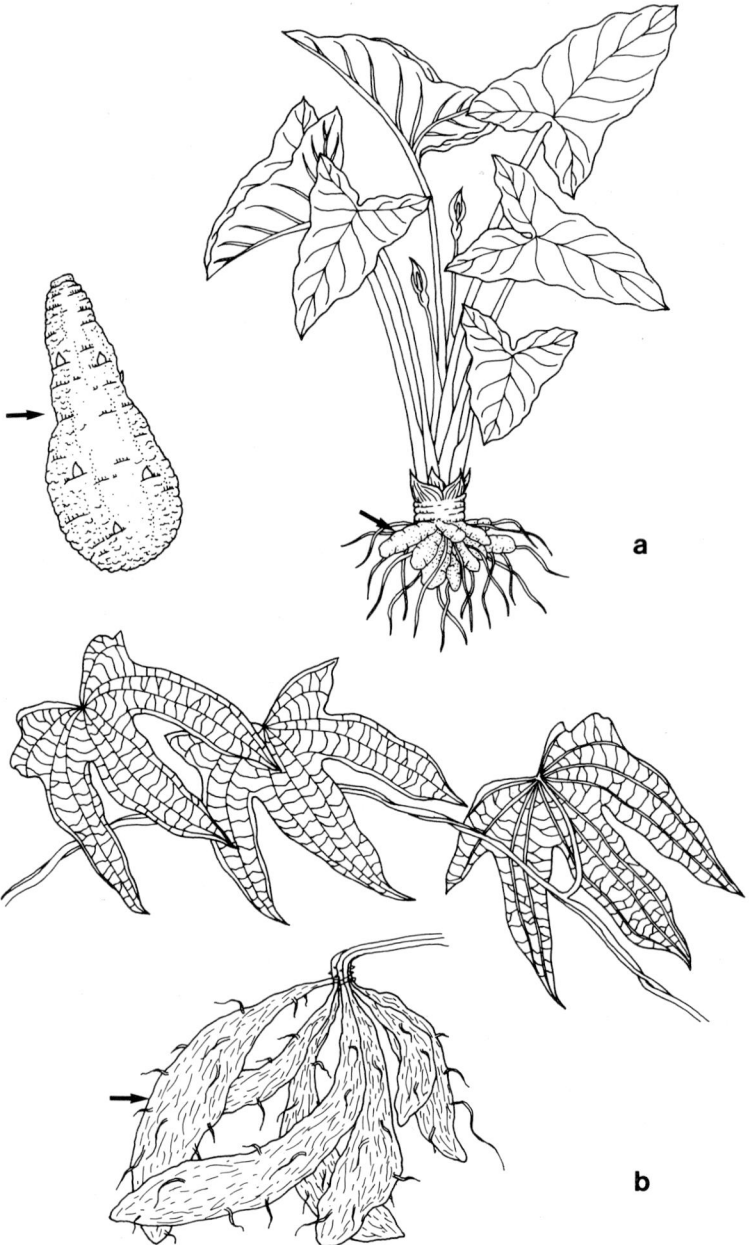

FIGURE 3.4   (a) *Xanthosoma saggitifolium*. Modified from Giacometti and León (1994), with permission from the Food and Agriculture Organization of the United Nations. (b) *Dioscorea trifida*. Modified from Purseglove (1972), with permission from Blackwell Science Ltd. Arrows indicate the utilized parts of the plants.

The multiplicity of lowland root/tuber plants, both monocots and dicots (discussed later), brought under domestication suggests a complementarity that is difficult to describe in detail because ecological information is scant for many species. Indeed, it is difficult to suggest specific areas of origin for any of these crops. Whether domestication occurred to the north or the south of the wetter Amazonian forests, in western Ecuador or northern Colombia, or in Central America is largely a matter of speculation for most species. Following Sauer (1952) and others, we are inclined to look in the seasonal formations of southern Central America and northern South America for evidence of domestication of many root/tuber species.

Arrowroot

Two New World genera within the Marantaceae have yielded plants cultivated for their edible tubers: *Calathea* and *Maranta*. Both are genera of robust herbs native to lowland forests; with 250 species, *Calathea* is the major genus of the family. There are approximately 30 species of *Maranta* (Gentry, 1993). A number of species of *Calathea* have edible roots; *C. allouia* (Aubl.) Lindl., leren, lairén, or topee tambu, is most frequently mentioned. It is cultivated in the Caribbean and northern South America (Purseglove, 1972). Other edible species are *C. latifolia* and *C. macrosepala* (NAS, 1975). Given the vast number of *Calathea* species indigenous to the Neotropics, however, it is likely more species were once grown than still exist in cultivation.

Maranta arundinacea L., the cultivated arrowroot, is also considered indigenous to northern South America and the Caribbean (Purseglove, 1972). Sturtevant (1969) reports that *M. arundinacea* occurs wild in Brazil, northern South America, and perhaps Central America (in fact, it has been observed in Panama), but insufficient information is known about wild related species to designate an area of origin.

Although early historical accounts demonstrate that *C. allouia* was grown for its edible tubers in the Caribbean prior to contact, Sturtevant (1969) argues that *M. arundinacea* was initially used there as an antidote for poisoned arrow wounds. *Maranta* rhizomes are very tough, requiring thorough grinding or maceration to release the starch. As we will discuss in Chapter 4, however, there is evidence for early use of *Maranta* as a food in mainland regions of Central and South America.

Arrowroot requires rainfall of 1500–2000 mm per year and grows best on rich, sandy loams. It cannot tolerate waterlogging (Purseglove, 1972). Propagation is by rhizome tips; because these are commonly broken off during harvest of the starchy rhizomes, the crop often selfpropagates. Planting occurs at the beginning of the rainy season, and mature rhizomes are produced in 10–12 months. The amount of undigestible fiber in arrowroot rhizomes varies among cultivars, but they are more fibrous than the edible portion of *C. allouia*—small tubers produced at the ends of fibrous roots. Leren is reproduced from the rootstock, the tubers lacking eyes (Hawkes, 1989). Tuber production requires approximately 12 months. The starch of both arrowroot and leren is of high nutrient value and easily digestible; processing is much easier for leren, however (Brücher, 1989).

Cocoyam

The Araceae have produced a number of widely cultivated crops, including several species of taro native to the Old World tropics (*Colocasia esculenta* (L.) Schott, *Alocasia macrorrhyza* (L.) Schott, and *Cyrtosperma chamissonis* (Schott) Merr.) (Plucknett, 1976; Pursesglove, 1972). The New World edible aroids (referred to as cocoyams, yautia, or malanga) belong to the genus *Xanthosoma*. There is disagreement about how many species are represented by the cultivated cocoyams, with most authors simply grouping them all under *X. sagittifolium* (L.) Schott. (Fig. 3.4a; Plate 3.1). Approximately 30–40 species occur in the Neotropics.

Cocoyams were cultivated in tropical Central and South America and the Caribbean at the time of European contact, with distinctive varieties (if not species) grown throughout the region (Purseglove, 1972). Too little is known to suggest an ancestral species or region. Of all the root crops considered here, cocoyams are adapted to the wettest growing conditions; 2000 mm or more rain gives the best yields, and water is required throughout the growing season (Onwueme, 1978). The literature is contradictory, however, about whether waterlogging can be tolerated. Cocoyams can be planted in alluvial soils too wet for sweet potatoes or yams, however (NAS, 1975). Cocoyams are not deep forest plants; although they can tolerate light shade, they go dormant with increasing shade (Purseglove, 1972). Useable cormels (small tubers surrounding the central corm, which is often replanted) are produced in 3–10 months, depending on the cultivar (NAS, 1975). Tubers are roasted or boiled; starch is extracted by grating and boiling. Because

PLATE 3.1   Yautia tubers from the Panamanian market.

the tubers contain raphides (microscopic pieces of hard, calcium oxalate crystals), they must be cooked before consumption. Harvested tubers can be stored 2 or 3 months if kept dry (NAS, 1975) and not damaged when dug up.

Yams

The Dioscoreaceae (yams) are a family of tropical plants, usually climbers, with rhizomes or tubers (Pursesglove, 1972; Hahn 1995). Species of *Dioscorea* occur in both the Old and New World tropics, and yams were domesticated independently in both hemispheres. Although the most thorough general discussions of yam cultivation and domestication remain those of Coursey and associates (Coursey, 1967, 1976; Alexander and Coursey, 1969; Alexander, 1971; Ayensu and Coursey, 1972), Chikwendue and Okezie (1989) present some interesting research on the impact of harvesting and planting on wild African yam. For example, they document, during an 8-year period, the transformation of the long, thin tubers of the wild species into uniform, rotund tubers much easier to harvest.

A number of species of *Dioscorea* and the closely related *Rajania* were used for food in the Neotropics at the time of European contact, but the status of most as cultivars is unclear. Among the utilized species are *D. adenocarpa* Mart., Brazil; *D. convolvulacea* Cham. et Achlecht., Central America; *D. dodecaneura* Vell., widespread in South and Central America, probably once an important food source; *D. piperifolia* Humb. et Bonpl. Brazil, now little cultivated; *D. trifoliata* Grisebach., northern South America; *Rajania cordata* L., used extensively in the Caribbean; and *D. trifida* L., the most commonly cultivated species (Coursey, 1967) (Fig. 3.4b). The tubers of several wild New World species contain chemical compounds called saponins to protect them from animal predation (a number of Old World species, but not New World species tested, also contain other compounds called alkaloids) (Coursey, 1967). Such bitter-tasting yams are considered inedible, and this characteristic is selected against in cultivated/utilized species.

Dioscorea trifida is native to northern South America and is grown there, in Central America, and throughout the Caribbean (Coursey, 1967). The border area between Brazil and Guyana is suggested as a center of origin of *D. trifida* domestication because many forms are known (Ayensu and Coursey, 1972; Alexander and Coursey, 1969). Wild yams are still used by hunter–gatherers in eastern Brazil and Venezuela. The widespread replacement of native yams by introduced, Old World varieties makes it difficult to determine which species were cultivated, where, and to what extent in the past. There is also the perception that yams were always less important than "easily" cultivated root crops such as manioc, cocoyam, or sweet potatoes (e.g., Alexander and Coursey, 1969). We think it equally likely that yams declined in importance, with the other crops we discuss in this section, as people began to focus on manioc cultivation.

Dioscorea trifida forms a group of small tubers, each 15–20 cm long (Coursey, 1967). The development of the tuber is an adaptation to a prolonged dry season, and cultivated yams grow best with a dry season of 2–5 months. However, during the growing season, approximately 1200–1500 mm of rain, evenly distributed, is required for optimal tuber production.

Yams can withstand drought during the growing season, but this affects the yield. Deep, sandy loam soil is best; yams cannot tolerate waterlogging. Propagation is by small tubers planted at the beginning of the rainy season. Harvested yams can be stored for as long as 6 months, if kept dry, because they are rarely attacked by insects or rodents (loss from natural metabolism and rotting increases after 3 months, however). Tubers may be roasted or boiled—small ones without peeling; flour can be produced by sun-drying sliced tubers (either fresh or after being boiled), which are then ground into a coarse meal (Coursey, 1967).

Achira

There are 10 (Gentry, 1993) to 55 (Purseglove, 1972) species of *Canna* distributed in the New World tropics and subtropics, including species native to the mid-elevation Andes, Brazil, and Florida. Gentry describes *Canna* as distributed mostly in disturbed or swampy areas at mid-elevation. The edible cultivated canna or achira (*C. edulis* Kerr.), is no longer commonly grown outside the Apurimac region of Peru but was once known from the Antilles to Argentina and through the Amazon (Gade, 1966) (Fig. 3.5). It can be cultivated at up to 2000 m elevation (León 1964). Other *Canna* species known to have been used for their fleshy rhizomes are *C. coccinea* Mill., *C. paniculata* R. et P., and *C. indica* L. (Purseglove, 1972).

Although there is insufficient information to suggest a precise area of origin for achira, open habitats within seasonally dry forest at the fringes of the tropics (altitudinally or latitudinally) are likely areas given the tuber-forming habit of the plant and its tolerance of cool conditions and disturbance. Gade (1966), following Sauer (1952), proposes the northern Andes.

Achira is propagated by planting small, corm-like rhizome segments at the beginning of the rainy season. Approximately 8 months is required for the crop to mature; it can be left in the ground for up to 2 years, however (León 1964). In the Apurimac region, the sweet rhizomes are baked, but they can also be boiled or eaten raw (Gade, 1966). A high-quality starch can also be extracted.

In summary, the suite of monocot root/tuber crops discussed here are likely the remnants of a large group of big-leafed, robust herbs domesticated by lowland peoples. We propose that a great diversity of root/tuber foods characterized the diet of early farming populations in the lowlands not only because a variety of these excellent carbohydrate sources were available but also because each type fits a slightly different ecological niche, enhancing the chances of successful propagation. Cocoyams, for example, thrive in the wettest, hottest, open areas, whereas achira ranges into the coolest. Yams, tolerant of drought and able to climb to reach sunlight, provide root/tuber resources in drier, more closed habitats. The large

FIGURE 3.5   *Canna edulis*. Modified from León (1964). Arrows indicate the utilized parts of the plant.

number of arrowroot, and especially leren, species native to the Neotropics ensured that root/tuber foods would be encountered in a wide array of habitats when foragers turned to these resources because of the decline of higher ranked foods at the end of the Pleistocene.

Because the nutritive values of all these monocot domesticates are quite similar (Table 3.1; note the percentage of carbohydrates), their relative importance in the early tropical forest agricultural system was likely a factor of local environmental conditions, at least initially. We think it likely that early cultivators used several species of roots and tubers. Later on, when population increased, the resulting pressure on arable land led to a focus on the most productive, and most labor-intensive, root/tuber crop of the Neotropics—manioc.

Manioc and Other Dicot Tubers

Two of the most important lowland root/tuber crops are dicots that are quite distinctive from each other and from the tuber resources previously discussed: *Manihot esculenta* Crantz (manioc, yuca, cassava) and *Ipomoea batata* (L.) Lam. (sweet potato) (Figs. 3.6, 3.7a, and 3.7b; Color Plate 3.2). Both manioc, a shrub, and sweet potato, a vine-like herb, are propagated by stem cuttings, i.e., a section of stem is stuck in soil and roots from the nodes or "eyes." Many of the wild manioc species that would have been encountered by tropical forest peoples have roots that are bad tasting or poisonous (manioc contains two chemical compounds called cyanogenic glucosides, which chemically decompose to liberate cyanide). *Ipomoea* is a genus with poisonous species as well. We think it likely that the eventual

TABLE 3.1   Nutritive Value of Root/Tuber Foods[a]

Root/tuber	Moisture	Carbohydrates	Protein	Source
		%		
Monocots				
Xanthosoma	70–77	17–26	1.3–3.7	Onwueme (1978)
Maranta	69–72	19.4–21.7	1.0–2.2	Purseglove (1972)
Dioscorea	70–80	25–28	1.0–2.8	Onwueme (1978)
Canna	73	24	1	León (1964)
Dicots				
Manihot	62	35	1	Purseglove (1968)
Ipomoea	70	27	1.5–2	Purseglove (1968)
Pachyrrhizus	87.1	10.6	1.2	Purseglove (1968)
Arracacia	74	23	0.7	León (1964)
Polymnia	70–93	4.7–10.5	0.3–2.2	León (1964)

[a] Data are given in percentage of fresh, edible portion by weight.

FIGURE 3.6   Depiction of manioc (*Manihot esculenta*); Cayman of the Water. Reproduced with permission from Lathrap (1973a).

emergence of manioc as the dietary staple of much of eastern South America was linked to its greater starch content, high productivity, and ability to grow in less than optimal soils.

Manioc

There are approximately 100 species of *Manihot,* ranging from southern Arizona to Argentina, with most being native to arid or seasonally dry regions, including

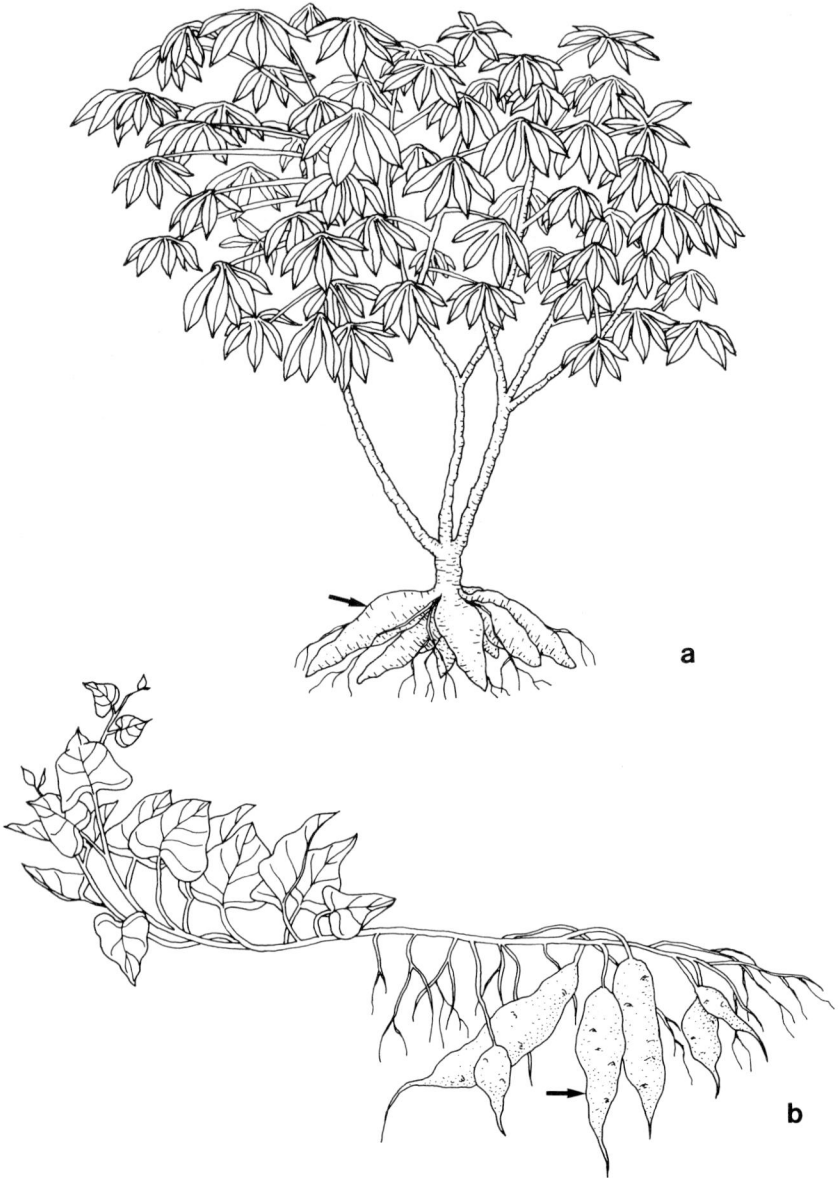

FIGURE 3.7 (a) *Manihot esculenta* (manioc). Modified from Hancock (1992). (b) *Ipomoea batatas* (sweet potato). Modified from "Plant Evolution and the Origin of Crop Species" by Hancock, © 1992. Adapted by permission of Prentice-Hall, Inc., Upper Saddle River, NJ. Arrows indicate the utilized parts of the plants.

deciduous tropical forest, and to open habitats. Most species are shrubs, with some vines and small trees; all produce milky latex (Gentry, 1993; Jennings, 1995; Rogers and Appan, 1973; Rogers and Fleming, 1973; Purseglove, 1968; Sauer, 1993; Nassar, 1978). The root system of many of the shrubby species is shallow and easy to uproot. Many produce enlarged roots with starchy reserves (Rogers, 1965). The most recent treatment of the genus is by Rogers and Appan (1973). These authors delineate two major concentrations of species—central and western Mexico and east-central Brazil, extending toward Paraguay. The genus is rather young, with many recent speciations. The large number of species in the Brazilian center may in fact be the result of human burning of vegetation (if this is the case, the concentration of species may reflect the later stages of use of manioc in this region rather than its domestication). All species are rather sporadic, never dominating the local vegetation; this pattern would have been accentuated by late Pleistocene climatic conditions. The Mesoamerican and Central American species basically do not overlap with the South American species. Most species (80 of 98) occur in South America.

An incredible amount of variability exists within cultivated manioc, some of which is likely the result of crossing among related species (hybridization) (Purseglove, 1968). Manioc is easily hybridized artificially with wild *Manihot* (especially if the cultiver is used as the female parent) (Nassar, 1980). Hybridization also occurs in nature if manioc is allowed to mature to the flowering stage (hybrid seedlings can then be selected and propagated as clones) (Sauer, 1993). There appear to be no interspecific barriers to hybridization in the genus (Fregene *et al.*, 1994). In nature the wild species are segregated ecologically and geographically (Sauer, 1993). In summary, it is possible that manioc varieties existing today are the result of crosses between the early domesticate and one or more wild species (as the incipient cultivar was spread) (McKey and Beckerman, 1993). However, in which region, or regions, of the seasonally dry lowlands did this process begin?

Based on morphological characteristics, Rogers and Appan (1973) proposed that *M. aesculifolia* (H. B. K.) Pohl (widely distributed in Mesoamerica and Central America) was the closest wild relative of manioc. Two Mexican species, *M. rubricaulis* I. M. Johnston (which includes *M. isoloba* as a subspecies in their treatment) and *M. pringlei* Watson, may also have contributed to manioc evolution. Based on these observations, a Mesoamerican origin for *M. esculenta* was proposed.

Recent research makes a strong case for a South American origin, however. *Manihot tristis* Muell-Arg. (southern Venezuela, southern Surinam, and northern Brazil) is closely related, if not the same species as the domesticate (Allem, 1987; Fregene *et al.,* 1994). Allem (1987) also reports observing forms of *M. esculenta* that are indistinguishable morphologically from cultivated manioc in the wild in central Brazil. This would place wild manioc in both the seasonal forest zone of northern South America and the cerrados of central Brazil. A study by Fregene *et al.* (1994) of genetic variability of DNA seguences in the chloroplast and nucleus in 21 manioc cultivars and 16 wild, related species, including *M. aesculifolia* and

M. rubricaulis (*M. pringlei* was not studied), revealed that *M. esculenta* subspp. *flabelli-folia, M. tristis* Muell-Arg., and *M. irwinii* Rogers and Appan (central Brazil) are most closely related to manioc. *Manihot esculenta* subspp. *flabellifolia* shares affinities with both of the major chloroplast DNA types characterizing manioc. Among those tested, these three South American species occupy the primary gene pool (i.e., are most closely related) with manioc; *M. aesculifolia* and *M. rubricaulis* do not. Fregene *et al.* (1994) note that manioc and the three closely related wild species could also have been derived from a common, unknown ancestral form.

The Fregene *et al.* (1994) study does not completely resolve the issue of the area of origin of manioc but does, in our view, weaken the case for a Mesoamerican origin by demonstrating (i) that the Mesoamerican species *M. aesculifolia* and *M. rubricaulis* are not as closely related to manioc as thought on morphological grounds, and (ii) that *M. tristis* and *M. irwinii,* two "good" wild South American species, are in the primary gene pool with the domesticated species. The ambiguous status (wild manioc or escape?) of *M. esculenta* subspp. *flabellifolia,* the most closely related form, makes the significance of its relationship difficult to assess. The distribution of this subspecies is also unclear. Study of the other Mesoamerican species considered as possible ancestors for manioc is obviously important, but we feel comfortable at this point in proposing an origin for manioc in either northern South America (Venezuela/Guianas/northern Brazil) or central Brazil with, following Sauer (1952), the former the most likely spot.

Two interrelated issues concerning manioc remain to be discussed: (i) why did manioc become the dominant root/tuber food of much of the eastern South American lowlands, outstripping all other root/tuber resources, and (ii) why did bitter, rather than sweet, strains of manioc largely fill this role? Although sweet manioc is widespread in South America, bitter is largely restricted to the eastern lowlands (Renvoize, 1972). Beckerman (1993) and McKey and Beckerman (1993) note that bitter forms of manioc in indigenous Amazonia were also clearly histori-cally correlated with major water courses, where fishing is probably a more produc-tive activity than hunting, and with more sedentary forms of lifestyles. Where only sweet manioc is grown (west of the Andes, mid-elevation Andean valleys, Central America, and Mexico), it tends to be a minor part of a diverse crop complex and of secondary importance to maize. Sweet manioc was, however, the principal food crop for some societies along the margins of Amazonia and in western Amazonia (McKey and Beckerman, 1993).

The answer to the first question, why manioc became a major staple, is largely given in Table 3.1. Manioc tubers are much higher in starch than are any of the other root crops of the lowlands; pound for pound, they provide a more efficient source of energy. Although yields are too variable for these lowland crops to admit meaningful comparisons, manioc is very productive and undemanding. It is often the last crop to be grown before a field is fallowed in traditional mixed horticulture. It can grow on almost any soil, providing it is not waterlogged or too shallow for root development (Purseglove, 1968). It can be grown with as little as 750 mm of rain per year but can also thrive under much wetter conditions, as long as soils are well drained (Sauer, 1993). We propose that this efficient, undemanding carbohydrate source became an increasingly attractive crop as human populations grew and pressure on land increased (i.e., shortening fallow periods and decreasing time for recovery of soil fertility). Its increasing importance may correlate with evidence for the development and intensification of swiddening (see discussion of the lake core data in Chapters 4 and 5.).

Manioc is probably the only example of a highly poisonous staple crop in the world (McKey and Beckerman, 1993). Why did bitter forms of manioc come to be widely used? Without getting into a long discussion of bitter versus sweet varieties, a wide range of toxicity exists in tubers (15–400 mg HCN/kg fresh weight), that varies both among clones and because of environmental conditions, such as drought and degree of shade (Lancaster *et al.*, 1982). Although there are many ways to process tubers that result in detoxification (see Lancaster *et al.*, 1982, for an overview), the key element is hydrolysis (chemical decomposition) of the cyanogenic glycosides (in the presence of the enzyme linamarase) and subsequent elimination of the liberated cyanide. Rupturing the cells by grating or pounding brings the enzyme and the glucosides together, and washing carries away the cyanide.

This is all much more work than that required for using "sweet" varieties (i.e., those low in glycosides), which can simply be boiled, roasted, or even eaten raw. However, if dried (i.e., storable) manioc products are desired, grating, washing, and squeezing to separate the starch from the pulp used for coarse flour are required, regardless of the glycoside content of the roots. Hence, no additional work is generated by using bitter varieties. Sweet varieties can (and were) selected and maintained under cultivation in the lowlands. There is nothing to suggest they are poorer yielding than bitter varieties. According to Lancaster *et al.* (1982) and McKey and Beckerman (1993) there are no concrete data to support the belief that bitter strains produce more starch, and are therefore superior for flour production, or that bitter varieties preserve better after harvest (i.e., unprocessed). Although long-season varieties can be stored "in the ground" for up to 2 years, harvested manioc roots begin to deteriorate within a few days (Purseglove, 1968).

Although we would not throw out folk wisdom concerning which manioc is better for what food products, this is clearly an area that could use research. There may, however, be a more important reason why bitter manioc would be favored by intensive manioc agriculturalists: protection of a valued crop from animal predators. Because most of the glycosides are concentrated in the peel and outer flesh of the tuber, this could provide a defense against herbivory (McKey and Beckerman, 1993; Sauer, 1993). As McKey and Beckerman (1993) indicate, the quality of the data on the matter is not satisfactory, but it is highly suggestive. In resource-poor areas, such as much of the eastern Amazon Basin, herbivory would be tolerated less well by plants with no chemical or mechanical defenses and result in lower growth rates than in resource-rich areas, necessitating the need for defensive compounds to maintain acceptable yields (Coley *et al.,* 1985; McKey and Beckerman, 1993).

Furthermore, the correlation of bitter manioc distribution with riverine areas, and with more sedentary fields, might be related to the advantages of toxicity in the situation in which fields are more prone to crop pest attacks, and in habitats in which higher densities of mammalian herbivores prey on crop plants (McKey and Beckerman, 1993). Also, fish contain significant amounts of amino acids that can detoxify the poisonous manioc compounds and reduce the costs of consuming manioc (Beckerman, 1993). In more permanent societies, food storage would be required, and women would have the time to carry out the costly processing techniques. In short, the maintenance and production of bitter varieties of manioc may have substantially related to the interactions between the plants, people, and their environments in the Amazonian lowlands.

Finally, there is evidence suggesting that both sweet and bitter varieties of manioc derived from wild ancestors of mixed (neither very bitter nor sweet) toxicity. Other alternatives—that sweet forms gave rise to bitter varieties, that bitter forms gave rise to sweet forms, or that each variety had its own ancestor—appear to be less likely (for a review see McKey and Beckerman, 1993). Natural and cultural selection then gave rise to both increased and decreased levels of chemical defenses.

Sweet Potato

The role of the second important dicot tuber of the lowlands, sweet potato (*Ipomoea batatas* (L.) Lam.), in the evolution of the tropical forest agricultural system is obscure, to say the least (Fig. 3.7b). There is more interest in the issue of the introduction of this crop into the Pacific, where it outproduces native aroids and is essential in pig husbandry, than in its area of origin and early evolutionary history. Where both manioc and sweet potato are grown today in the Neotropics, sweet potato tends to be in a secondary role. However, Contact-period accounts from places such as Panama, interior South America, and around the Caribbean indicate it was a more important crop in the past (Sauer 1952). Until recently, for example, sweet potato was the staple crop for large numbers of people in central Brazil who now rely on manioc (Beckerman, 1993).

There are approximately 500 species of the pantropical genus *Ipomoea* (Hancock, 1992; Bohack *et al.,* 1995). It displays a polyploid series (i.e., more than two sets of homologous chromosomes) of $2n = 30$, 60, and 90. The domesticated sweet potato is the only hexaploid (i.e., 3 × the diploid number, or $2n = 90$). Tetraploid (i.e., 2 × the diploid number, or $2n = 60$) *I. batatas* is also known; it tends to be weedy both in wild and in cultivated settings (Austin *et al.,* 1992). Approximately 15 wild species and the forms just discussed make up the *Ipomoea* section *batatas* complex. The mechanism for the origin of the domesticate has been considered alternately as hybridization among species or as an origin from a single species through chromosome doubling. Another option, by analogy with bread wheat, is an origin of the hexaploid cultigen from an earlier tetraploid form and a weedy diploid (Sauer, 1993). The issue cannot currently be resolved; it is easy to envision any of these processes taking place in the Neotropical house garden.

Ipomoea trifida (H. B. K.) G. Don is considered by several authors to be the closest relative of sweet potato (Hancock, 1992). This seemingly straightforward statement hides a wealth of taxonomic confusion, however. A number of important accessions of hexaploid *I. trifida,* used to argue for its position ancestral to the sweet potato (i.e., Nishiyama, 1971), are rejected as that species by Austin (1978), who considers them to be in part feral (i.e., escaped from cultivation) sweet potato and in part hybrids with sweet potato. Sweet potatoes apparently escape easily, and when allowed to breed freely (i.e., from seed rather than cloned from stem cuttings) they produce many wild-looking forms (Brücher, 1989). Disagreement about how species are defined makes it difficult to evaluate recent research on relationships among species (e.g., Orjeda *et al.,* 1990; Shiotani and Kawase, 1989) because it is unclear whether the "disputed" strains are included in the research. Little is known of the ecology of wild *Ipomoea* populations, which is especially vital for distinguishing among wild, weedy, and feral forms (Sauer, 1993).

Ipomoea trifida as defined, conservatively, by Austin (1978) is distributed in Mexico, Central America, northern South America, and the Caribbean. We have found no discussion of the habitat preferences of the species, but in southern Mexico it has been collected from Veracruz, Chiapas, Oaxaca, Guerrero, and Michoacan states at elevations ranging from sea level to 1250 m (Contreras *et al.,* 1995). This suggests a wide habitat tolerance. Although describing the distribution as circum-Caribbean, the map by Austin (1978) also shows a west coastal distribution in Mexico and Central America, a distribution into the Venezuelan interior, and a single collection from coastal Brazil.

Sweet potato is the only species in *Ipomoea* section *batatas* to form tubers. Although a number of wild *Ipomoea* species form thickened, fibrous roots, none of these species are closely related to the domesticate (Martin *et al.,* 1974). The source of the enlarged root of the cultivated species, as well as the anthocyanin and carotenoid pigments found in it, remains obscure. Austin (1978) notes, however, that the extreme "wild" type of this highly variable domesticate has a simple thickened root almost identical to those of *I. trifida*. Ploidy level also influences root thickening (Díaz *et al.,* 1992). Selection under domestication for tetraploid plants with thicker roots, or perhaps hybridization with an as yet undiscovered tuber-forming related species, is the source of this trait.

Whether or not *I. trifida* gave rise to *I. batatas* (alone or with other species) or evolved with it from an unknown but common ancestor, the distribution of the former species suggests an origin for sweet potato somewhere in northern South America or Central America. Given the confusion about the relationships among species in *Ipomoea* section *batatas,* even this broad statement must be considered tentative, however. The lack of ecological information on most species in the complex is another complication; we can say little except that forms producing underground storage organs would tend to be adapted to areas with long dry seasons (i.e., Hawkes, 1989).

At European contact, sweet potatoes were grown throughout the Neotropics (Purseglove, 1968). There are a large number of cultivars (approximately 88 in the Caribbean alone), varying in texture and color of the tuber flesh, tuber size and shape, and various vegetative characteristics. Sweet potatoes grow best in well-drained soil with plenty of sun and rainfall of 750–1250 mm distributed throughout the growing season. Propagation is by stem cuttings, usually planted on ridges or mounds. In the tropics, plants often flower and fruit, but because most cultivars are self-sterile, viable seeds are produced only if cross-compatible cultivars are grown together. Tubers can be harvested in 3–6 months, depending on the cultivar. They store poorly. No special processing is required.

Before concluding our discussion of the dicot tubers, we should mention four, unrelated root/tuber-producing species that are grown to a limited extent today in the mid-elevation Andes and were likely brought under domestication there: *Arracacia xanthorrhiza* Bancroft (arracacha), *Polymnia sonchifolia* Peoppig & Endlicher (yacón), *Pachyrrhizus* spp. (jícama; there are three cultivated species, including a Central American one), and *Mirabilis expansa* R. et P. (mauka) (Hawkes, 1989). None of these crops has been studied systematically. We think it likely that, like achira (*Canna edulis*), these were once commonly cultivated above the altitude where the lowland tubers thrived and below the potato zone. This complex of root/tuber foods, which likely varied in composition and relative importance of species by region, was largely replaced throughout the temperate Andes (and, in the case of jícama and achira, on the coast) by maize in precolombian times.

In summary, we interpret the depiction of achira and manioc on the Obelisk Tello as representing the diverse array of root/tuber foods that was responsible for, and fueled the expansion of, the tropical forest agricultural system throughout the lowlands and into the mid-elevation Andes and fringes of the tropics. The eventual dominance of bitter manioc, consumed in the form of flour or starch, in the South American lowlands east of the Andes was probably a relatively late development—a response to the need for a portable, storable carbohydrate that could be intensely cropped on marginal land. We believe it is likely that the transition to bitter manioc production, and a narrowing of the subsistence base, is correlated with the expansion of swiddening in some areas of the lowlands. In other areas, including the western lowlands of Ecuador, a non-root/tuber food took on the role of a storable, tradeable carbohydrate source.

Peanut and Other Legumes

The Cayman of the Water brings an interesting plant that produces an edible underground portion that is a fruit, not a tuber: the peanut or maní, *Arachis hypogaea*

L. (Figs. 3.8 and 3.9a). An excellent source of both protein (25–30%) and oil (45–50%) (Sauer, 1993), the peanut fills a vital dietary role in the lowland tropics. As a nitrogen-fixing plant (i.e., a plant capable of utilizing atmospheric nitrogen) peanuts do well interplanted with other crops or grown late in the crop rotation (Purseglove, 1968). The geocarpic habit—fruits are produced on the ends of stalks known as pegs, which elongate and push the fertilized fruit 2–7 cm into the soil, where it finishes developing—is likely a survival mechanism for seasonal drought, a type of *in situ* seed bank (Sauer, 1993). Seeds are dispersed by being washed out of the soil and redeposited elsewhere; water dispersal clearly played a role in the distribution of wild species related to the peanut, as discussed later, and the domesticate is at home in alluvial soils. The geocarpic habit also means that seeds do not require protection by other means, i.e., chemical compounds found in other legumes, including *Phaseolus,* that must be removed before they can be eaten. All these positive qualities suggest that peanut should be among the earliest Neotropical forest domesticates.

Peanut

Arachis is a small genus of approximately 40 species, all distributed in South America east of the Andes from the Amazon River south to approximately 35°S (Purseglove, 1968). The domesticated species is an allotetraploid, the result of hybridization between two distinctive genomes or sets of chromosomes, referred to as the A and the B genomes (producing an AB-genome form). The domesticate is part of section *Arachis,* which contains one wild allotetraploid species, *A. monticola* Krap. et Rig. (also AB genome) (the likely progenitor of the peanut), and approximately 20 wild diploid species (Stalker, 1990).

FIGURE 3.8   Depiction of peanut (*Arachis hypogaea*); Cayman of the Water. Reproduced with permission from Lathrap (1973a).

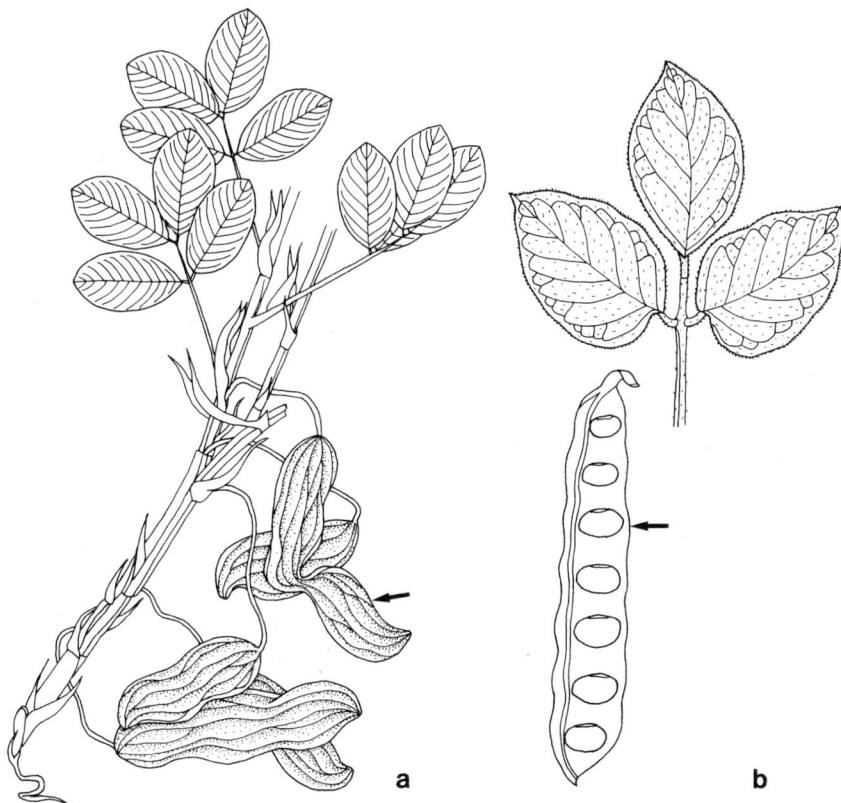

FIGURE 3.9   (a) *Arachis hypogaea* (peanut). Modified from *Useful Plants of Neotropical Origin and Their Wild Relatives*, Brücher, H., pp. 105–266, 1989, with permission from Springer-Verlag. (b) *Canavalia ensiformis* (jack bean). Modified from Purseglove (1968), with permission from Blackwell Science Ltd. Arrows indicate the utilized parts of the plants.

The systematics of the genus are still being worked out, but we know more about the geographic distribution of species related to the peanut than is the case for most Neotropical crops. This is due both to interest in using wild germ plasm to improve peanut cultivars and to the relatively restricted distribution of the genus. The center of origin of *Arachis* is the Mato Grosso region of central Brazil, where representatives of all the sections are found (Krapovickas, 1969). Species of section *Arachis* are mostly distributed there and westward, in the Paraná/Paraguay river drainage, including the upper reaches of the western tributaries that arise in northwest Argentina and southern Bolivia and in an upper tributary of the Madeira river, the Mamoré, which arises just to the north of this region and flows northward into the Amazon (Stalker, 1990; Wynne and Halward, 1989). Scattered occurrences of species in this section are known from other south Amazonian tributaries. *Arachis* species thrive in a variety of habitats, with most preferring more open habitats, such as rock outcroppings, open forest, forest–grassland margins, and disturbed areas (Stalker and Moss, 1987).

The wild allotetraploid species, *A. monticola,* is known from Jujuy state, northwest Argentina, where it occurs between 1400 and 2800 m elevation (Krapovickas, 1969; Stalker, 1990). This is the only known tetraploid, besides the peanut, and the only wild species that gives fully fertile hybrids with the domesticate. *Arachis monticola* is thus the best candidate for the immediate ancestor of the peanut, barring discovery of another wild tetraploid that could have given rise to both. A number of lines of evidence suggest that allotetraploidy arose only once in section *Arachis* (and again in section *Rhizomatosae*) (Halward *et al.,* 1991). The peanut is self-pollinating, with a narrow genetic base; much of the morphological variation observed in peanut cultivars is controlled by only a few genes (Halward *et al.,* 1992).

Which wild diploid species gave rise to the tetraploid line leading to the peanut is not entirely settled. The B-genome parent is *A. batizocoi* Krap. et Greg, the only known B-genome diploid species (Stalker and Moss, 1987). It is distributed along the upper Mamoré (Santa Cruz state, Bolivia) (Stalker, 1990). Various species have been suggested for the A-genome parent, including *A. villosa* Benth, *A. duranensis* Krap. et Greg., and *A. cardenasii* Krap. et Greg. A species distributed in the south Bolivia/northwest Argentina region is most likely because this is where the B genome parent and *A. monticola* occur. *Arachis duranensis* and *A. cardenasii* both fit this pattern, but there are many other possibilities, including recently collected species that have not even been named. In the cooler climate of the late Pleistocene, species adapted to the upper-elevation range of the genus (e.g., *A. monticola* and a number of diploids) could have ranged into the lower elevations of the western Paraná/Paraguay tributaries.

The peanut has undergone considerable diversification under domestication. Two subspecies, each with two botanical varieties, have been defined; these are the Virginia and Peruvian varieties (both *A. hypogaea* ssp. *hypogaea*) and the Valencia and Spanish varieties (both *A. hypogaea* ssp. *fastigiata*). Despite the common names, all originated in precolombian times in South America (Krapovickas, 1969). Five or six centers of variation of peanuts are recognized, some of which are associated the extant ethnic groups (Krapovickas, 1969; Gregory and Gregory, 1976). *Arachis hypogaea* ssp. *fastigiata* cultivars are distributed mostly to the south and east of the primary center of peanut distribution (northwest Argentina/south Bolivia); A. *hypogaea* ssp. *hypogaea* varieties are distributed mostly to the north and northwest. Traditional peanut cultivars of Peru and Ecuador are of the Peruvian, or "runner" type (*A. hypogaea* spp. *hypogaea,* var. *hirsuta*); this is the type found archeologically on the Peruvian coast (Krapovickas, 1969). This primitive cultivar is rarely grown in this region today (Banks, 1990).

By European contact, peanut cultivation had spread throughout the South American lowlands, much of the mid-elevation Andes, and the Caribbean but apparently not into coastal Colombia or Central America (Sauer, 1993). Whether the peanut was introduced by the Spanish into Mexico (from the Caribbean) or arrived late in prehistory (from Peru) is undetermined.

Peanuts require at least 500 mm of rain during the growing season; most are grown with 1000 mm (Purseglove, 1968). Dry weather is required for ripening and harvesting. Soil should be well drained and loose (sandy soils are preferable). Fruits mature in 3.5–5 months or more, depending on the variety.

Like the root/tuber resources discussed previously peanuts are adapted to seasonal environments; the ecology of the wild species and traditional cultivars suggests the ancestor species of the peanut occupied riverine habitats in open environments, such as dry forest or savanna. We think it is likely that these oil- and protein-rich seeds became increasingly important in diet over time, contributing the protein needed in a manioc-based diet and aiding in the recovery of soils.

Jack Beans

Much less is known of the origin of a second group of lowland pulses, the jack beans, *Canavalia ensiformis* (L.) DC. and *C. plagiosperma* Piper. (Fig. 3.9b). In comparison to peanut, common bean, and lima bean, jack beans are uncommon in cultivation today in the Neotropics, except as green manures and fodder (Purseglove, 1968). Indeed, *C. plagiosperma* is close to extinction (NAS, 1979). We know, however, from the archeological record of the west coast of South America that *Canavalia* beans enter the archeological record relatively early (see Chapter 5).

Canavalia is a genus of slender vines to erect herbs with trifoliate leaves. Approximately 20 species occur in the New World; they are most common along coasts (Gentry, 1993) and are often found in secondary vegetation and along river banks (Sauer, 1964). *Canavalia ensiformis* is considered native to Central America and the Caribbean; *C. plagiosperma* is considered native to South America (Purseglove, 1968). Plants are deep rooted and drought resistant and will tolerate shade. Jack beans thrive, however, under a variety of rainfall regimes, and are reported to grow well under from 700 to 4200 mm of rain (NAS, 1979). The deep root system allows plants to survive dry conditions; jack beans will produce a crop when common bean fails. They will also tolerate a wide range of soil textures and fertility, including depleted lowland soils, and are less affected by waterlogging and salinity than other pulses. Dry, mature jack beans are very resistant to insect damage. Although jack beans can be grown up to an elevation of 1800 m (NAS, 1979), this approaches the range where *Phaseolus* beans are better adapted.

Mature jack beans contain 23% protein, 55% carbohydrate, and 1% fat (Purseglove, 1968). The flat, straight pods, among the largest of any domesticated legume, mature in 3 or 4 months (NAS, 1979). Pods can be picked green (either before seeds enlarge, i.e., used like snap beans, or when seeds are still soft) but must be cooked. Also, the dry seeds are only edible after extensive boiling, with changes of water and peeling of the seed coat. The reason is that the same chemical compounds that protect the seeds from predation are toxic to humans. These are growth-inhibiting compounds and include the protein con-canavalin A, which reduces the body's ability to absorb nutrients (NAS, 1979).

Little research has been done concerning the evolution of the jack beans. Neither domesticated species is known in the truly wild state (Sauer, 1964). There are four New World wild species closely related (i.e., in the same genus section) to the jack beans. *Canavalia brasiliensis* Mart. ex Benth., a wild species widespread in the drier regions of lowland South America, Central America, Mexico, and the Carribbean, is considered on morphological and geographic grounds to be the most likely ancestor of *C. ensiformis*. *Canavalia piperi* Killip & Macbride, a species that is distributed in the Mata Grosso area of Brazil into lowland Bolivia and northwest Argentina (with an outlier in southern Peru), is considered by Sauer (1964) to be the ancestor of *C. plagiosperma*. As will be reviewed in Chapter 5, archeological remains place *C. ensiformis* in Mexico, and *C. plagiosperma* in coastal Ecuador and Peru; the record is silent on which species was used east of the Andes [both species occur under cultivation or as escapes in eastern South America (Sauer, 1964)]. Purseglove (1968) reports that a form that is probably *C. plagiosperma* crosses with *C. ensiformis* on Trinidad, producing fertile offspring.

The tolerance of jack beans to drought, salinity, and low-fertility soils suggests to us that both species were domesticated in dry coastal settings. Shade tolerance and the ability to thrive under much higher rainfall and soil moisture made them adaptable to inland, riverine environments, as well as to wetter coastal areas. In the absence of any molecular or biochemical data on the relationship between the two domesticated species, or between them and related wild forms, we suggest (i) that *C. ensiformis* was domesticated somewhere in the circum-Caribbean region (either the dry coast of northern South America or the gulf region of Mexico seem likely), with a spread into the interior; and (ii) that *C. plagiosperma* was domesticated in South America. Although the current distribution of *C. piperi* suggests northwest Argentina/southern Bolivia for the latter development, it makes little sense for jack beans to be brought under domestication in the heartland of the peanut. A dry coastal setting seems more likely; perhaps *C. brasiliensis,* which ranges into coastal Ecuador, was involved with the emergence of both domesticates. Whether the two species share a common ancestor at some point in their evolution or have introgressed in regions where both are grown (i.e., the case of fertile crosses reported in Purseglove, 1968) is unknown.

This scenario, although nothing more than a model for future testing, brings us to a puzzling question regarding domestication of the *Canavalia* beans: Why grow a scarsely edible vegetable protein in the first place? In the absence of pottery, it is difficult to see how dried jack beans could have been detoxified and cooked easily enough to make the effort worthwhile for coastal populations with ready access to marine protein sources. Although soaking beans in a net bag in a stream or fermentation would have removed toxins (Brücher, 1989; NAS, 1979), the difficulty of extended cooking times remains. Was drought resistance and the unattractiveness to pests of the dried seeds sufficient to attract early agriculturalists? We think it is more likely that the first use of jack beans was the gathering of immature pods as a seasonal vegetable in dry coastal regions of the Neotropics, with mature seeds, along with other low-ranking resources, used as a carbohydrate/protein source only in times of nutritional stress. The transformation from vegetable to dry carbohydrate/protein resource is perhaps linked to the introduction of new cooking technology (i.e., ceramics). At this point, ease of storage may also have made *Canavalia* beans a more desirable resource.

A second question has to do with the relationship of jack beans and peanut. As will be discussed in Chapter 4 and 5, available data indicate that peanut was introduced into western South America prior to the appearance (local domestication?) of jack bean. Later, both were grown in some areas (i.e., coastal Peru), whereas jack beans faded away in other areas (i.e., coastal Ecuador). Whether these divergent histories of use are due to differing ecological requirements of the crops or cultural preferences is unknown. Documenting the early history of pulse use is hindered by the apparent lack of well-silicified, diagnostic phytoliths in this group.

Phaseolus Beans

The best known of the New World pulses, common bean (*Phaseolus vulgaris* L.) and lima bean (*P. lunatus* L.), by European contact had spread from homelands in the mid-elevation Andes and highlands of Central America to throughout the tropics (excluding the wettest regions), including the west coast of South America, where common bean also contributed to declining use of *Canavalia* (Figs. 3.10 and 3.11).

The case of common bean is better understood. Wild common beans grow in relatively dry environments with intermediate temperatures (i.e., dry montane forest and thorny scrub vegetation) from Mexico to the southern Andes of Peru, Bolivia, and Argentina (Gepts, 1991; Debouck *et al.,* 1989a, 1993; Brücher, 1988). In Central America, the most common elevation range is 1500–1900 m, with seasonal rainfall of 550–1000 mm (Delgado *et al.,* 1988). Plants mature during the dry season and are often found in open and disturbed habitats. In the Andean region, wild bean populations tend to be distributed on the eastern slopes of the Andes in Venezuela and Colombia, on the western slopes in Ecuador and northern Peru, and on the eastern slopes again from central Peru southward (Debouck *et al.,* 1989a, 1993). Wild beans are found in subhumid montane to dry forests in the Andes at elevations ranging from 900 to 2600 m, with 1400–1900 m being a common range (Freyre *et al.,* 1996). Koenig *et al.* (1990) hypothesize that the larger seed size of wild Andean forms (in relation to Mesoamerican wild beans) is an adaptation to the more forested habitat because populations living in such environments tend to have larger seeds than those in open habitats. In some areas the range of wild beans is expanding because wild beans thrive in secondary scrub vegetation. In other areas, populations are nearly extinct.

FIGURE 3.10   *Phaseolus vulgaris* (common bean). Arrow indicates the utilized part of the plant. (Modified from Bailey, 1947).

Multiple lines of evidence, including reproductive isolation, differences in morphology, distinctive biochemical markers [phaseolin (seed-storage proteins) and isozymes (different molecular forms of an enzyme)] and molecular markers (nuclear and mitochondrial DNA), demonstrate that there are two primary gene pools of domesticated common beans, the Middle American (or Mesoamerican) group of small-seeded cultivars and the Andean group of large-seeded cultivars (Gepts *et al.*, 1986; Gepts, 1988, 1990; Koinange and Gepts, 1992; Hamann *et al.*, 1995; Vargas *et al.*, 1990; Freyre *et al.*, 1996; Koenig *et al.*, 1990). The Middle American group is distributed in Mexico, Central America, and Colombia; the Andean group

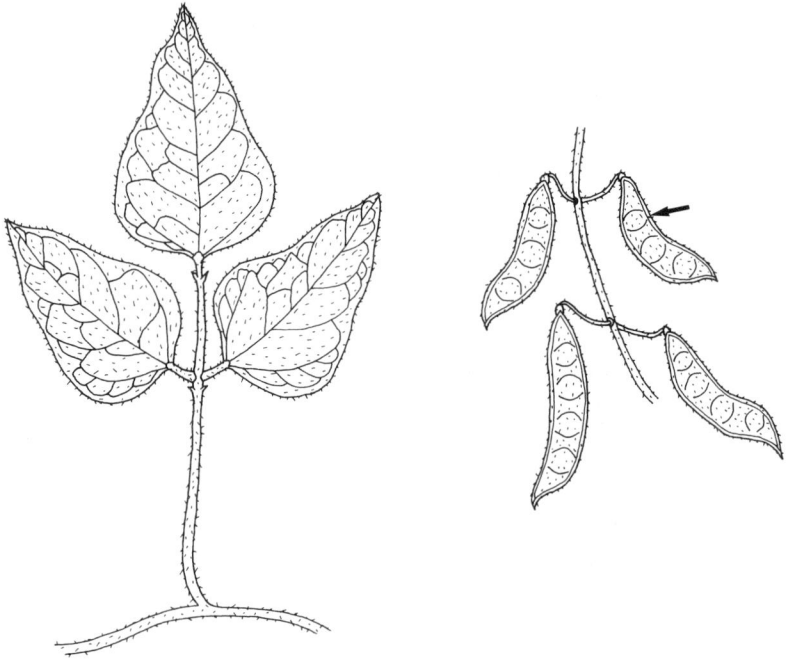

FIGURE 3.11   *Phaseolus lunatus* (lima bean). Arrow indicates the utilized part of the plant. (Modified from Lackey and D'Arcy, 1980).

is distributed in the southern Andes. Lowland South American and Caribbean beans largely cluster with the Middle American group and share phaseolin affinities (i.e., the "Sb" seed protein) (Castiñeiras *et al.*, 1994). Traditional cultivars of northern Peru and Ecuador, as well as some in Colombia, share traits with both primary gene pools (Gepts and Bliss, 1986) but cluster slightly closer to the Andean (Freyre *et al.*, 1996).

Evidence indicates that the primary gene pools separated prior to domestication because, within each gene pool, wild and domesticated beans cluster together by region (Koinange and Gepts, 1992; Sonnante *et al.*, 1994; Becerra Valásquez and Gepts, 1994; Debouck *et al.*, 1993). Similarities in mitochondrial DNA are especially strong between Middle American cultivated beans and wild bean accessions from Guatemala and between Andean cultivars and wild accessions from southern Peru/Bolivia (Khairallah *et al.*, 1992). Phaseolin data indicate a west-central Mexican origin in or near Jalisco state for the common bean in Mesoamerica, however (Gepts *et al.*, 1986; Gepts, 1990). An origin in this region might make more sense because maize was domesticated nearby. Phaseolin markers agree that the southern Andes was the major domestication center in South America (Gepts, 1990).

The eastern Andean region of Colombia is a "meeting place" for the two primary gene pools because both the "S" (Mesoamerican) and "T" (Andean) phaseolin types occur in wild beans from this region (Gepts and Bliss, 1986). Also, "B" phaseolin occurs only in wild beans from this region, and it is also distributed in cultivars from Colombia, Central America, and Ecuador. This factor suggests that northern Colombia, or perhaps Central America, may also be a minor domestication center (cultivated genotypes have a lower frequency of B phaseolin types) (Gepts and Bliss, 1986; Gepts, 1990; Koenig *et al.,* 1990).

Recently discovered populations of wild beans of western Ecuador and northern Peru, which were nearly extinct due to human habitat destruction, are distinctive from those of the other regions. They are characterized by a combination of genetic traits from both primary gene pools (Debouck *et al.,* 1993) and a type of phaseolin ("I" seed protein) not found in other wild or cultivated beans and considered ancestral to the species as a whole (Kami *et al.,* 1995). Because cultivars lack the I phaseolin type, this region is not a likely area of origin for common bean but perhaps a center for early divergence of the wild species.

Within each primary gene pool, three races of domesticated beans can be defined based on morphological, agronomic, and molecular data (Singh *et al.,* 1991). Each evolved under human selection to fit specific ecological conditions. For example, race Nueva Granada is adapted to intermediate elevations in the Andes, whereas race Mesoamerica is suited to lowland environments. Common beans never adapted to the wettest areas of the Neotropics, however (the distribution map of bean races in Singh *et al.,* 1991, is blank for the Amazon and Orinoco basins). Despite high phenotypic variability within cultivated bean, there is a low level of genetic diversity in comparison to that which exists in wild bean populations (Sonnante *et al.,* 1994; but see Beccera Valásquez *et al.,* 1994, for data that suggest less change in diversity). Reduced diversity implies that domestication events involved circumscribed populations of wild beans, representing only a fraction of the overall diversity of the wild species (Gepts, 1991).

In summary, evidence to date suggests two or three domestications of common bean (from local wild beans in west-central Mexico or Guatemala, northern Colombia, and southern Peru/Bolivia) from an already diverged wild species that emerged in western Ecuador/northern Peru. Domesticated beans were spread widely and continued to evolve through human selection and crossing with wild forms.

We now discuss the lima bean. Wild lima beans grow in low to mid-elevations in seasonally dry shrubland and savanna formations in Central America, the Caribbean, and the Andes from Colombia to Argentina (Sauer, 1993; Baudoin, 1988; Esquivel *et al.,* 1993; Debouck *et al.,* 1989b). There are two forms of the wild species; a small-seeded form (the Mesoamerican, or sieva type) that occurs from southern Mexico to Argentina (eastern Andes slopes), generally below an elevation of 1600 m, and in the Caribbean, and a large-seeded form (the Andean, or big lima type) that occurs in the western Andean foothills of Ecuador and northern Peru, between 320 and 2030 m elevation (Debouck, 1994; Esquivel *et al.,* 1993;

Debouck *et al.,* 1989b; Gutiérrez *et al., * 1995). Thus, as was the case for common bean, there are two forms of wild lima beans in the Andean region. As might be expected from the wide range of the wild forms, under cultivation lima beans can tolerate a broad range of ecological conditions. Although able to tolerate wetter conditions than common bean, limas also require dry weather for seed maturation (Purseglove, 1968).

The large-seeded, or Andean, lima beans were likely domesticated at mid-elevation in the western Andes of Ecuador or northern Peru, where larger seeded, truly wild forms occur (Debouck *et al.,* 1987, 1989b; Gutiérrez *et al.,* 1995). Seed protein (phaseolin) studies of wild and cultivated limas from this region show at least three families of cultivars, suggesting that multiple domestications occurred within the large-seeded lima gene pool (Gutiérrez *et al.,* 1995) or that hybridization between cultivated and wild forms was occurring as the early domesticated form spread. The large-seeded form, the pallar, was widely cultivated in precolombian times along the coast of Peru (Kaplan and Kaplan, 1988; see Chapter 5).

The location of initial domestication of the small-seeded, Mesoamerican form is more difficult to determine because the wild species is widely distributed in Central and South America, as is the most common phaseolin pattern (Gutiérrez *et al.,* 1995). The wide genetic distance between the two gene pools, as revealed by DNA analysis (Nienhuis *et al.,* 1995), the presence of two distinct phaseolin types (Debouck *et al.,* 1989b; Maquet *et al.,* 1990; Gutiérrez *et al.,* 1995), and the existence of both two wild and domesticated forms, supports a predomestication separation of the Mesoamerican and Andean genepools, as is the case for common bean. It is possible that the small-seeded limas were domesticated more than once. Esquivel *et al.* (1993), for example, argue for domestication in Cuba from local wild limas. The distribution of wild small-seeded forms indeed seems to be discontinuous—split into northern (Central America/northern South America) and southern (south Andes) groups in northern Peru/Ecuador by the large-seeded forms. However, given that the center of diversity of small-seeded cultivars is Central America (Sauer, 1993), and that all archeological finds in South America are of the large form prior to 4000 B.P. (Kaplan and Kaplan, 1988), an origin in the northern tropics for the small limas seems likely.

Wild lima beans of both types have high levels of a chemical compound called glycoside linamarin or phaseolunatin (usually more than 1500 ppm; cultivated limas have values lower than 200 ppm), which produces cyanide through chemical decomposition (hydrolysis) (Baudoin *et al.,* 1991). Most cyanide is released from the seeds. Hydrolysis occurs rapidly when macerated seeds are cooked in water, as the liberated cyanide is lost through vaporization. However, even extensive soaking or boiling with changes of water will not get rid of all the cyanide if levels of cyanogenic glycosides are high. This is because cooking eventually destroys the enzyme that releases the cyanide (Baudoin *et al.,* 1991). In other words, the high cyanide levels in wild lima beans render them inedible. Wild common bean seeds are edible but require a longer cooking time (6 hr) than cultivated bean because of hard seed coats (Delgado *et al.,* 1988). So why did people domesticate these pulses?

Although Debouck (1989) suggests that early cultivation of beans (both common and lima) may have been for their aesthetic value, allowing domestication to begin before selection yielded seeds that could be rendered edible, we think it is more likely that there was immediate selection of wild beans with edible green fruits. In the case of lima beans, for example, cyanide is concentrated in the seeds, suggesting foragers could safely use fruits of wild limas picked in the snap bean stage (i.e., before seeds begin to develop). This could be tested by measuring cyanide levels in steamed fruits. With selective pressure for high toxicity levels released (i.e., by humans protecting the plants from predation), once the technology for rendering dried beans edible was available, levels of toxins were already substantially reduced.

In summary, pulses are important sources of protein and carbohydrates, and, in the case of the peanut, oil in Neotropical diets. The fact that species were brought into domestication to fit every environment in which agriculture could be practiced [in addition to those discussed previously, the following are also included: lupine (*Lupinus mutabilis* Sweet) for the high elevation Andes, scarlet runner bean (*Phaseolus coccineus* L.) for the cool uplands of Central America and tepary (*P. acutifolius* Gray.) for the arid areas of the northern tropics and subtropics] is an indication of their critical role as a storable carbohydrate and vegetable protein source.

We propose, however, that this role did not emerge at the same time for these individual pulses. Peanut may have been the earliest cultivated for its mature seeds. Peanuts are nontoxic and no special processing is required. It makes more sense for the others, especially the jack beans and lima bean, to have first been used as vegetables in the green stage (i.e., before seed development). This use may be ancient, but it is difficult to document. Mature seeds (planting stock) would not be cooked and therefore would rarely enter the macroremain record of early sites. As we will see in Chapter 5, most *Canavalia* and *Phaseolus* bean finds occur at ceramic-age sites. We argue that the development of appropriate cooking technology is the key to common and lima bean becoming widely cultivated. Increased use of pulses may also correlate with the advent of short fallow swiddening because they contribute to soil fertility.

Gourds, Squashes, and Cotton: Containers, Edible Seeds and Vegetables, and Fishing Nets

Gourd

The Cayman of the Sky also brought useful plants to the people. One of these gifts, the bottle gourd (*Lagenaria siceraria* [Molina] Standl.), was widely distributed around the globe in ancient times (Figs. 3.12 and 3.13a). Truly wild populations of gourd occur today only in South and East Africa. Bottle gourd is considered native to the semidry tropical lowlands of Africa south of the equator (Whitaker and Bemis, 1964, 1976; Sauer, 1993; Purseglove, 1968). Gourds are adapted to riparian habitats. The seeds are dispersed by fruits floating away from the parent plant and eventually becoming deposited on banks and broken open. Wild African gourds washed out to sea in the south Atlantic would be carried west by the south equatorial current, ending up on the coast of Brazil or the northern South American coast (i.e., swept north across the equator by the Caribbean current). Because experiments have shown that seeds in gourds afloat in seawater 224 days are still viable, this is the likely mechanism for their dispersal to the New World (Purseglove, 1968).

For gourds washed on shore, humans took over as dispersal agents; gourds are not adapted to the seashore, and no free-living gourd populations have been found in the New World (Sauer, 1993) (as might be expected if a few were tossed far enough inland by a storm to reach hospitable soils). Tropical lowland peoples were likely familiar with the tree gourd (*Crescentia cujete*) (an unrelated plant) and native wild *Cucurbita* squashes with hard, thin rinds. This new gourd would look familiar and would soon be found to be superior as a container and float. Bottle gourds have thicker, tougher, and more durable rinds than squashes, and they produce a wider variety of shapes than the tree gourd. Bottle gourds can be grown successfully throughout the tropics and subtropics and into the temperate zone. The flesh of most cultivers is too bitter to eat, but the oily seeds are edible, as are young shoots and leaves (Purseglove, 1968).

FIGURE 3.12   Depiction of bottle gourd (*Lagenaria siceraria*); Cayman of the Sky. Reproduced with permission from Lathrap (1973a).

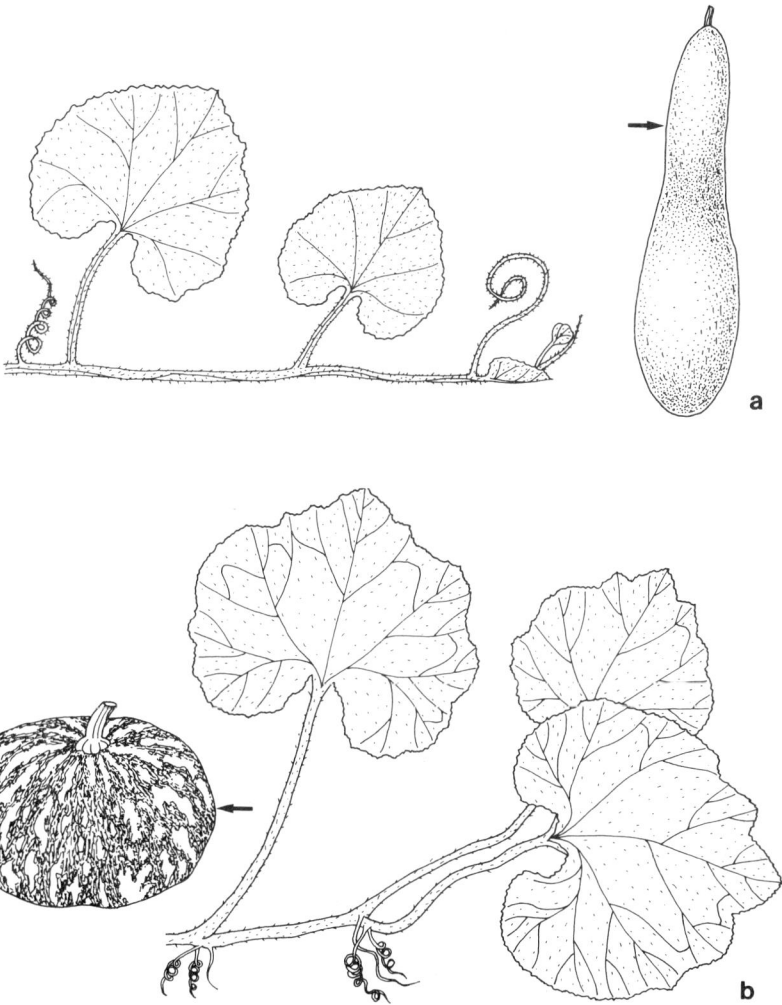

FIGURE 3.13   (a) *Lagenaria siceraria* (bottle gourd). Modified from Brücher (1989). (b) *Cucurbita moschata* (crookneck, butter nut squash). Modified from Purseglove (1968), with permission from Blackwell Science Ltd. Arrows indicate the utilized parts of the plants.

As the archeological data show, bottle gourds were taken into the house garden during the early Holocene. This step would have been most successful in a semidry, tropical habitat such as that of the gourd's homeland (i.e., riparian habitats in deciduous tropical forest). Not every population that had gourds grew them, but it is likely that those populations that used them routinely as net floats did grow the gourds.

Squashes

Bottle gourds and squashes (*Cucurbita* spp.), all in the Cucurbitaceae family, are much alike in growth habit and the appearance of the flowers and fruits. Five species of native squashes were brought under domestication in the New World: *Cucurbita pepo* L., *C. argyrosperma* Huber (= *C. mixta* Pang.), *C. moschata* (Lam.) Poir, *C. maxima* Duch., and *C. ficifolia* Bouché (Figs. 3.13b and 3.14a–d). Although squashes of modern commerce are grown for their flesh, initial cultivation was likely for the edible seeds and usefulness as containers because the flesh of wild squashes is stringy and bitter (because of secondary chemical compounds called cucurbitacins) (Whitaker and Bemis, 1976). As a group, the domesticated squashes and their allied wild forms do not do well in the ever-wet tropics. Rather, they prefer moderate rainfall and are often grown in the dry season (Purseglove, 1968). *Cucurbita pepo, C. maxima,* and *C. ficifolia* are adapted to cool growing conditions and extend into the temperate (*C. pepo*) and Andean (*C. maxima* and *C. ficifolia*) zones.

The five domesticated species are relatively isolated genetically. Although they might hybridize with related/ancestral wild species, they do not cross readily with each other (Andres, 1990; Nee, 1990; Merrick, 1995). Some researchers feel that species grown together for centuries (e.g., *C. moschata, C. mixta,* and *C. pepo* grown in the American Southwest and northwestern Mexico) show evidence of infrequent introgression or the gradual infiltration of the germ plasm of one species into another (Decker-Walters *et al.,* 1990). The domesticated squashes are thus not derived from a common ancestor; each domesticate was derived from a local wild form(s) and spread within its optimal habitat. Some of these habitats overlapped.

Cucurbita moschata is the most likely species to have been domesticated by lowland tropical forest groups. At contact, *C. moschata* was grown from the American Southwest into northern South America, but it was the predominant form grown from Mexico City southward (Whitaker, 1968). It is adapted to the lowlands. It prefers high temperatures and high humidity and is the least cold tolerant of the five domesticated species. The ancestor has not been identified (none of the known wild species are closely related), but possible wild occurrences have been reported in northern Colombia (Nee, 1990).

A wild species of squash related to *C. moschata* and/or *C. argyrosperma* has recently been discovered in Panama (Andres and Piperno, 1995) (Plate 3.3). Populations of this squash are present in areas along the central Pacific coast of Panama that once supported a tropical deciduous forest (Plate 3.4). *Cucurbita argyrosperma* is not grown in Panama today, where *C. moschata* is dominant and exhibits some of the highest diversity of fruit morphology seen today (Plate 3.5). As will be discussed in Chapter 4, it is possible that an ancestral relationship exists between the wild Panamanian

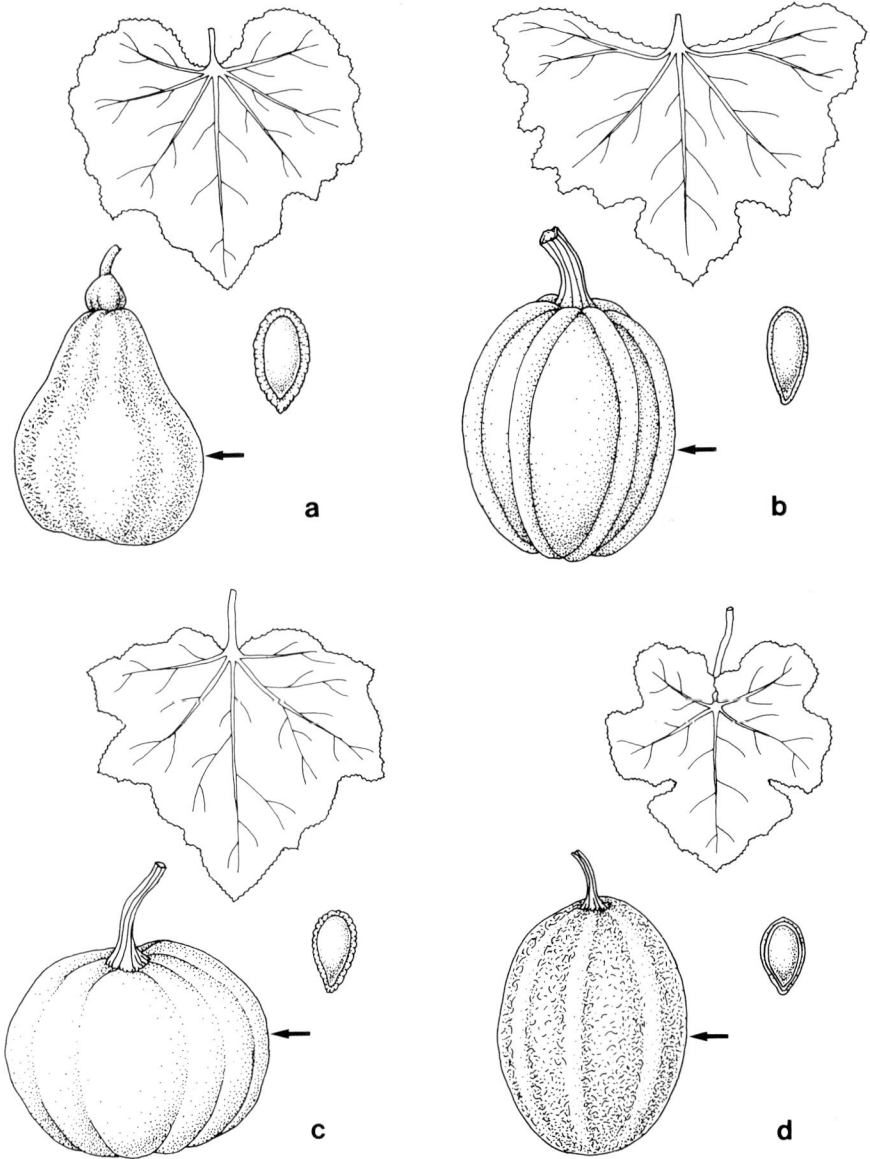

FIGURE 3.14   (a) *Cucurbita argyrosperma* (silverseed gourd, cushaw). (b) *Cucurbita pepo* (summer pumpkin, summer squash, marrow.). (c) *Cucurbita moschata*. (d) *Cucurbita ficifolia* (Malabar gourd). Modified from Lira Saade and Montes Hernández (1994), with permission from the Food and Agriculture Organization of the United Nations. Arrows indicate the utilized parts of the plants.

PLATE 3.3   Fruits of the wild squash *Cucurbita argyrosperma* ssp. *sororia* recently discovered in Central Pacific Panama.

squash population and *C. moschata*. Merrick (1995) believes that *C. moschata* was domesticated in southern Central America or the northwestern tip of South America.

Both allozyme (different molecular forms of an enzyme) (Decker-Walters *et al.*, 1990) and chloroplast DNA data (Wilson *et al.*, 1992) suggest that *C. moschata* and *C. argyrosperma* are "sister" species. The contact period distribution of *C. argyrosperma* is from the American Southwest to the border of Mexico and Guatemala, and the likely ancestral species, *C. sororia* Baily, is native to the lowland thorn–scrub and deciduous forest vegetation of the Pacific coast from Mexico (including the Balsas Valley) to Nicaragua (Nee, 1990; Merrick, 1990). Thus, the distributions of these two lowland squashes, one (*C. moschata*) adapted to hot, humid conditions and the other (*C. argyrosperma*) adapted to more cool conditions and drought tolerant, overlap at the northern edge of the tropics.

Cucurbita ecuadorensis Cutler and Whitaker is a lowland squash species that may have been domesticated and then "lost." Known today only from southwest Ecuador, this species has larger fruits than other wild squashes, and some are not bitter (Nee, 1990) (Plates 3.6 and 3.7). The possibility of a local domestication of squash in the dry tropical forest zone of western South America is supported by phytolith data from the Vegas site (see Chapter 4).

PLATE 3.4   The current habitat of *Curcurbita argyrosperma* ssp. *sororia* in Panama. Before human destruction of the environment, the habitat would have been a tropical deciduous forest.

PLATE 3.5   Some of the many varieties of *Cucurbita moschata* grown by Panamanian campesinos today.

PLATE 3.7   A fruit of *Cucurbita ecuadorensis*. The scale is 10 cm long.

The relationship of *C. ecuadorensis* to other wild and domesticated species is a matter of debate. *Cucurbita ecuadorensis* groups with *C. andreana* and *C. maxima* Naud by chloroplast DNA to form a South American group of allied species (Wilson *et al.*, 1992). *Cucurbita andreana*, a wild squash now found in Uruguay and Argentina (warm temperate zone), is considered to be either ancestral (Nee, 1990) to *C. maxima*, the domesticate grown throughout the western Andean slopes, or derived from a common ancestor (Wilson *et al.*, 1992). In the latter case, *C. ecuadorensis* was likely derived from the same ancestor as *C. maxima* and *C. andreana* (Wilson *et al.*, 1992). We believe it is likely that a cool-tolerant, ancestral species to the South American group occurred widely in the mid-elevation Andes and the warm south temperate zone and was brought under domestication more than once, becoming *C. maxima* in the western mid-elevation zone and *C. ecuadorensis* on the west coast.

At still higher elevations in the Andes, people domesticated *C. ficifolia*, the high-altitude, cool-tolerant species that ranges from the Mexican Plateau to Bolivia (Whitaker, 1968; Andres, 1990). Wild *C. ficifolia* is reported only from Bolivia, despite extensive searches for it in Mexico (Nee, 1990). Cold-tolerant *C. pepo* may have been brought under domestication in the subtropics of northeast Mexico (Andres, 1995).

To summarize, the five extant domesticated squashes of the New World, domesticated initially for their edible, protein-rich seeds, are adapted to a wide variety of habitats in the Neotropics, from the humid lowlands to the cool Andean and temperate zones. This was not the result of the spread and diversification of a single ancestral population but of several domestication processes. The history of two lowland species, *C. moschata* and *C. argyrosperma*, is linked (with common ancestry in the northern tropics, perhaps) as is that of one living (*C. maxima*) and one "lost" (*C. ecuadorensis*) South American domesticate (with common ancestry in the southern tropics).

Cotton

It seems appropriate in a discussion that began with the bottle gourd in its role as a container to end with the plant most closely linked to gourd in its role as a net float: cotton.

There are two domesticated species of cotton in the New World, both of which are allotetraploids (AADD genome): *Gossypium barbadense* L. and *G. hirsutum* L. (Stephens, 1973; Purseglove, 1968) (Figs. 3.15a and 3.15b). In addition to wild populations of these species, there are 4 other wild tetraploid species (one each endemic to Hawaii, Brazil, the Galapagos, and Mexico) and approximately 13 diploid, D-genome species (most in Mexico, with 1 Peruvian and 1 Galapagos species) (Percival and Kohel, 1990). Only the tetraploid species produce spinnable fiber. No A-genome diploid species occur in the New World.

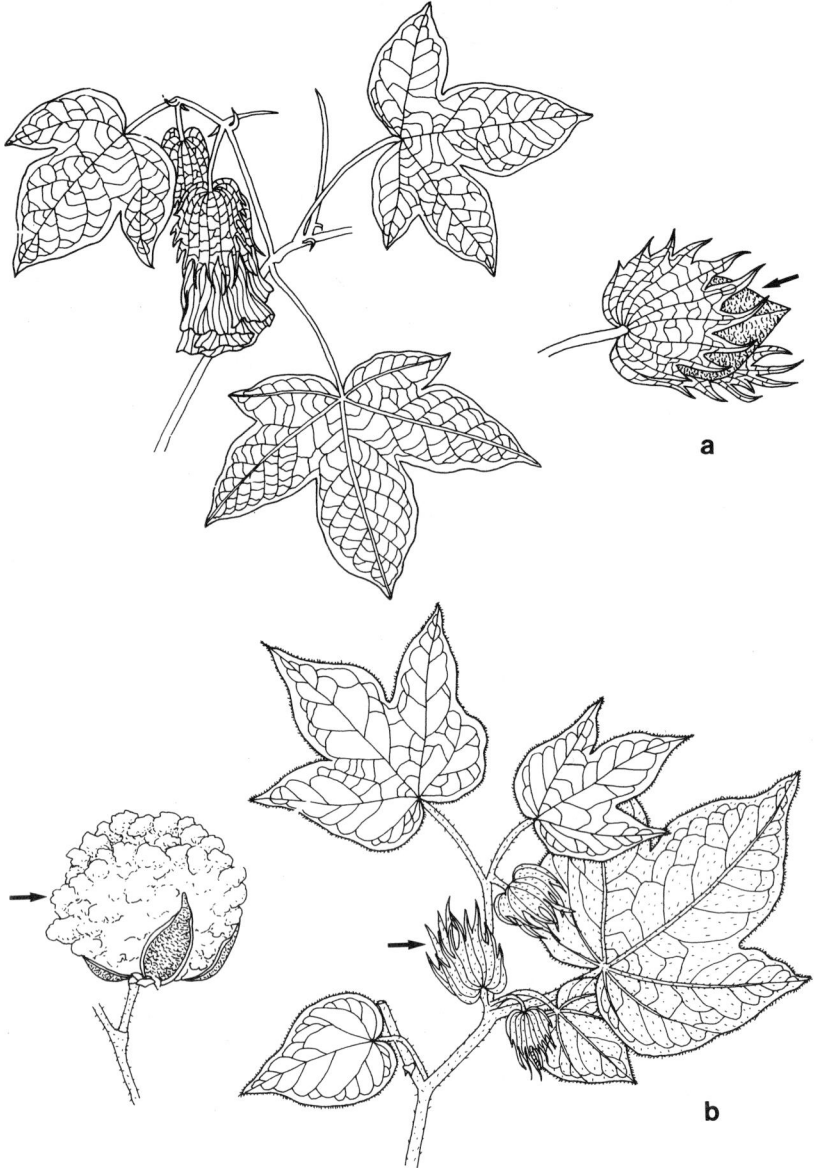

FIGURE 3.15 (a) *Gossypium barbadense* (cotton). Modified from Purseglove (1968), with permission from Blackwell Science Ltd. (b) *Gossypium hirstum* (cotton). Modified from Purseglove (1968), with permission from Blackwell Science Ltd. Arrows indicate the utilized parts of the plants.

Gossypium species are tropical and subtropical in distribution; they are sun-loving, xerophytic plants that do not thrive under shade, heavy rainfall, or cool temperatures. Although cultivated cotton has been adapted to more mesic conditions (landraces of G. *barbadense* exist in the Choco region of Colombia and the central Amazon, for example), it prefers hot, dry conditions (Purseglove, 1968).

The ancient, predomestication history of *Gossypium* has been revealed by chloroplast DNA study of New and Old World cottons (Wendel and Albert, 1992). Two introductions of *Gossypium* to the New World took place. There was a very early dispersal from Africa, leading to the evolution of the D-genome diploids, and a second, later dispersal of the A-genome ancestor of the tetraploids, either from Africa or Asia. In each case oceanic drift of fruits is the likely mechanism of dispersal, and speciation occurred prior to human occupation of the New World.

Wild diploid cottons probably evolved in northwestern Mexico and radiated outwards (Wendel and Albert, 1992). The wild tetraploid species possibly evolved on the Pacific coast of either Mesoamerica or South America. *Gossypium raimondii* (Peru) and G. *gossypoides* (southern Mexico) are from the same stock and are considered the closest living models for the D parent of the tetraploids. The hybridization of the D-genome and A-genome parents to produce the AADD genome allotetraploids is thought to have occurred only once, with subsequent radiation and speciation. Wild G. *barbadense* and G. *hirsutum* emerged from this common tetraploid stock; domestication occurred from the respective wild species at a much later date.

Wild G. *barbadense* is a perennial shrub; the annual habit of modern cultivars is a late development (Purseglove, 1968). Noncultivated G. *barbadense* occurs in dry habitats from the western South American coast to the intermontane valleys of the northern Andes to riverine northern South America. In addition to the previously mentioned regions, the precolombian range of domesticated G. *barbadense* may include Central America and the Caribbean (Percy and Wendel, 1990); G. *barbadense* var. *brasiliense* is widely grown in Amazonia (Phillips, 1976). *Gossypium hirsutum* is also perennial, with some varieties, such as G. *hirsutum marie-galante,* growing to be small trees. *Gossypium hirsutum* occurs throughout the drier regions of the circum-Caribbean area into northeast Brazil and into the West Indies (Purseglove, 1968). Wild populations are documented in the Yucatan, Caribbean, and coastal Venezuela (Percival and Kohel, 1990). The two domesticated species are both grown in Central America, parts of the Caribbean, and northeast Brazil. Crossing between landraces of G. *barbadense* and G. *hirsutum* where they are sympatric is rare; several reproductive isolating mechanisms exist (Percy and Wendel, 1990). However, G. *hirsutum marie-galante* populations in northeast Brazil show characteristics suggesting introgression from G. *barbadense* and G. *mustelinum,* an endemic tetraploid (Percival and Kohel, 1990).

By the time of European contact, distinctive varieties of each species were established that varied in fruit size, time of fruiting, length of fiber (seed coat hairs or lint), and "freedom" of seeds (i.e., amount of fuzz on the seeds) (Purseglove,

1968). Local cultivated varieties (or landraces) are still widely grown as "dooryard" crops, and feral plants of these and of improved cotton varieties can be difficult to distinguish from wild forms (Percy and Wendel, 1990).

An allozyme study by Percy and Wendel (1990) of a diverse group of *G. barbadense* accessions, representing the range of wild, domesticated, and feral populations discussed previously reveals that northwestern South America (Peru, Ecuador, and Colombia) west of the Andes is the center of genetic diversity of *G. barbadense*. This is the likely home of the species. Closely clustered with this group of accessions were inter-Andean specimens. Diffusion and differentiation in the species occurred along two separate pathways: south into Argentina–Paraguay (at probably post-European contact) and into eastern and northern South America east of the Andes (precontact), then from the latter area into the Caribbean and Central America.

The clustering of Amazonian (i.e., *G. barbadense* var. *brasiliense*), Caribbean, and Central American *G. barbadense* by allozymes parallels morphological traits, such as "kidney seed" and long, slender bolls, linking those areas (Percy and Wendel, 1990). Interestingly, the Central American accessions are nested within a Caribbean cluster, which is in turn embedded within the larger east-of-the-Andes cluster. Diffusion of *G. barbadense* into the Caribbean may have occurred after contact (Percy and Wendel, 1990), making it likely that *G. hirsutum* was already present in the region. Unfortunately, the archeological record is silent on the timing of these introductions.

It seems clear from the allozyme data discussed previously that *G. barbadense* emerged as a species in a dry, tropical region of northern South America west of the Andes. The coastal plain of southern Ecuador/northern Peru is a strong possibility given the habit of ocean dispersal noted for the genus (and the possibility that the original hybridization to produce the tetraploids occurred in this region). Was this also where domestication occurred? We think this is likely. Large populations of endemic wild *G. barbadense* occur in Guayas and Los Rios provinces of Ecuador (Percival and Kohel, 1990). Cotton occurs in early Valdivia strata at the Real Alto site in southwest Ecuador (see Chapter 5); Stephens and Moseley (1974) document that cotton from preceramic sites in the Ancon–Chillon area of coastal Peru show features intermediate between present-day wild forms and local cultivars. We think it is likely that the radiation of the species, as documented by Percy and Wendel (1990), occurred after domestication via the house gardens of tropical forest horticulturalists. Although cotton would have been abundant in its preferred habitat— the xerophytic areas of the outer coastal plain—groups occupying better watered inland valleys would have needed to grow this useful plant to ensure a supply. A rapid spread into the interior and the low-elevation Andes is likely; a delay in adaptation of *G. barbadense* to the more humid areas of the Amazonian lowlands might be expected. Eventually, *G. barbadense* would have been spread to groups already growing cotton, in this case, *G. hirsutum*.

There are two major cultivated forms of *G. hirsutum;* shrubby cotton (modern upland cotton cultivars and various landraces) and tree cotton (*G. hirsutum* var.

marie-galante). The best documented wild population is *G. hirsutum* var. *yucatanense,* which is a prominent feature of natural beach-ridge scrub vegetation along the north coast of the Mexican Yucatan Peninsula (Sauer, 1993). Similar wild *G. hirsutum* has been collected in the Caribbean, Baja California, on islands off the Pacific coast of Mexico, and on the dry coast of northern South America (Sauer, 1993). As discussed earlier, the species likely originally emerged along the west coast of either southern Mexico or southern Ecuador/northern Peru from a common tetraploid ancestor with *G. barbadense.*

Sauer (1993) believes *G. hirsutum* was probably rare before domestication, restricted to the narrow littoral zone and to semidesert climates because it cannot compete in dense coastal thickets in humid climates. Following Stephens (1973), he proposes [as does Lee (1984) and Phillips (1976)] that the two cultivated varieties were independently domesticated; the shrubby variety was domesticated on the Gulf coast of Mexico, and then taken inland and northward, and the tree type was domesticated on the Caribbean coast of South America and then spread northwest along both sides of Central America, into the Antilles, and finally down the coast of Brazil in recent (i.e., post-Colombian) times. Percival and Kohel (1990) report considerable variability in *G. hirsutum* collections in Venezuela (including along the north coast), with wild cotton found in large but locally restricted populations. This lends support to domestication somewhere in this region. However, these authors also confirm that truly wild *G. hirsutum* occurs in the Caribbean (along with feral and dooryard stands).

Recent allozyme (Wendel *et al.,* 1992) and DNA (Brubaker and Wendel, 1994) analyzes shed some light on the issue. The allozyme study included 538 accessions of feral, wild, or dooryard *G. hirsutum,* representing broad geographic coverage (but no northern South American material) and morphological diversity (both shrubby and tree types) and 50 modern upland cultivars. The DNA study was based on a subset of these accessions (65 Mesoamerican feral, wild, or dooryard cottons and 23 upland). The accessions studied were found to fall into three lineages: a basal Yucatan Peninsula lineage, a "sister" lineage derived from the Yucatan group encompassing the Mexican accessions (palmeri and latifolium shrubby races), and a Central American lineage composed mostly of marie-galante accessions. Most of the material in the Yucatan group was wild (i.e., race yucatenese) or showed minimal human selection (i.e., race punctatum). Based on this, and the basal position of the Yucatan material, Brubaker and Wendel (1994) suggest that the coastal Yucatan Peninsula was the location of the initial domestication of cotton in Mesoamerica, and that races latifolium and palmeri diverged in southern Mexico and Guatemala from the early cultivar, as did race punctatum. Race punctatum also apparently spread into the western Caribbean. The origin of marie-galante is still obscure, however. There was apparently little gene flow between the Central American lineage and the Mesoamerican races (Brubaker and Wendel, 1994). Introgression between marie-galante and *G. barbadense* has been documented (Wendel *et al.,* 1992).

In light of the information summarized previously, although we accept that wild G. *hirsutum* was brought under domestication on the Yucatan Peninsula (perhaps by populations living in the interior, where wild stands would not have been abundant), a process that gave rise to the shrubby cotton races of Mesoamerica, the possibility of a second domestication in Central America or northern South America has not been eliminated. Wild, feral, and dooryard G. *hirsutum* from the latter area have not been subjected to genetic analysis. We think it is likely that cotton was abundant enough on the dry islands of the Caribbean to meet the needs of the earliest immigrants and that cultivated varieties of cotton were introduced in concert with other lowland domesticates.

Ají: The Spice of Life

It is hard to imagine Latin American cuisine without ají, the *Capsicum* chili peppers. Four species were brought under domestication: *C. annuum* L., *C. frutescens* L. (includes *C. chinense* Joeg.), *C. baccatum* L., and *C. pubescence* Ruiz et Pav. Each possesses an incredible array of forms (Fig. 3.16). Approximately 25 additional wild ají species are native to the New World tropics (Sauer, 1993). Feral, weedy, and intermediate cultivars are common within each species, adding to the taxonomic complexity (Eshbaugh *et al.,* 1983; Pickersgill, 1984). At European contact, peppers were cultivated throughout the Neotropics; valued for the pungent flavor of their fruits (resulting from the chemical capsaicin), peppers are also a good source of vitamins A and C (Purseglove, 1968). Peppers can be grown on a wide variety of soils but require good drainage; waterlogging causes leaf shedding. Too heavy

FIGURE 3.16   Depiction of aji (*Capsicum*); Cayman of the sky. Reproduced with permission from Lathrap (1973a).

rainfall is also detrimental, leading to poor fruit set (Purseglove, 1968). In general, habitat descriptions of wild collections emphasize the wide habitat tolerance of *Capsicum*. Wild Mexican peppers, for example, grow in both xeric areas and inside dense tropical forests (Loaiza-Figueroa *et al.,* 1989). Seed dispersal is by water (plants are often found near streams or steep hills) and through the activity of birds, which favor the pungent fruits.

Capsicum annuum was grown from the southern United States into northern South America; wild forms occur throughout this range (Heiser, 1976). *Capsicum annuum* overlaps in the southernmost part of its traditional range with *C. frutescens,* the most commonly cultivated pepper of the Amazon Basin and Caribbean. Formerly, only wild, weedy, and dooryard varieties of this pepper were called *C. frutescens,* whereas the more advanced cultivars were assigned to *C. chinense*. Botanists now place them all in a single species (Sauer, 1993). Wild forms occur throughout the range of cultivation.

The second lowland South American species, *C. baccatum,* was cultivated along the eastern Andean foothills and adjoining lowlands and into the Gran Chaco, La Plata Basin, and southern Brazil. It was also grown throughout the western South American coast as far north as Ecuador (Heiser, 1976). Wild forms are largely restricted to lowland Bolivia. The final domesticated species, *C. pubescens,* was cultivated widely in the mid-elevation Andes (typically 1500–3000 m) but it can be grown at lower elevations on dry, sandy soil (Eshbaugh, 1979). Its fruits and seeds are somewhat distinctive from those of the other three cultivated peppers, and it does not hybridize easily with them. *Capsicum pubescens* is not known from the wild but shows affinities to two wild peppers native to mid-elevations in Bolivia and Argentina.

The evolution of the three lowland domesticates, *C. annuum, C. frutescens,* and *C. baccatum,* appears to be interrelated (*C. baccatum* more distantly, as discussed later), whereas *C. pubescens* is genetically isolated. South-central Bolivia (McLeod *et al.,* 1982) and the mountains of southern Brazil (Pickersgill, 1984) have been proposed as the center of evolution within the genus, with subsequent migration into the Andes, the Amazon, and Middle America, accompanied by radiation and speciation. Both humans and birds may have played roles in the spread of chili peppers. For example, both utilized species and species with no history of use by humans were transported to Middle America. As Pickersgill (1984) notes, it is difficult to separate the relative importance of these two dispersal agents (and the role of water transport), but we feel the role of humans in the initial radiation of the genus may be underestimated.

A recent allozyme study of *Capsicum* (Loaiza-Figueroa *et al.,* 1989) has somewhat altered our understanding of relationships among the species derived from earlier crossing studies (e.g., Egawa and Tanaka, 1984; Eshbaugh *et al.,* 1983; Eshbaugh, 1976; Pickersgill, 1984). Based on an analysis of genetic distance among *Capsicum* populations, Loaiza-Figueroa *et al.* (1989) demonstrated (i) that *C. pubescens* (Andean) is the most widely diverged of the domesticated species and related wild forms; (ii) that the South American wild species *C. chacoense* and *C. praetemlisum* and domesticated and wild *C. baccatum* (southern South American lowlands) form a second group; and (iii) that *C. annuum* and *C. frutescens* (including *C. chinese*) (Mesoamerica to northern South American lowlands) form a complex of related domesticated, semidomesticated, and wild forms. Within the *C. annuum–frutescens–chinense* complex there are two clusters, corresponding to *C. chinense/C. frutescens* and *C. annuum*. Within *C. annuum* there are three, including a cluster in eastern Mexico that is considered basal to domesticated *C. annuum*. A possible second center of domestication is western Mexico. *Capsicum annuum* accessions from other countries were not included in the Loaiza-Figueroa *et al.* (1989) study, however, making it impossible to exclude *C. annuum* domestication futher south. In Mexico, wild *C. annuum* and *C. frutescens* appear to occupy nonoverlapping regions, but ecological differences between the two are not entirely clear (e.g., wild *C. annuum* appears to have been collected both from "tropical" and "xeric" regions in eastern Mexico). Wild *C. frutescens* appears to prefer moister habitats, as suggested by its distribution in the Amazon Basin and the fact that the center of diversity of the domesticated forms is in Amazonia (Pickersgill, 1984).

In summary, the domestication of *C. baccatum* appears to have occurred in relative isolation from events in the northern area. In the northern South American/Central American area, wild *C. frutescens* appears to have evolved to thrive in the most mesic conditions, whereas wild *C. annuum* adapted to mesic and xeric habitats. Domesticated *C. annuum* eventually emerged in northern Central America (perhaps eastern Mexico, as suggested by Loaiza-Figuero *et al.,* 1989) and *C. frutescens* in the northern South American lowlands. *Capsicum pubescens* likely originated in Bolivia from wild related species (such as *C. eximium* Hunz) native to mid-elevations (Eshbaugh, 1979; Eshbaugh *et al.*, 1983).

Domestication of the *Capsicum* peppers is a complex issue. The distributions of three of the four species are not obviously explained by ecological factors; there is considerable overlap between the Amazonian species, *C. frutescens,* and *C. annuum* and *C. baccatum,* the species occupying the northern and southern tropics, respectively. More research is needed on the ecological preferences of these species, however. Perhaps cultural preference for a diverse array of these spicy, vitamin-rich foods explains the overlapping ranges.

Summary

As a starting point for this discussion of the phytogeography of Neotropical crops, we have used the Obelisk Tello, which we argue (following Lathrap, 1973a) depicted foods of symbolic importance to Chavin culture/belief. These are key plants for Neotropical horticulturalists and fisher folk, constituting reliable carbohydrate sources, vegetable protein and oil sources, net floats and containers, and spices and medicinals. Depiction of two tuber crops highlights the importance of this type of cropping to the tropical forest horticultural system. Most of the crops domesticated in the lowlands and adjacent mid–elevation zones are related to this core group, either taxonomically or by propagation practice. A few interesting ones are missing, however.

PLANTS THE CAYMAN NEGLECTED

Palms and Other Tree Fruits

Descriptions of Neotropical house gardens always note that useful trees form a broken, upper canopy within the garden, re-creating the microenvironment of the tropical forest. Palms, the source of roofing thatch and fruits valuable for oil, papaya, avocado, and various trees in the Sapotaceae family, are among the diverse array either planted in the garden or left when the area is cleared. Useful tree and shrub taxa are also often left when chacras are cleared; families will retain the right to harvest the fruits long after the field is in fallow. Indeed, one way to identify former settlements and swidden areas is to look for clusters of such useful trees, many of which otherwise would be widely dispersed in the forest.

Smith *et al.* (1992) list more than 100 domesticated perennial species of the forests of the Neotropics (and this is exclusive of strictly medicinal and ornamental taxa) (Plate 3.8). Table 3.2 summarizes what we know of the area of origin and use of selected taxa. Given how useful tree taxa are today, and the archeological evidence for the use of many prehistorically, why did the Great Cayman neglect to give these to his people? Perhaps because, appearances aside, tree crop use was not central to the evolution of the tropical forest agricultural system.

We think it is likely that many perennial forest taxa were low-ranking resources for foragers in the early Holocene. This is due to the dispersed nature of many of these taxa and the facts that they are largely not carbohydrate sources and that humans faced stiff competition with other mammals and birds for many of these fruits. With a few exceptions, notably the oil-rich avocado and palms (Arecaceae), most of the trees in Table 3.2 (selected because of their importance and/or occurrence in the archeological record) contribute mostly vitamins and minerals to the diet. They produce sweet fruits favored by nonhuman foragers. Most follow the overall lowland forest pattern of occurring as widely dispersed individuals. Again, some palms are exceptions; batana or seje (*Jessenia bataua*), for example, occurs in monospecific stands in inundated areas, often in densities of 450 trees/ha (Balick, 1989; Schultes, 1989). Species of *Astrocaryum* and *Euterpe* palms can also cover hundreds of kilometers in inundated savannas (Braun, 1968). For the most part, however, return rates for foragers would be low in terms of calories gained.

If, as we propose, people began planting to increase their foraging efficiency, long-generation forest species would not be good candidates for selection because the return would be low in the short run. For example, sapote (*Calocarpum mammo-*

TABLE 3.2   Selected Tree Crops of the Neotropics

Crop	Area(s) of origin; dietary contribution
Anacardiaceae	
Anacardium occidentale L. (cashew)	Semiarid coasts of Venezuela or Brazil; "apples" rich in vitamin C, nuts in fat and protein
Spondias mombin L. (yellow mombin)	Lowland Neotropics; bittersweet fruit
S. purpurea L. (red mombin)	Mexico to Colombia; bittersweet fruit, creamy flesh, vitamins A and C
Annonaceae	
Annona cherimolia Mill. (cherimoya)	Andean South America; one of nine domesticated species; sweet fruit, good source of thiamine, riboflavin, and niacin
A. diversifolia Saff. (ilama)	Pacific lowlands Mesoamerica; sweet fruit, flavor superior to cherimoya
A. glabra L. (anona)	Mesoamerica; sweet fruit
A. muricata L. (guanábana)	Caribbean, northern South America, or Brazil; sweet fruit, good vitamins B_1, B_2, and C
A. reticulata L. (anona)	Mesoamerica; sweet fruit
A. squamosa L. (sweet sop)	Mesoamerica or Caribbean; sweet fruit
Arecaceae	
Acrocomia mexicana Karw. (coyole)	Central America; one of approximately five utilized species; palm wine from cuttings, oil-rich fruits
Bactris gasipaes H.B.K. (peach palm)	Lower Central America to South America; domesticated from several wild populations; starchy fruits, heart of palm, vitamin A, protein, oil
Elaeis oleifera H.B.K. (corozo)	Lower Central America to South America; oil-rich fruits
Jessenia bataua (Mart.) Burr. (batana, seje)	Lower Central America, northern half of South America; a widespread species in the *Oenocarpus–Jessenia* complex of palms; oil-rich and protein-rich fruits
Orbignya phalerata Mart. (= *O. martiana*) (babassu)	Southern Amazonia, related species found throughout lowland Central and South America; oil-rich and starch-rich fruits
Bromeliaceae	
Ananas comosus (L.) Merrill (pineapple)	Southern Brazil, Paraguay; fruits vitamin C, carotene, riboflavin
Caricaceae	
Carica papaya L. (papaya)	Caribbean coast, Central America; one of six domesticated species; fruits vitamin C and A.
C. pubescens Linne and Koch (papaya de monte)	Andean South America; ripe fruits eaten raw, green fruits boiled or baked

Ebenaceae	
Diospyros ebenaster Retz (black sapote)	Mexico–Guatemala; sweet fruit pulp
Lauraceae	
Persea americana Mill. (avocado)	Central America to northern Andes; hybridization between two wild Guatemalan species produced one of the three varieties; creamy fruits high in oil, minerals, and vitamins A, B, and E
Leguminosae	
Inga spp.	Approximately 10 species, in various parts of Neotropics; spongy white pulp is eaten
Prosopis spp. (algarrobo)	Semiarid to arid conditions; fruits milled into flour, protein rich and sweet
Malpighiaceae	
Bunchosia armeniaca (Cav.) Rich (ciruela de fraile)	Amazonia; bittersweet fruits
Byrsonima crassifolia (L.) C.C. (nance)	Lower Central America to Amazonia; bittersweet fruits, good vitamin C
Malpighia punicifolia L. (acerola)	Central America or Caribbean; bittersweet fruits, very high vitamin C
Moraceae	
Brosimum alicastrum Swartz (ramón)	Mexico to Colombia, Brazil; seed kernels have high percentage essential amino acids
Myrtaceae	
Psidium guajava L. (guava)	Neotropics; one of five domesticated species; high vitamins C and A, calcium, iron, and phosphorous
Sapotaceae	
Calocarpum mammosum Pierre (= *C. sapota*, *Pouteria sapota*) (sapote)	Central America; one of approximately four species domesticated in lowland Neotropics; sweet fruit, high in vitamins A and C
Manilkara achras (Mill.) Fosberg (= *Achras zapota*) (sapodilla)	Central America to northern South America; one of approximately five domesticated species; fruits and latex (chicle)
Pouteria lucuma (= *Lucuma bifera*, *P. caimito*, *P. obovata*) (lucuma)	Colombia to Chile; one of 4–10 domesticated species
Sterculiaceae	
Theobroma cacao L.	Amazonia and Mexico, Central America; one of six domesticated species; sweet pulp around seeds, seeds as stimulant

Note. Sources consulted: Balick (1986, 1989), Brücher (1989), Cuevas (1994), Clement (1989, 1994), Furnier *et al.* (1990), Harlan (1992), León (1994), Mahdeem (1994), Mora–Urpi (1994), Morera (1994), Morishidi (1996), Scora and Bergh (1990), Smith *et al.* (1992), Storey *et al.* (1986), and Schultes (1989).

sum) trees take 7 or 8 years to produce fruits (Mahdeem, 1994). However, even casual collection and use of low-ranking tree fruits would have led to the dispersal of seeds around settlements and eventually to locally dense clumps of wild species. In the case of batana, for example, the practice of soaking fruits to soften the mesocarp, making it easier to remove and process (this is the oil-rich portion of the fruit), stimulates germination of the discarded seeds (Balick, 1989). This process— alteration of densities of wild stands—would have accelerated with human environmental disturbance, which, by creating open habitats, permitted invading species to grow in larger clumps. It is no coincidence that most of the species in Table 3.2 thrive in secondary growth and open habitats; some, such as guava (*Psidium guajava*) and babassu (*Orbignya phalerata*), thrive to the point of invasiveness.

We believe the process described previously happened around every settlement in the Neotropics. In other words, we see tree crops following the "dump heap" or incidental route to domestication and propose that this process was secondary, essentially derived from the primary focus on carbohydrate-rich root/tuber resources. Most of the species in Table 3.2 are essentially unaltered from the "wild" state and thus were not the subject of selection beyond range expansion. We are not saying that tree fruits do not contribute to the diet—they are a prime contributor to the breadth that characterizes lowland diet—but rather that they are not the key to understanding the emergence and spread of the tropical forest agricultural systems.

Clement (1989) argues that northwestern Amazonia (northeast Peru and the adjoining areas of Colombia and Brazil) was a center of "domestication" of perennial fruit species. He bases this view primarily on the distribution of large-fruited landraces of peach palm (*Bactris gasipaes*) but argues that six other unrelated perennial taxa whose distributions are centered in this region also show significant modifications due to domestication (increased fruit size and percentage of pulp). All are terra firme species. Floodplain species occurring in continuous distribution along river courses do not show evidence of selection. The six trees could also have been modified by incidental selection pressures by people practicing food production during the early Holocene.

In this, the least seasonal area of the basin and the largest expanse of unbroken moist forest during the Pleistocene (see Fig. 2.4), perhaps return rates on forest taxa, particularly the peach palm, were high enough to lead foragers to focus on these plants to a greater extent than elsewhere in the lowlands. This suggestion depends on whether this area of the basin was well penetrated by humans who were not practicing food production.

The Role of Maize in Tropical Forest Agriculture

The plant most conspicuous by its absence on the Obelisk Tello is, of course, maize (*Zea mays* L. ssp. *mays*). As we argue here, and supported by the data presented in Chapter 4 and 5, this is because maize was relatively unimportant in the Neotropical diet prior to, and in some areas through, the period represented by the Chavin horizon.

The story of the origin and evolution of maize is a complex one that has been the topic of book-length treatments. Briefly, we are convinced by the large body of research that corn originated from teosinte (wild *Zea* native of southern Mexico and Guatemala) rather than from an extinct, wild pod corn (Figs. 3.17a and 3.17b). For recent reviews of each side of this debate, which began in the 1930s, see Beadle (1972, 1977, 1980), Galinat (1975, 1988, 1992, 1995), and Doebley (1990) (teosinte side) and Mangelsdorf (1974, 1986; Mangelsdorf and Reeves, 1939) (pod corn side).

Zea is a small genus of monoecious grasses (i.e., bearing male and female flowers on the same plant) native to the Neotropics. Most researchers follow the taxonomic scheme of Iltis and Doebley (Iltis and Doebley, 1980; Doebley, 1990) and recognize the following species and subspecies: *Zea diploperennis* Iltis, Doebley, & Guzmán;

FIGURE 3.17   (a) *Zea mays* ssp. *parviglumis* (Balsas teosinte), from Doebley *et al.* (1990). *Proc. Acad. Sci.* **87**, 9888–9892, with permission. (b) *Zea mays* (maize). Modified from "Plant Evolution and the Origin of Crop Species" by Hancock, © 1992. Adapted by permission of Prentice-Hall, Inc., Upper Saddle River, NJ. Arrows indicate the utilized parts of the plants.

Z. perennis (Hitchc.) Reeves & Mangelsdorf; *Z. luxurians* (Durieu & Ascherson) Bird; and *Z. mays* L. subsp. *mexicana* (Schrader) Iltis, subsp. *parviglumis* Iltis & Doebley, subsp. *huehuetenangensis* (Iltis & Doebley) Doebley, and subsp. *mays* (maize). The genus most closely related to *Zea* is *Tripsacum;* there are natural hybrids between *Zea* and *Tripsacum* species (Talbert *et al.,* 1990).

Isozyme and chloroplast DNA studies indicate that *Z. mays* subsp. *parviglumis* is the teosinte most closely related to maize; the two share a recent common ancestor (Doebley, 1990). Subspecies *parviglumis* grows today from 400 to 1700 m on the upper slopes of river valleys in the tropical deciduous forest zone of southern and western Mexico, where rainfall ranges from 1250 to 2000 mm per year. There are three natural populations within subspecies *parviglumis*. Of these, maize most resembles the central Balsas population. Thus, the events leading to the divergence of maize from an ancestral teosinte population likely occurred in this region or a similar well-watered, low-elevation habitat in west or south Mexico.

Ribosomal internal transcriber spacer evidence further clarifies the divergence of the species (Buckler and Holtsford, 1996). Divergence of *Zea* species occurred very recently in comparison to the divergence of *Zea* from *Tripsacum* (contra Eubanks, 1995, there is no molecular support for *Tripsacum* playing a role in maize domestication). Subspecies *huehuetenangensis* is established as basal to the group (i.e., it diverged earliest), with subspecies *mays, parviglumis,* and *mexicana* diverging at about the same time, or subspecies *mays* diverging earlier. The early divergence of maize and its large effective population size suggests a terminal Pleistocene/ early Holocene domestication (Buckler *et al.,* 1995; Buckler and Holtsford, 1996). The molecular evidence offers no support for multiple domestications (Doebley, 1990).

The ears (female inflorescences) of maize and teosinte are morphologically very distinctive (Plates 3.9 and 3.10). Five regions of the genome control most of the differences in the infloresences (Doebley and Stec, 1991; Doebley, 1992). These differences probably did not start to arise from sudden phenotypic changes unrelated to human intervention, as proposed by Iltis (1983). Rather, the fixation of a series of stepwise mutations at these few genetic loci under human selection probably produced the changes, which, nevertheless, represent the most dramatic alteration of morphology seen between a major domesticated plant and its wild ancestor (i.e., Galinat, 1975; Doebley, 1990, 1994a,b; Dorweiler *et al.,* 1993).

There is little archeological evidence that *Zea* spp.were commonly used in the wild state. This is possibly because *Setaria,* which lacks the hard fruit case of the teosintes, could be easily harvested from large stands in valleys such as the Balsas and in drier regions, such as Tehuacán, where it grows in association with mesquite and was heavily exploited during the early Holocene (Buckler *et al.,* 1997; Pearsall, 1995). Although teosinte grains can be ground (or popped), *Setaria* may have been easier to process (Pearsall favors this view). Because the probable hearth of maize domestication (the Balsas River valley) has not seen archeological study, the lack of evidence for wild *Zea* exploitation may be purely a sampling error. Also, data

PLATE 3.9   An ear of Balsas teosinte.

on the return rates of these grasses are not yet available, making the two alternatives difficult to evaluate (Piperno's view).

Pearsall believes it is likely that lowland peoples using teosinte for its green fruits or sweet stems encountered a plant with a mutation for a four-ranked spike and collected some of the seeds for planting. Piperno believes that teosinte was, from the beginning, exploited for its seeds, which were either popped or ground before consumption. Whatever the case, successive harvesting and planting of seeds from these plants led to the emergence of the nonshattering maize cob with its naked seeds. Glume softening and kernel nakedness may have been among the final changes to occur (see discussion of the central Panamanian archeological data in Chapter 4).

Many local landraces of maize developed as a result of the spread and diversification of this initial domesticated population. Isozyme evidence indicates that a number of maize races grown in the higher elevations of central and northern Mexico contain isozyme alleles transferred from teosinte subspecies *mexicana* by introgression (Doebley, 1990). Overall, introgression played a minor role in maize evolution. Subspecies *mexicana* is adapted to higher elevations (1800–2500 m) and lower rainfall (500–1000 mm) than subspecies *parviglumis*. The introgression of maize derived from teosinte adapted to moist, warm habitats with a highland teosinte may have been crucial for the adaptation of maize to higher elevations and drier growing conditions in Mexico. This process likely accounts for the delay

PLATE 3.10   An ear of modern maize from Panama.

in the spread of early maize northwards into the central Mexico plateau (see discussion of Tehuacan, Guila Naquitz data, in Chapter 5).

If, as seems likely, the maize that was spread southward from the Balsas region into Central America lacked introgression with subspecies *mexicana* (i.e., maize radiated out from the Balsas, with the populations that became Central American races not passing through the Mexican highlands), then the process of adapting to higher elevations was repeated in the highlands of Central America and again in the higher, and much drier, central Andes. The need for the crop to adapt to conditions very different from its Balsas homeland likely contributed to the very long delay in the establishment of maize agriculture in Peru (see Chapter 5).

As isozyme studies of maize races from Mexico, Guatemala, and Bolivia have shown, genetic variation in maize tends to correlate with altitude, suggesting that factors such as length of growing season, temperature, and available moisture limit crossing among varieties adapted to different elevations (Doebley, 1990, 1994a). Hybridization among distinctive races thus played less of a role in maize evolution than previously thought. The Bolivian study showed that the morphologically diverse Andean races share a common isozyme signature and likely a common origin (Goodman and Stuber, 1983). We think it is likely that early maize was spread throughout the seasonally dry South American lowlands fairly rapidly, and that its introduction into higher elevations probably began in northern Colombia. An isozyme study of Caribbean maize races (Bretting *et al.,* 1987) revealed affinities between those races and maize from both northern South America and the Central American lowlands.

The early maize that was spread by tropical forest agriculturalists south from the Balsas River region had small, hard kernels that would not have been a highly desirable carbohydrate source. The earliest maize certainly would not have competed with root crops, such as *Calathea, Canna,* or manioc, in terms of productivity. Maize kernels are a better source of protein, however, and can be stored without processing into flour. Outside the range of introgression with teosinte, kernel size increased, epidermal thickness decreased, and changes in endosperm hardness occurred, leading to increases in productivity and ease of use of maize. Whether maize replaced tuber crops or remained supplementary to them likely depended on ecological factors (i.e., abundance and timing of rainfall, soil fertility, and length of growing season), dietary considerations (i.e., alternative protein resources), and cultural factors, especially the importance of having a storable, easily portable carbohydrate source.

SUMMARY OF THE PHYTOGEOGRAPHY OF CROP PLANTS

As should be clear by now, our knowledge concerning the area of origin and ancestral species of the crops of the Neotropics varies considerably among crops. The distributions of related species (and likely the species themselves) were probably altered in the late Pleistocene/early Holocene from what is seen today. Based on the best available knowledge of the crops and their related species, and our understanding of their ecology, we delimit the following 10 discrete areas (Fig. 3.18) and three macroregions (Fig. 3.19) of the lowlands as possible areas of domestications:

1. Yucatan, Mexico, area: *Gossypium hirsutum* (cotton)
2. Guatemala area: *Phaseolus vulgaris* (common bean, Middle American type; one of two possible areas)
3. West-central Mexico/Balsas River valley area: *Zea mays* (maize), *Cucurbita argyrosperma* (squash), *Phaseolus vulgaris* (common bean, Middle American type; one of two possible areas)
    Middle America macroregion: *Capsicum annuum* (ají)
4. Central Pacific Panama area: *Cucurbita moschata* (squash, one of two possible areas)
5. Northern Colombia area: *Cucurbita moschata* (squash, one of two possible areas), *Phaseolus vulgaris* (common bean, B phaseolin type)
6. Guiana area (southern Venezuela, southern Guianas, and northern Brazil): *Manihot esculenta* (manioc, one of two areas), *Dioscorea trifida* (yam)
    Southern Central America/northern South America macroregion: *Capsicum frutescens* (ají), *Xanthosoma sagittifolium* (yautia, one of two regions), *Calathea allouia* (leren), *Maranta arundinaceae* (arrowroot), *Canna edulis* (achira), *Ipomoea batatas*

1. Yucatan:
Gossypium hirsutum

2. Guatemala:
Phaseolus vulgaris (Meso.) (1)

TROPIC OF CANCER

Atlantic Ocean

3. West-central Mexico / Balsas:
Zea mays
Cucurbita argyrosperma
Phaseolus vulgaris (Meso.) (2)

Mouth of the Amazon

4. Central Pacific Panama:
Cucurbita moschata (1)

Lagenaria siceraria

5. N. Colombia:
Cucurbita moschata (2)
Phaseolus vulgaris ("B" phaseolin)

6. Guiana:
Manihot esculenta (1)
Dioscorea trifida

7. SW Ecuador / NW Peru:
Gossypium barbadense
Phaseolus lunatus (Andean)
Cucurbita ecuadorensis
Canavalia plagiosperma (1)

8. Western Amazon:
Bactris gasipaes

9. E. Andes / lowland Bolivia / W. Brazil:
Phaseolus vulgaris (Andean)
Arachis hypogaea (low)
Capsicum baccatum (low)
Capsicum pubescens (mid)
Canavalia plagiosperma (2)

10. Central Brazil:
Manihot esculenta (2)

Subtropics & high elevation species:

Cucurbita pepo	Mexican subtropics
C. maxima	Southern Andes, mid-elevation
C. ficifolia	Bolivian Andes, high elevation

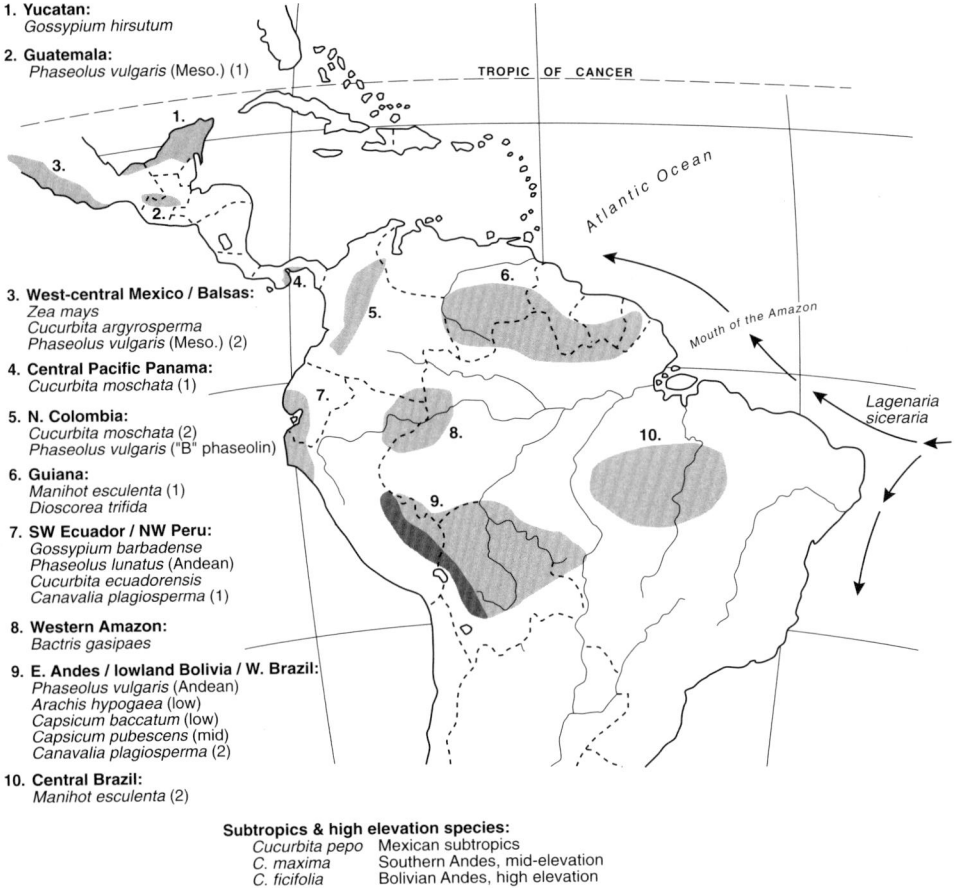

FIGURE 3.18   Map showing likely domestication areas for various crop plants. Crops for which there are two equally likely areas are designated (1) and (2).

(sweet potato), *Phaseolus lunatus* (sieva-type lima bean), *Canavalia ensiformis* (jack bean, probably from dry circum-Caribbean zone)

7. Southwest Ecuador/northwest Peru area: *Gossypium barbadense* (cotton), *Phaseolus lunatus* (Andean lima bean), *Cucurbita ecuadorensis* (squash), *Canavalia plagiosperma* (jack bean, one of two areas)

8. Western Amazonia area: *Bactris gasipaes* (peach palm)

9. Eastern Andes/lowland Bolivia/western Brazil area: *Phaseolus vulgaris* (common bean, Andean), *Arachis hypogaea* (peanut), *Capsicum baccatum* and *Capsicum pubescens* (ajis), *Canavalia plagiosperma* (jack bean, one of two areas)

10. Central Brazil area: *Manihot esculenta* (manioc, one of two areas)

Southern South American macroregion: *Xanthosoma sagittifolium* (yautia, one of two regions).

Middle America
Capsicum annuum

TROPIC OF CANCER

Atlantic Ocean

20°N

Dry Circum - Caribbean
Canavalia ensiformis

Mouth of the Amazon

0°

C. America / N. South America
Capsicum frutescens
Xanthosoma sagittifolium (1)
Calathea allouia
Maranta arundinacea
Canna edulis (N. Andes)
Ipomoea batatas
Phaseolus lunatus (sieva)

20°S

Southern South America
X. sagittifolium (2)

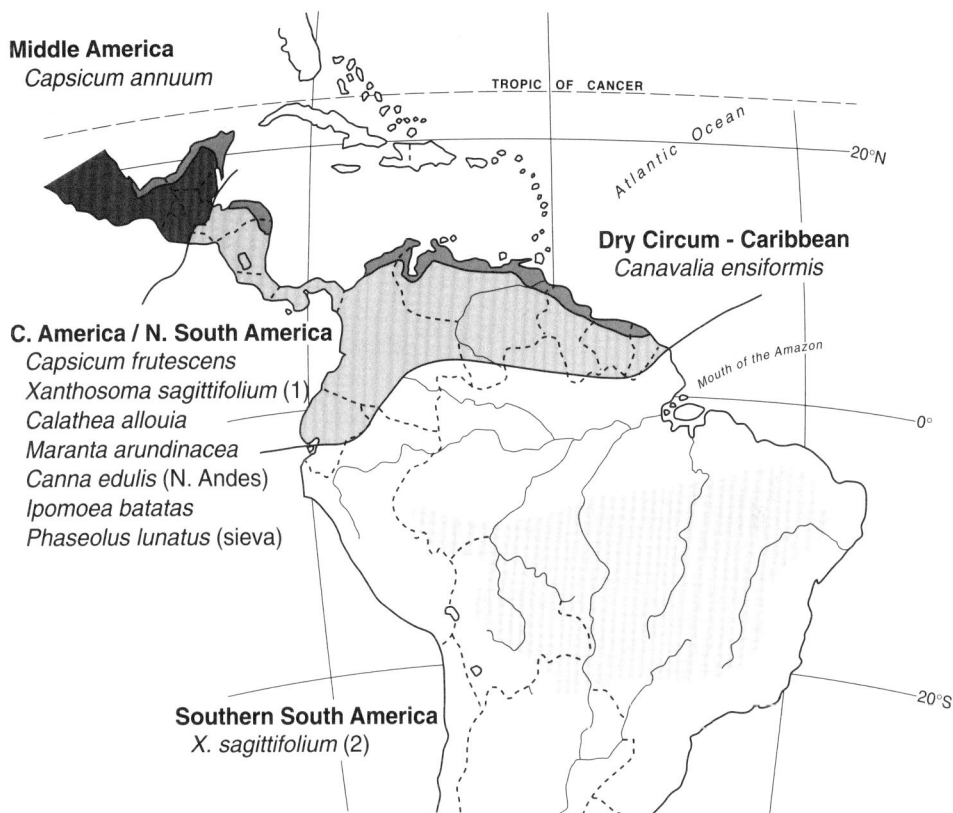

FIGURE 3.19   Map showing likely domestication macro-regions for various crop plants. Crops for which there are two equally likely regions are designated (1) and (2).

Figures 3.18 and 3.19 make clear the broad geographic spread of areas of the potential origin of the crops important in the tropical forest agricultural system (also refer to Table 3.2 for a sample of the tree crops). As discussed in Chapter 2, several independent domestication events in the tropics are likely because the influential environmental processes that favored food production were felt over very wide areas of Central America and South America at about the same time. As will be discussed in detail in Chapter 4, evidence for the beginning of the domestication process appears during the early Holocene in a few of the areas we have delimited previously. All these considerations persuade us that the cultivation of plants began independently in a number of areas in the lowland Neotropics in the early Holocene. We add, however, that we do not take the data to mean 10 independent domestication developments in tropical America during the early Holocene. Some regions and crops discussed may have seen later domestications, which were influenced by interregional contacts and interaction spheres operating during the early and middle Holocene and other factors. Also, future molecular and botanical data may force revisions to the zones of likely or possible domestication events that we have presented here.

The alternative scenario, that food production began in one area and spread rapidly because of the success of the adaptation, with new species of familiar crops brought under cultivation along the way to fit changing ecological conditions (i.e., Lathrap, 1976, 1977a), is not supported by the current data. Pearsall would add that this position is subjective to revision because the early history of plant domestication in many areas east of the Andes is virtually unknown, and early Holocene data points remain few in number.

The Evolution of Foraging and Food Production

INTRODUCTION

In this chapter we present evidence for occupation of the Neotropics by late Pleistocene hunters and gatherers and the subsequent development of food production by some of these groups during the early Holocene. The discussion can be summarized as follows. Between 11,000 and 10,000 years ago[1] foraging populations were living in tropical forests and were both exploiting and modifying the plants of the forest. Between 10,000 and 8600 years ago, intensification of economic practices oriented toward the plant world and human interference with the environment resulted in forms of tropical forest horticulture that emphasized both native tubers and seed plants and that almost certainly involved the deliberate planting and/or management of tree species.

The settlement patterns of the horticulturists of this period were largely characterized by dispersed single or multihouse units that were positioned to exploit small parcels of alluvium along secondary and seasonal streams. Site size and density of refuse lenses suggest that some of these settlements were not occupied year-round or that they were used for no longer than a few years at a time.

---

[1] Years B.P. and years ago refer to radiocarbon years.

Early Holocene settlements that were exploiting estuarine resources along with practicing plant husbandry, and that may have been occupied on a year-round basis, are also apparent in coastal Ecuador. In places such as central Pacific Panama, where there is a close juxtaposition of coastal and terrestrial resources, both kinds of resources were exploited from an early time by populations who were living part of the year in the "interior" locations of the country but within 60 km of the sea.

Domesticated plants were present on the landscape by at least 9000 – 8000 years ago. By 7000 years ago systems of slash-and-burn (or swidden) cultivation similar to those that can still be observed in the modern forest had emerged. For the first time, exogenous (introduced) crops played significant roles in the subsistence cycle of some regions, leading to greater levels of human interference with the environment, which, unlike the earlier food-producing systems, involved the progressive removal of primary forest trees over relatively large areas.

EARLY HUMAN OCCUPATION OF THE NEOTROPICS: WHEN AND UNDER WHAT CONDITIONS DID ADAPTATIONS TO THE TROPICAL FOREST BEGIN?

> *Here human ecology must begin with limnology as known and used by primitive man.*
>
> Carl O. Sauer (1958)

The absolute number of localities in the Neotropics with convincing evidence of human occupation dating to the late and terminal Pleistocene periods (between ca. 13,000 and 10,000 years ago) is small but growing. They are situated on both the Pacific and Atlantic slopes, in the lowlands as well as in the mountains, and have recently been identified from the interior of the Amazon Basin itself (Roosevelt *et al.,* 1996). Many of these sites have ^{14}C dates and/or lithic implements that suggest occupation shortly before or within 1000 years after 11,000 B.P., although whether all of these are Paleoindian (Clovis) in cultural affinity is a matter of debate (e.g., Fiedel, 1996; Gibbons, 1996). Clovis is the first stage of the Paleoindian period, representing the first certain human culture identified in the New World (Bonnichsen and Fladmark, 1991; Stanford, 1991). Clovis sites in the United States are well dated to between 11,200 and ca. 10,600 B.P. (Haynes, 1991). Sites from the Central American tropics, such as Los Tapiales, Guatemala, Turrialba, Costa Rica, and La Mula, Panama, contain projectile points and other types of stone tools virtually identical to those found in Clovis sites from the United States, indicating a similar time frame for a tropical occupation by these groups (Ranere and Cooke, 1991).

Some sites identified in the tropics (i.e., Taima-Taima and El Abra) are probably 1000–3000 years earlier than those associated with the Clovis tradition. The recent acceptance of a ca. 13,000 B.P. occupation at the site of Monte Verde in southern South America (Dillehay, 1989) by a consensus of early man experts (Meltzer, 1997) makes it ever more likely that sites such as Taima-Taima and El Abra date to approximately this age (for recent overviews of early human populations in the tropics, see Cooke, 1997; Dillehay *et al.*, 1992; Lynch, 1983, 1990; Meltzer *et al.*, 1994; and Ranere and Cooke 1991, 1995, 1996).

A discussion of late and terminal Pleistocene human adaptations in the Neotropics (ca. 13,000–10,000 B.P.) is needed here because it is becoming clear that, as in other areas of the world such as the southern Levant and the Far East, food production emerged shortly after this time. Thus, people were developing increasingly close relationships with certain elements of their plant repertoire during that period when, as discussed in Chapter 2, the world's climate and vegetation were in a state of considerable perturbation after the final retreat of the glaciers and were moving rapidly toward their modern forms.

A discussion of the protracted controversy over the antiquity of a human presence in Middle and South America is beyond the scope of this book. We feel obliged to point out, however, that we have divergent points of view. Pearsall accepts entry into the New World by approximately 20,000 B.P. Until recently, Piperno identified most strongly with the conservative group that espouses late entry by peoples of Clovis age. Monte Verde's acceptance as a site of the pre-Clovis cultural tradition leads her to believe that tropical America was probably colonized by ca. 13,000 B.P. However, Piperno does not accept that sites such as Pikimachay Cave in Peru (MacNeish, 1979; MacNeish *et al.*, 1981) or the Pedra Furada rock shelter in Brazil (Guidon and Delibrias, 1986; Guidon, 1989) have produced a valid cultural sequence older than 13,000 B.P.

Our divergent thoughts on this matter make little difference to the arguments we develop in this book because, in either scenario, humans would have been occupying the Neotropics during the latter stages of the last glacial cycle when climate and vegetation were still at full-glacial modes. By both scenarios, humans would have encountered profoundly changing environments after initial colonization. Clovis-age groups would have had to deal with these environmental perturbations shortly after their first penetration of the tropics.

A significant issue regarding early humans and the emergence of food production is the antiquity of human adaptations to the tropical forest. Early scholars assumed that the first colonizers of the New World were hunters and gatherers exploiting mainly now-extinct large game. They felt that limited faunal availability in tropical forest made this biome unfit for habitation and a barrier for human migration from north to south (e.g., Sauer, 1944, Lothrop, 1961). Consequently, they proposed that early human adjustments to the tropical biome took place largely in open landscapes on the drier Pacific watershed of Central America, in northern South America, and in the intermontane valleys of the Andes.

Contemporary scholars have largely agreed with this assessment (e.g., Lynch, 1990; but see Bryan, 1989 and Roosevelt, 1989). Dillehay (1991) took the argument beyond subsistence issues and considered that tropical forest would have presented early populations with unacceptable health problems, largely in the form of new pathogens and diseases, which made occupation of forest undesirable and generally disastrous if attempted. Ranere, on the other hand, has long argued (Ranere, 1980a; Ranere and Cooke, 1991, 1996) that early human (Paleoindian) populations moving through the Central American Isthmus would have traveled through and exploited tropical forest habitats. He felt that movement through a wide range of habitats was possible because hunting, at least initially, was their primary subsistence strategy rather than plant exploitation. Ranere argued that for populations entering a new region, plant resources would be harder to obtain and process than faunal resources. Reduction in game species and increasing human population densities during the latest parts of the Pleistocene possibly led to increased emphasis on plant exploitation.

Ranere's hypothesis rested on the fact that high annual precipitation, characteristic of such Paleoindian localities as Turrialba (Snarskis, 1979) and Madden Lake, Panama (Bird and Cooke, 1978), indicated that they were still covered by forest during the late Pleistocene, an argument supported by recent paleoecological data. Ranere's approach of correlating adequately reconstructed paleoenvironments with late Pleistocene archeological sites still provides the best insight into the question, given the rarity of archeological subsistence information.

The past 10 years have seen a significant increase in archeological and paleoecological data from the Neotropics, placing us in a stronger position to correlate site location and paleovegetation. (For a detailed discussion of the nature and timing of environmental change during the late and terminal Pleistocene, see Chapter 2). If Paleoindian and earlier sites are largely confined to open areas, the implication is strong that forest was avoided. If, however, settlement before approximately 10,000 years ago can be consistently documented in forests, then this biome played more important roles in early subsistence than originally conceived and some subsistence orientations must almost certainly have been broader than specialized big-game hunting. When archeological site locations are plotted against the reconstructed late-glacial vegetation presented in Chapter 2 (Figs. 2.4a and 2.4b), the patterns shown in Figs. 4.1a and 4.1b emerge.[2]

The sites where a pre-Clovis presence may be indicated are located in open deserts/woodlands/grasslands at low elevations in northern South America (Taima-

---

[2] We do not pretend to be exhaustive in our coverage of early archeological sites in the tropics. Because many sites are controversial and not well dated, we offer an assessment of what we feel is a reasonable sampling of the most well-known and well-documented sites. In choosing which sites from the Neotropics identified as pre-Clovis to include in Fig. 4.1a and 4.1b, we relied partly on the assessments of Dillehay et al. (1992) and Meltzer et al. (1994) concerning the reliability of cultural materials and dates for individual sites.

Taima and Río Pedregal) and open environments (páramo) of the northern Andes (Tibitó and El Abra). We think that the number of sites is currently too few to enable firm conclusions concerning settlement and subsistence, although now-extinct game species were hunted at Taima-Taima and Tibitó.

On the other hand, Paleoindian-period sites appear to have been located in a diverse array of environments, including alpine meadow (Los Tapiales, Guatemala), high- (Turrialba, Costa Rica, and La Elvira, Colombia) and low-elevation forest (Corona and Madden Lake, Panama, and the middle Magdalena Valley, Colombia), and open thorny scrub/savanna types of vegetation (La Mula, Panama; Lowe Ranch, Belize; and El Cayude, Venezuela). [For purposes of simplicity, we call all sites occupied between 11,000 and 10,000 B.P. Paleoindian, even though some of them (e.g., Caverna de Pedra Pintada, Brazil; El Abra, Colombia; and Talara, Peru) do not contain characteristic Clovis-type tools.] In some cases (e.g., Arenal and Turrialba, Costa Rica, and Madden Lake and Corona, Panama) paleoecological sites occur so near to the archeological sites as to permit good documentation of the environmental setting for human occupation. The reconstructed environment is tropical forest.

The Panamanian forests that were occupied at an early time were drier and more open than those that exist today (Fig. 4.2). For example, the Madden Lake forest was an arboreal association that held many more deciduous trees than the semi-evergreen forests currently typical of the area, and the same was probably true of the forest surrounding the Corona site (Piperno et al., 1992). Hence, large and medium-sized game availability may have been greater than characteristic of forest today. At the Caverna de Pedra Pintada near Monte Alegre, Brazil, in the lower Amazon about 10 km north of the current Amazon River channel (Roosevelt et al., 1996), settlement between ca. 11,000 and 10,500 B.P. was in an area that we reconstructed as having supported a tropical forest during the Pleistocene. Numerous ^{14}C dates on charcoal and carbonized plant remains clearly document occupation of the cave and use of both tropical forest and riverine resources between approximately 10,800 and 10,000 B.P. Plants exploited included palms and other tree fruits, such as the Brazil nut. Faunal remains were not well preserved in the Paleoindian levels of the site but, importantly, they included fish, shellfish, and river turtles, as well as rodents and large land animals of a size consistent with the forest tapir. Hence, it is fairly clear that the earliest inhabitants of Pedra Pintada were exploiting the considerable freshwater faunal resources of the nearby Amazon River, including the larger fauna such as turtles.

Roosevelt and co-workers suggest that the presence of certain trees near the site argue for an Amazonian forest largely unchanged by cooling and drying during the Pleistocene. We disagree. A very narrow subset of plants from an archeological site may not be useful for a general environmental reconstruction, especially in an area that may have supported a forest before the onset of the Holocene. Also, the radiocarbon dates on tree seeds from Pedra Pintada and their error ranges suggest

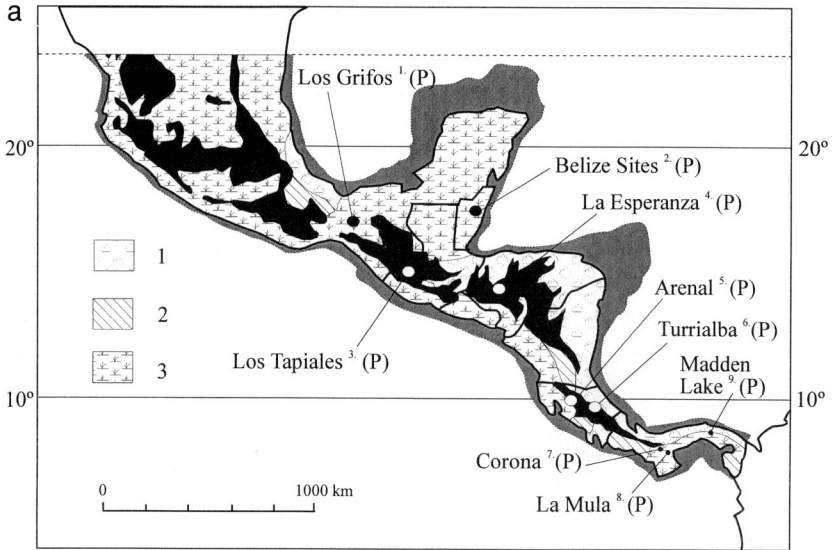

FIGURE 4.1   (a) Location of Paleoindian sites in Middle and Central America plotted against the reconstructed late Pleistocene vegetation. 1, García-Bárcena (1982); 2, MacNeish et al. (1980) and MacNeish and Nelken-Terner (1983); 3, Gruhn et al. (1977); 4, Bullen and Plowden (1963); 5, Sheets and McKee (1994); 6, Snarskis (1979); 7, Ranere and Cooke (1991) and Cooke and Ranere (1992a); 8, Ranere and Cooke (1991, 1996); and 9, Bird and Cooke (1978). P, Paleoindian site; Gray area, land exposed by sea level drop. In most cases, exposed land contains vegetation similar to that of adjacent terrestrial zones. (b) Location of pre-Clovis and Paleoindian sites in South America plotted against the reconstructed late Pleistocene vegetation. 10, Cruxent (1970); 11, Oliver and Alexander (1990); 12, Gruhn and Bryan (1984); 13, López Castaño (1993, 1995a,b); 14, Correal Urrego (1986) and Correal Urrego et al. (1966–1969); 15, Correal Urrego and Van der Hammen (1977); 16, Correal Urrego (1981); 17, Gnecco Valencia (1994, 1995); 18, Roosevelt et al. (1996); 19, Richardson (1978); 20, Lynch (1980); 21, MacNeish (1979) and MacNeish et al. (1981); and 22, Dillehay et al. (1992). PC, pre-Clovis site; P/P, pre-Clovis and Paleoindian site; P, Paleoindian site; Gray area, land exposed by sea level drop. 1, Largely unbroken moist forest, often with a mixture of currently high-elevation and lowland forest elements. In some area, montane forest elements (e.g., *Podocarpus, Quercus, Alnus,* and *Ilex*) are conspicuous. Annual precipitation was lower than it is today, but sufficient precipitation exists to support a forest. 2, Forest containing drier elements than characteristic today. High-elevation forest elements occur, especially in moister areas of the zone. Areas near the 2000-mm precipitation isohyet and areas with sandy soils may contain savanna woodland. The vegetation may be patchy. 3, Mostly undifferentiated thorn woodland, low scrub, and wooded savanna vegetation. Some regions (e.g., Guatemala) have temperate elements (e.g., *Juniperus*). Areas receiving greater than 2000 mm of rainfall today may still support a drier forest, as in No. 2. River- and stream-side locations support a forest. 4, Probably substantially drier vegetational formations than No. 5, with fewer trees and more open-land cerrado and caatinga taxa. Paleoecological data are lacking for the zone. 5, An open and fairly humid forest containing many currently high-elevation taxa (e.g., *Ilex, Podocarpus, Rapanea,* and *Symplocus*) combined with elements of the modern semi-evergreen forest and cerrado. Northward shifts in the southern polar fronts and other factors ameliorate the general precipitation reduction experienced elsewhere. The modern, seasonal forest–cerrado vegetational formations of the region do not appear until approximately 10,000 B.P. 6, Desert/cactus scrub.

b

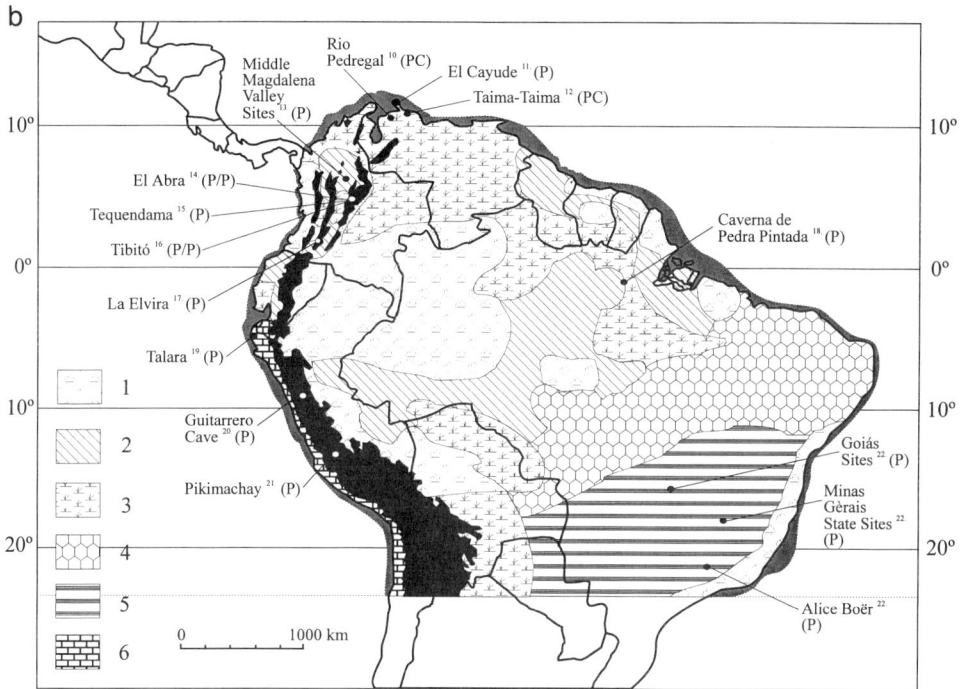

FIGURE 4.1   (*Continued*)

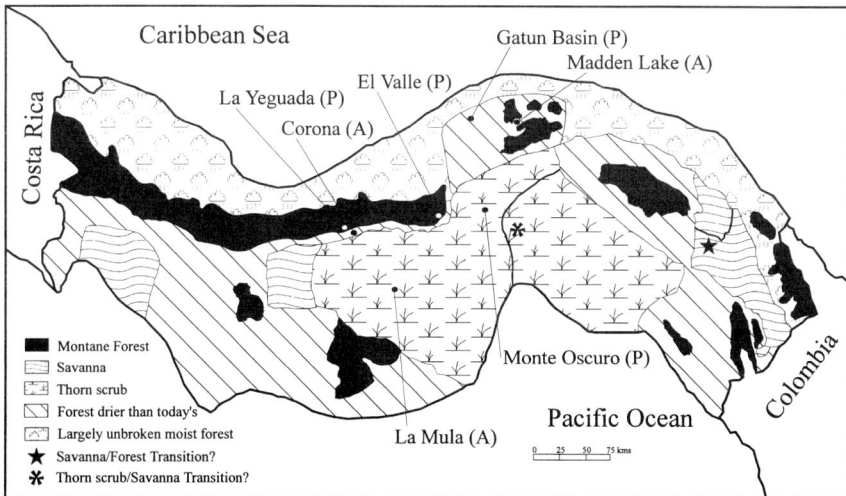

FIGURE 4.2   The environments of late Pleistocene Panama. P, Paleoecological site; A, Archeological site.

to us that they were exploited beginning ca. 10,800–10,500 B.P., when the post-Pleistocene floral changes were surely already under way.

Ranere and Cooke (1991), in playing their own devil's advocate, noted that Paleoindian site finds in forested areas in Panama do not necessarily mean that people were spending significant intervals in the forest and exploiting its subsistence resources. It could be argued that early sites, which have yielded very few direct subsistence data, simply represent forays into the forests to hunt and/or gather and/or to obtain other resources (e.g., lithic raw materials) from a home territory with a more open savanna setting.

In Panama, where four paleoecological records spanning the late Pleistocene and early Holocene periods have been retrieved and three Paleoindian sites identified, the late-glacial vegetation of the Pacific slopes appear to have been heterogeneous and patchy (Fig. 4.2). Lake Madden and Corona are clearly located in forest but were only a short distance from land holding open types of vegetation, making the open-terrain home territory with short forays into the forest scenario credible. However, Panamanian paleoecological data (discussed later) indicate that forest settlement around a lake, La Yeguada, between 11,000 and 10,000 B.P. probably cannot be characterized as short and casual forays, although this settlement, like that at Caverna de Pedra Pintada, may initially have been substantially oriented around the exploitation of *aquatic,* in this case, *lacustrine* resources.

Caverna de Pedra Pintada may also have been located in an ecotone near open terrain. Our reconstruction of the Pleistocene Amazonian vegetation predicted that a wide corridor of considerably drier, open land existed just to the east of the site. Today, patches of savanna and drier forest on sandy soils can still be found near the site (Bush *et al.,* 1997). In fact, early settlement in this part of the Amazon might be explained by the fact that people would have been able to move quickly down the eastern part of northern South America through an open corridor into the area, a route that would have been easier than slogging through the wet forests that grew over large areas of the basin to the northwest of the site.

Overall, the evidence linking early archeological sites and environmental context is strong in indicating some kind of early adjustments to the tropical forest. Out of 22 Paleoindian localities noted in Figs. 4.1a and Fig. 4.1b, 9 were probably located in some kind of forest. Humans could not easily have passed through southern Central America and entered South America without living in a tropical forest some of the time. It seems that Paleoindians may have been exploiting a number of different vegetation types and subsistence resources. To characterize Paleoindians either as narrow megafaunal specialists or as generalized hunters and gatherers is probably inaccurate. This is also true because many Paleoindian sites have not yielded materials suitable for dating, and we cannot follow possible trends in Paleoindian habitat preference and subsistence from 11,000 to 10,000 B.P. (a point to which we return later).

In addition to the correlation of archeological site location and paleoenvironment, paleoecological records from Panama are providing insights into the permanency of early settlements in forest and are helping to clarify the nature of early human interactions with the tropical forest.

Paleoecological Evidence for Tropical Forest Occupation and Modification

Lake La Yeguada is a large (1.5 × .75 km) body of water located at an elevation of 650 m in the central Pacific watershed of Panama (Fig. 4.2; Plate 4.1) (Piperno *et al.*, 1990, 1991a,b; Bush *et al.*, 1992). The lake's watershed was covered by tropical evergreen forest until it was removed by humans between 7000 and 4000 years ago. Archeological signals of human occupation of the watershed are evident by 10,000 B.P., from lithic implements characteristic of that time (Weiland, 1984). A point recovered on the lakeshore may be Paleoindian in age, although it is broken at the base and not unequivocally identifiable. If not Paleoindian, it certainly dates to the earliest part of the Holocene, however (Ranere, 1992).

A disturbance horizon indicating frequent firing of the forest and small-scale land clearing begins shortly after the climatic snap that represents the termination of the Pleistocene at ca. 11,000 B.P., after being absent from the record during the first 3000 years that the lake was in existence and accumulating sediment (Piperno *et al.*, 1990) (Fig. 4.3). It is manifested by a sudden and large rise of microscopic charcoal along with plants typical of forest gaps such as *Heliconia* and grasses. More than 70% of the *Heliconia* phytoliths and 5% of the grass phytoliths were charred, providing direct evidence that they had been fired. The interpretation of charcoal in lake sediments can be a tricky undertaking. Natural fires, although they are rare and of minimal intensity, sometimes occur today in tropical evergreen forest (e.g., Uhl *et al.*, 1988). They might contribute something in way of measurable charcoal to a paleoecological profile.

However, it is highly unlikely that the La Yeguada disturbance patterns were the result of natural perturbations for several reasons. Charcoal suddenly increased by more than several orders of magnitude at ca. 11,000 B.P. after being virtually absent during the late Pleistocene, when the climate was much drier. High levels of charcoal were sustained across the Pleistocene–Holocene boundary of increasing precipitation, and they are consistent with levels later in time when human influence is certain.

Charcoal influx can also be compared to that at the modern surface of La Yeguada as well as in another lake in Panama where human burning is not currently undertaken in the watershed forest (Piperno, 1994). The amount of charcoal injected into these sediments annually by natural fire is far lower than that recorded at 11,000 B.P. at La Yeguada.

Although the charcoal evidence is persuasive, what is perhaps a more compelling line of evidence is found in the frequencies of plant disturbance indicators in the phytolith record and the proportions of these that show evidence of having been

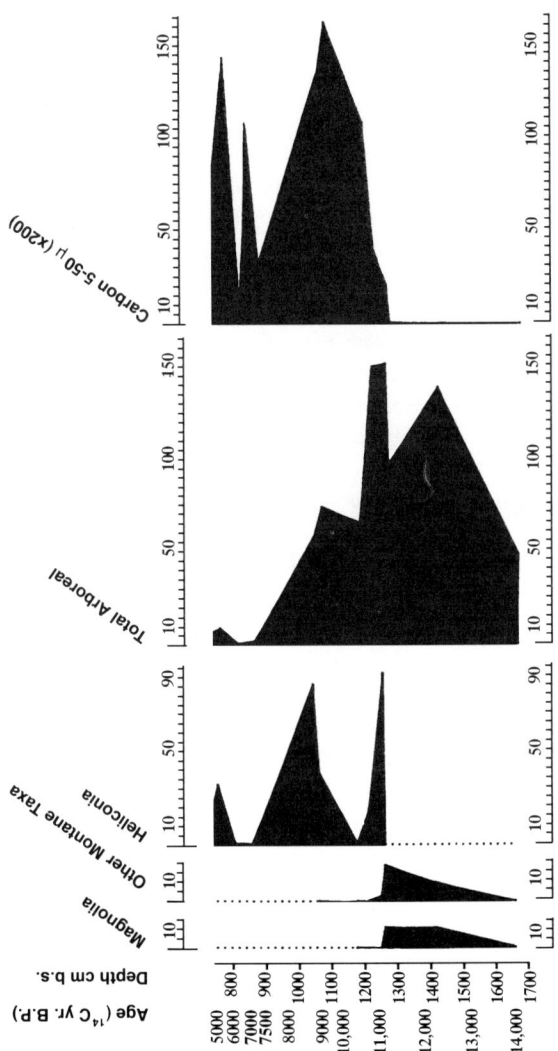

FIGURE 4.3 A summary phytolith and charcoal profile from Lake La Yeguada showing the Pleistocene disturbance horizon and its continuation and intensification during the early Holocene. Phytolith and charcoal values are expressed in absolute rates of deposition (20 cm^2 $^{-1}$ yr^{-1}). Total arboreal = the sum total of all arboreal and arboreal-associated phytoliths, which are largely from the primary forest. Carbon = charcoal.

burned (Fig. 4.4). When a plant is burned it does not just leave behind charred organic debris. If the plant holds phytoliths, then these are also likely to have a black coating indicating charring, while the morphology of the phytolith remains intact. Thus, knowledge of precisely which plants have been subjected to burning and at what frequency is possible through quantification of the percentages of burnt phytoliths in any taxon.

Figure 4.4 shows that percentages of phytoliths from such plants as *Heliconia* and the Gramineae, plus the proportion of phytoliths from these plants that were

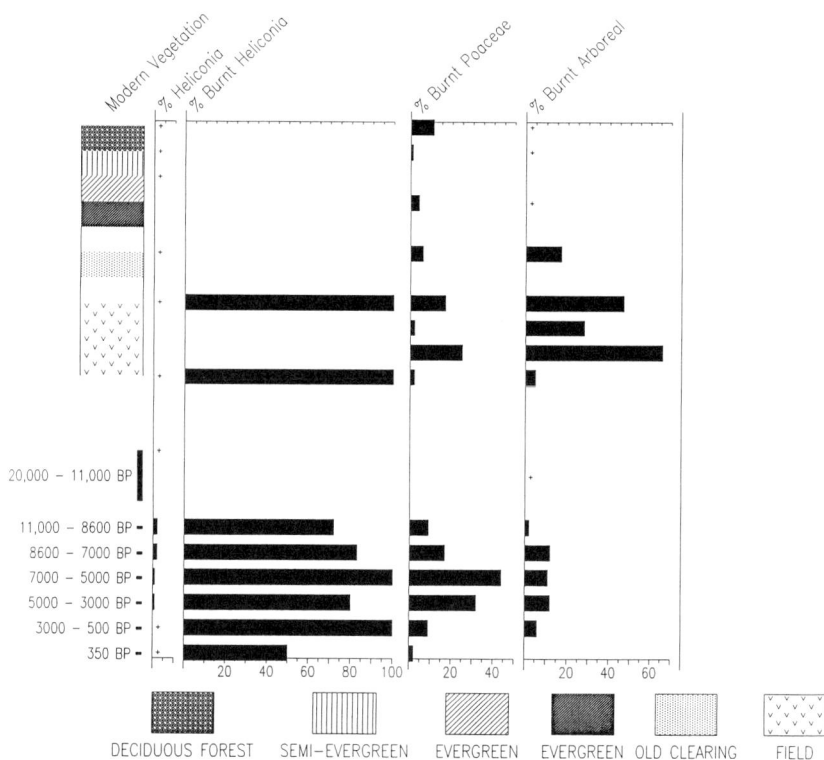

FIGURE 4.4   The frequencies of successional phytoliths and burnt successional and arboreal phytoliths in modern forests and through time at Lakes La Yeguada and El Valle. Modern Vegetation: Evergreen forests are from El Cope, Panama, and north of Manaus, Brazil. Semi-evergreen forest is Barro Colorado, Panama. Deciduous forest is Guanacaste Province, Costa Rica. Old clearing (cleared from forest and planted with banana 50 years ago) is from Guanacaste, Costa Rica. Fields are present day slash-and-burn agricultural plots from Panama planted in manioc and maize. Phytolith frequencies for each site are averages from a series of soil transects or "pinch samples" taken from the upper soil surface at the sites (see Piperno 1988a, 1995a, for details). Records for 20,000–11,000 B.P. are from El Valle and La Yeguada. Records for 11,000–350 B.P. are from La Yeguada. +, observed at a frequency of less than one percent; B.P., uncorrected radiocarbon years before the present; Poaceae, Gramineae.

burnt, are much higher at La Yeguada beginning at 11,000 B.P. than in soil samples taken from underneath modern Panamanian forests where the sole causes of alteration are from natural processes. On the other hand, they are consistent with phytolith frequencies in modern vegetation under active modification by humans (see also Piperno, 1995a, pp. 147–148). *Heliconia,* an important indicator taxon for early successional herbaceous growth, invades the edges of human clearings much more frequently than it does the limited open area typically created by natural tree-fall gaps (D. Piperno, personal observation). It is then quickly replaced by woody successional taxa in 1–3 years after the clearing or gap begins to regenerate.

In order to maintain a large proportion of these plants, which are also burned, as is characteristic of late Pleistocene and early Holocene deposits at La Yeguada, it seems that forest openings of a substantial size (when compared to natural tree-fall gaps) must be made frequently and then also subjected to further disturbance, including fire, a short time later. It can be seen in Fig. 4.4 that *Heliconia* is nearly invisible in modern phytolith assemblages from various types of forest in Panama, Costa Rica, and Brazil not currently under human pressure. Also, when present in these forest records the phytoliths are never burned. Only in modern anthropogenic vegetation do frequencies of burnt *Heliconia* approach those seen at La Yeguada.

Thus, it seems that beginning approximately 11,000 years ago, humans repeatedly subjected the forest around La Yeguada to fire and small-scale clearings, strongly suggesting that settlement there involved more than short-term excursions into the forest from base camps located elsewhere to procure raw materials. In addition to the evidence for frequent fires, there is a measureable vegetation response that likely came from persistent interference with the primary forest by people living there on a regular basis and exploiting its resources. It is to be expected that early humans would have sought habitats such as lakes to camp around because the edges of these water bodies would have offered a considerable supply of wild fauna and useful wild plants to exploit. During the driest parts of the late Pleistocene year, other sources of water may have been difficult to find, and the mammalian fauna and humans alike may have sought the edges of lakes and other permanent freshwater bodies. Semisedentary occupations by small groups of people would probably have been possible. It is very possible that the resources of the lake edge and not the forest were the primary attraction for the first occupants of the La Yeguada watershed.

Few other late-glacial/early Holocene paleoecological records have been examined in as much detail as have those of Lake La Yeguada. Hence, it is not surprising that similar trends have not been noted at all of them. However, some lake paleoecological sequences from the tropics are demonstrating frequent forest fires during the early Holocene after showing little evidence of fire during the drier late-glacial period (e.g., Colinvaux *et al.,* 1997; Behling, 1996). At Lake Curuca, located 100 km northeast of Belem at the mouth of the Amazon River (Behling, 1996), frequencies of burnt grass and palm phytoliths at 9500 B.P., when charcoal levels rise dramatically, are consistent with human and not with natural disturbances

(D. Piperno, unpublished data). Phytolith data are needed to confirm the anthropo-genic nature of the other disturbances, but in light of the La Yeguada data this interpretation seems increasingly likely.

Summary: Implications of the Archeological and Paleoecological Data for Late Pleistocene Subsistence and for the Transition to the Holocene

The general dearth of subsistence information from the earliest archeological sites occupied in the tropics precludes a general discussion of diet drawn directly from animals or plants used by people. However, assuming that energetic efficiency was a primary variable conditioning settlement and subsistence strategies, and drawing from existing archeological and paleoecological data, we can ask some questions and make some suggestions about subsistence and possible subsistence changes during the late and terminal Pleistocene periods. Operating under conditions of energetic constraints, humans should have first exploited habitats in which return rates were relatively high and then moved into more "unfavorable" habitats after those first occupied began to experience drops in return rates. Habitats with the least costly resources should, for the most part, have been those in which the now-extinct bigger game were to be found in the highest numbers. We add that in a patchy habitat, variation in habitat exploitation could well occur during the course of a single year (Simms, 1987).

We think that the following picture of late Pleistocene and early Holocene human adaptation is most relevant to areas of the tropical lowlands that today are subject to highly seasonal precipitation regimes and support a potential vegetation of semi-evergreen and, especially, deciduous forest on fairly fertile soils, as described in Chapter 2. In Central America, they include much of the Pacific watershed from Mexico to northwestern Costa Rica, including the crucial Balsas River Valley of Mexico, in addition to Pacific-side Panama. In northern South America, they also comprise a considerable area, including the Colombian Cauca and Magdalena valleys, Venezuela, the southern Guianas, northeastern Brazil, and southwestern Ecuador, where important tubers and species of squash were domesticated.

The following question arises: Why were the occupants of the La Yeguada watershed firing and disturbing the vegetation? Perhaps to increase the abundance of animal game and the biomass of useful plants, which thrive under conditions of human interference. Creation of campsites for living spaces must also have involved some land clearance and maintenance of open ground. We refer again to our discussion in Chapter 2 that stated that the late-glacial forests possibly supported a higher biomass of big game than do forests today because their canopies may not have been as closed. The number of large game animals occurring in such forests would have increased following an increase of suitable plant food resulting from human interference.

Whatever the proximate motives for the habitat modification, a likely and important outcome was to increase substantially the reproductive fitness of many plants beneficial in the human diet, such as tubers and palms, as well as various other herbaceous and arboreal plants important in supplying materials for construction, tools, baskets, mats, and other economic items. Such increases of wild resources also contributed to the opportunity for increased settlement permanency in the tropical forest before the beginning of food production.

The fact that humans entered the New World tropical forest armed with the intent and capacity to manage the landscape represented a signal event for the subsequent history of this biome. Carl Sauer (1958) argued that humankind's successful adaptation to the tropical forest was facilitated by the use of fire and the creation of successional habitats within the primary forest, where useful plants are concentrated and animal densities are greater than they are elsewhere. Neotropical trees are thin barked and, therefore, poorly insulated, having apparently had a short evolutionary history with fire, thus making fire a highly effective tool for mastery of the forest both by foragers and by early food producers (Sauer, 1958). Because trees were so easily killed by fire and the creation of open spaces and admission of sunlight to the forest floor required no more than rubbing flint against flint, the use of stone axes would have been irrelevant during the early days of food production in areas having long and marked dry seasons.

Other important factors to consider in assessing early subsistence and subsistence changes are the patchiness and heterogeneity of the late Pleistocene landscape. Edge habitats, where successional forest growth met the open terrain and where an abundance of browse was available for game to feed on, were more extensive on the late Pleistocene landscape. Paleoindians operating out of strategically located base camps could have exploited the richest resources of these varied and productive vegetation zones without being highly mobile.

Simms (1987) notes that early subsistence in the Great Basin, where a patchy landscape similarly held a variety of closely ranked and low-cost resources, likely was oriented around a diet with some considerable variability. He adds that a restricted focus on very large game animals never necessarily occurred in that region despite their probable very high rank in the optimal diet set because very large "package size" may not always be associated with lower handling times and higher return rates. For example, killing some of the larger and more ferocious game animals after they were encountered may have been difficult for hunters using simple spear-wielding technologies, effectively raising the costs of exploiting them.

Given these factors and the occupations documented in tropical forest, it seems that Paleoindian subsistence in such areas cannot be described as either generalized, connoting "broad spectrum" with continuous year-round attention to many varied plants and smaller game, or as specialized, with a narrow focus on the pursuit of very large game. Certainly, use of optimal foraging theories to model subsistence results in no prediction that megafauna will be exploited to the exclusion of medium-sized and smaller game and some plants or that short-term dietary variability will not occur.

Also important in relation to this issue, and that the data cannot address at this time because of lack of subsistence information and dates on many Paleoindian sites, are possible differences in subsistence between early Paleoindian (ca. 11,000–10,500 B.P.) and late Paleoindian (10,500–10,000 B.P.) populations (see Ranere and Cooke, 1996, for a similar argument). Not long after these groups entered tropical America at ca. 11,000 B.P., the vegetation began changing and some large and medium-sized fauna probably started to decline and disappear due to environmental change, human predation, or both. Such changes may have negatively altered the costs of exploiting some pre-10,000 B.P. environments, if less severely than when forests virtually closed the landscapes at 10,000 B.P.

When high-ranked resources start to become less common in the environment they may contribute smaller proportions of food to the diet, although they still should be exploited whenever encountered. If such changes were occurring, they may have induced more subtle changes in subsistence (greater emphasis on smaller mammals and higher ranked plants not requiring much processing or other energetic investments) that are not immediately obvious from the archeological lithic records. One hallmark of these records is that Paleoindian fluted projectile points, with little doubt used to hunt large-sized game, probably continued to be produced until the beginning of the Holocene (Cooke, 1997; Ranere and Cooke, 1991). Another hallmark of the pre-10,000 B.P. lithic assemblages, whether or not they bear Clovis-type tools, is their apparent absence of plant processing and soil-tilling implements, indicating that plant foods were not yet a primary focus of subsistence. However, subsistence may have been increasingly more "generalized" toward 10,000 B.P. These are clearly considerations subject to empirical verification, but the current correlation of habitat reconstruction and site location suggests that there was no single, monolithic "Paleoindian" subsistence orientation in the tropics.

Approximately 10,000 years ago, when forests claimed the landscape, human populations were compelled to face an environment they could not have conceived of when they first entered low latitudes. The removal of many mega- and large- to medium-sized fauna from a landscape would immediately force subsistence options in the direction of lower ranked resources and very substantially broaden the diet breadth. The evidence indicates that successful adjustments to some tropical forest resources had already taken place during the terminal Pleistocene so that the reforestation of the landscape per se did not present a significant hurdle to continued human occupation by these sophisticated and flexible hunters and gatherers. Rather, the various negative factors associated with the marked decline of foraging return rates discussed in Chapter 2 presented the most important challenges.

We have proposed that in those areas experiencing dramatic declines of foraging return rates and housing suitable wild plants, a period of intense experimentation and manipulation of the tropical forest flora immediately followed glacial time, and that the cultivation of root crop and seed plants, such as teosinte and squash, soon followed during the early Holocene. In the following section, we explore the archeological and paleoecological evidence bearing on these propositions.

ADAPTATIONS AND SUBSISTENCE DURING THE EARLY HOLOCENE (10,000–7000 B.P.) IN AREAS WITNESSING EARLY FOOD PRODUCTION

> *Plants change with time as they diffuse through space. If one wishes to be specific about origins, one must inquire as to the origins of what. What kind of wheat or what kind of maize is under discussion?*
>
> Jack R. Harlan (1986)

There is increasing evidence for human occupation of the lowland Neotropics during the early Holocene. More archeological sites dating to this period than to earlier times have been discovered and excavated, and there is, for the first time, substantial recovery of plant and animal remains. The data indicate that sophisticated adaptations to the tropical forest were developing over large areas early in the tenth millennium B.P. Tropical forest tubers, seed plants, and tree fruits are well represented in occupation sites dating to between 10,000 and 9000 B.P., in association with a grinding stone technology that was used to process roots and tubers. Also present in lithic assemblages during the earliest Holocene are implements that appear to be designed to till soil.

In addition, sites are now identified near the coast, with the fish, shellfish, and terrestrial fauna of the intertidal, mangrove, and estuarine zone being well represented in these occupations. By 9000–8000 B.P. both local domestication of crops and acceptance of exogenous domesticates from tropical forest elsewhere are indicated in some regions.

Another important change in settlement pattern that appears to have taken place at approximately 10,000 B.P. is the occupation of rock shelters for the first time in regions such as the interior of central Pacific Panama. This may also connote a shift away from a focus on more mobile animal game toward an emphasis on plant foods. (Cooke, 1997, argues that rock shelter occupations during the early Holocene may also reflect a reduction of group size consequent to loss of the abundant and clumped Pleistocene resources.) Along with the data from archeological excavations, important results have been obtained from lake core sediments, which largely parallel those obtained from the archeological settings in indicating the evolution of food production between 10,000 and 8600 years ago.

In contrast to previous summaries, we start from South America and make our way north, ending with a brief review of the well-known Tehuacán and Oaxaca sequences of highland Mesoamerica. Much of the most compelling evidence for early Neotropical food production currently comes from coastal Ecuador, the low, western Andean slopes of northern Peru, the mid-elevational regions of southwestern Colombia, the Colombian Amazon, and the lowlands of the central Pacific watershed of Panama.

Coastal Ecuador

An early Holocene culture, called Las Vegas, has been defined from the identification of 31 sites located on the Santa Elena Peninsula of southwest Ecuador (Figs. 4.5 and 4.6) (Stothert, 1985, 1988). The environment of the immediate area is semiarid, owing to a peculiar interaction between the cooler than usual ocean surface and adjacent land that leads to little rainfall. The modern vegetation is sparse and largely represented by perennial grasses, cacti, and thorn scrub plants. In prehistoric times, however, there were probably many more large trees, especially along the watercourses, which have been cut during the past 50 years by human populations for fuel. More humid environments that supported a tropical deciduous

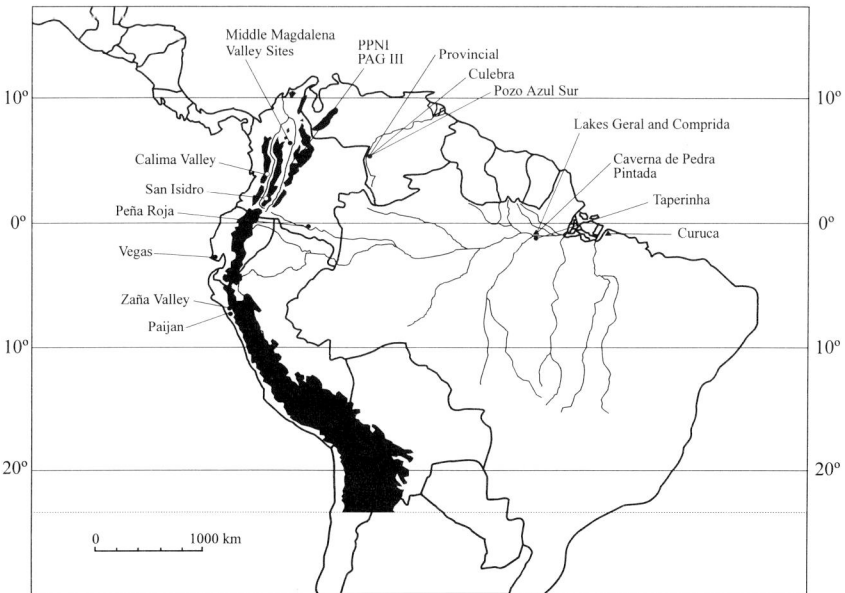

FIGURE 4.5   Map showing the location of the principal early Holocene sites in South America discussed in the text. Triangles are paleoecological sites; circles are archeological sites.

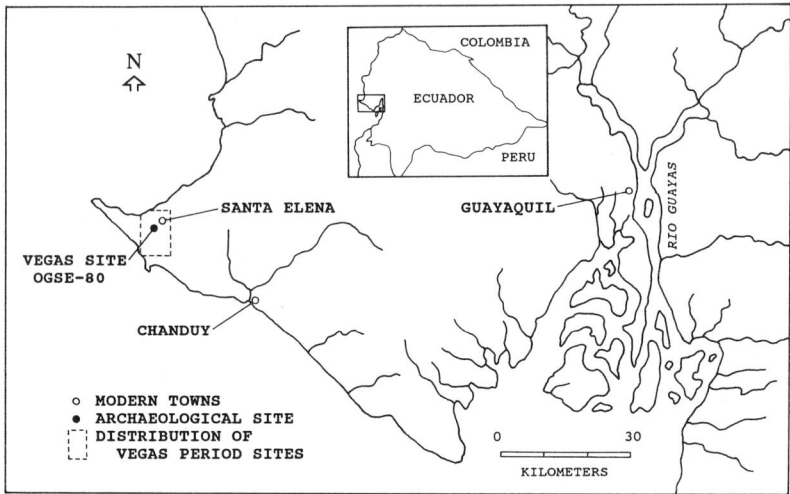

FIGURE 4.6   Map showing the distribution of the preceramic Vegas-period sites on the Santa Elena Peninsula, Ecuador.

forest occur less than 30 km from this part of the Santa Elena Peninsula, which lies on the boundary between the dry, desertic region to the south and forested, higher rainfall regions to the north and east.

The environment of the area during the early Holocene is open to debate. We believe that although conditions may have been somewhat moister between 10,000 and 8000 B.P. than they are today, they were not moist enough to move the tropical forest into the area of the peninsula where the Vegas sites lie. Phytoliths from such taxa as the Marantaceae and Palmae that inhabit the humid tropical forest today, and other typical tropical forest plants (Piperno, 1993), are not present in archeological soils dating to this period (with an exception related to the introduction of a crop plant, as discussed later).

Use of such plants by forest people is common and widespread today. If they were present on the landscape, it would be reasonable to expect some evidence for them. Also, the archeological faunal assemblages lacked animals of the tropical forest and pointed to the exploitation of thorny scrub environments and thickets; the latter would have been found at the edges of watercourses. Such evidence suggests early Holocene environments were not much different from those of today.

The Vegas type site, OGSE-80 (Site 80), was excavated by Karen Stothert in 1971 and 1977–1981. Located 4 km from the present-day coastline, it is a large (once 13,000 m^2) and deep midden from which numerous remains of stone artifacts, shells, and terrestrial fauna such as deer, rabbit, and squirrel were recovered (Plate 4.2). Seeds and nuts were not preserved and pollen grains were recovered in low quantity. However, phytoliths were present in very large numbers. Stothert believes

PLATE 4.2   General view of the excavations at Site OGSE-80 with the sparsely vegetated Santa
Elena Peninsula in the background.

that the size of OGSE-80, the density of materials found there, and the juxtaposition
of diverse and productive marine, estuarine, and terrestrial environments within
5 km of the Vegas sites indicate the likelihood of sedentary occupations.

A series of radiocarbon determinations on charcoal, shell, and human bone
indicate that human activity at Site 80 began approximately 10,800 years ago and
ended approximately 6600 years ago. A few sherds belonging to the early ceramic
Valdivia period (5500–3000 B.P.) were found on the surface of the site, but no
intrusive material was recovered in the excavated contexts considered here. The
Las Vegas culture, which left behind most of the cultural materials discussed here,
is considered by Stothert to have started approximately 9800 years ago (Table 4.1)
(Here and throughout, all radiocarbon ages are in uncalibrated years). A small area
of the site holding deeper deposits but few cultural remains yielded ^{14}C dates
ranging between 11,000 and 10,000 years ago, which Stothert believes represents
a poorly understood, terminal Pleistocene pre-Las Vegas occupation. There are
some limited data on plant exploitation during this period that will be discussed later.

Several other aceramic sites near Site 80 on the Santa Elena Peninsula yielded
radiocarbon determinations consistent with those from the type site, reinforc-
ing the early to mid-Holocene age of the Las Vegas culture. Stothert further
divides the Las Vegas occupation of OGSE-80 into an early Las Vegas phase, lasting from ca. 9800 to ca. 8000 B.P., and a late Las Vegas phase, from ca. 8000 to

TABLE 4.1   Radiocarbon Dates on Phytoliths, Charcoal, Shell, and Human Bone from Vegas Site OGSE-80 (Year B.P.)

[14]C Phytolith age	Provenience (cm)	Laboratory number
12,130 ± 70	I5 Feature 52 (80–100)	UCR-3281
		CAMS No. 14215
9,740 ± 60	F8-9 (110–120)	UCR-3284
		CAMS No. 14218
9,080 ± 60	E8-9 (110–120)	UCR-3461
		CAMS No. 27729
7,960 ± 60	G10-11 (130–140)	UCR-3285
		CAMS No. 14219
7,170 ± 60	GH8-9 (105–110)	UCR-3282
		CAMS No. 14216
5,780 ± 60	Feature 1 (112)	UCR-3283
		CAMS No. 14217
[14]C Dates on shell and charcoal[a]		
7,440 ± 100 (shell)	GH8-9 (90–95)	
7,150 ± 70 (shell)	GH9 (95–100)	
8,170 ± 70 (shell)	GH8-9 (105–110)	
9,550 ± 120 (shell)	GH8 (140)	
8,810 ± 395 (composite charcoal)	GH1-5 (90–100); FH8-11 (100–140)	
[14]C Dates on human bone from burials[a]		
7710 ± 240, 8250 ± 120, 6750 ± 150, 6600 ± 150		

Also: 10,100 ± 130 (shell), 10,300 ± 240 (charcoal), and 10,840 ± 410 (charcoal) from between 150 and 300 cm[a]

[a] From Stothert (1985, 1988).

ca. 6600 B.P., based on a stratigraphic break detected in one of the large, deep excavation cuts (F-H/8-11).

Subsistence data relating to the earliest, pre-Las Vegas-phase occupation of Site 80 are sparse. Stothert identified the outline of a fruit in a pre-Vegas-phase stratum that she tentatively identified as bottle gourd, while acknowledging that she could not rule out the possibility that it was the remains of *Crescentia cujete,* the tree gourd. A soil sample taken from in and around this feature contained two *Cucurbita* phytoliths but no evidence for *Lagenaria* [detailed discussions of the phytolith results from Vegas can be found in Piperno (1988a,b) and Piperno and Holst (1996b)]. However, because bottle gourd produces few phytoliths, this finding does not necessarily rule out *Lagenaria*. The possibility that the tree gourd is represented cannot be evaluated because this species does not produce phytoliths. The *Cucurbita* remains probably derive from the general midden soil of the period and indicate that people were using squash by approximately 10,000 B.P. at the site. The small sample of phytoliths recovered (with a mean length of 63 μm) does not allow a comparison of fruit size with phytoliths of post-10,000 B.P. contexts (discussed later), but it appears that some kind of exploitation of squash was occurring in terminal Pleistocene times.

Faunal and floral data from the Vegas-phase occupation of OGSE-80 indicate a broad-spectrum subsistence strategy. Remains of deer, fox, rabbit, squirrel, peccary, and opossum, along with clams, fish, and crabs, point to the exploitation of diverse and productive terrestrial, estuarine, and mangrove environments. The lithic inventory included plant-grinding implements, called "edge ground" cobbles because the grinding facet is on the narrow edge of the tool (Ranere, 1980b) (Plates 4.3a and b). Edge ground cobbles are the most typical plant-processing lithic tools found in early and middle Holocene sites in the humid tropics. Replicative experiments by Ranere (1975) suggest that tubers were a primary plant material being prepared with these implements, a finding supported by starch grain studies from a Colombian site (San Isidro) and from sites in Panama (discussed later).

Direct evidence for plant cultivation and domestication by the Las Vegas peoples comes from the phytolith record, recently restudied by Piperno, which revealed numerous *Cucurbita* spp. and maize phytoliths as well as phytoliths of *Calathea allouia* (leren, and topee tambu) and bottle gourd. Today, leren is a minor domesticated root crop grown mainly in northern South America and the Antilles (see Chapter 3). The phytoliths from these plants come mainly from a large stratigraphic cut, F-H/8-11, away from any burials, where the midden was deep and undisturbed. A soil sample from a deep Vegas level in cut E8-9, an undisturbed context away from burials, was also analyzed.

In order to evaluate the contention (Piperno, 1988a,b) that the *Cucurbita* and *Zea* phytoliths are of preceramic age, direct AMS dating of phytolith assemblages that contained numerous remains of these genera was undertaken (Table 4.1).[3] Results indicated that the phytoliths were deposited close to the time periods bracketed by the shell and charcoal ^{14}C dates, confirming their Las Vegas associations. The two phytolith contexts in which maize had been identified (Piperno, 1988a) yielded dates of 7170 and 5780 B.P. The former determination is consistent with the late Las Vegas-phase age assigned by the excavator to this level. It is clear that maize is first present at the site shortly before the end of the late Las Vegas-phase occupation, which probably dates between 7000 and 6700 B.P. after correction for the reservoir effect (determinations on shell from the same stratigraphic cut yielding the maize).

The maize phytoliths dated to 5780 B.P. are from the bottom of a large secondary burial that Stothert did not date but thought might belong to a somewhat later

---

[3] Phytoliths are recovered for dating purposes using the same procedures as when routinely retrieving them for identification and interpretation. All the phytolith dates reported here come from the University of California, Riverside, radiocarbon facility, where the feasibility of dating phytoliths by AMS was first demonstrated (Mulholland and Prior, 1993).

PLATE 4.3   (a and b) Edge ground cobbles from (a) Vegas Site OGSE-80 and (b) the Aguadulce rock shelter and Monagrillo (early ceramic) sites in Panama. The grinding facets of the tools are along the sides and the ends. The ends of the tools were also used as pounders.

time period (K. Stothert, personal communication, 1996). The phytolith date supports this contention and provides evidence for a later preceramic occupation near the peninsula that immediately preceded the earliest ceramic component in the area, belonging to the Valdivia culture and dating to approximately 5500 B.P. (see Chapter 5).

The 12,130 ± 70 B.P. phytolith date is older than the earliest date of 10,800 ± 400 B.P. on cultural material from the site. It is from a feature in a superficial level consisting of a large concentration of stones. In addition to the stones, much dirt appears to have been moved into and around the area and the soils appear to have been thoroughly mixed (Stothert, 1988). We cannot provide a firm explanation for its antiquity.

Cucurbita phytoliths are common and continuously present in phytolith assemblages directly dated from 9740 to 7170 B.P., sampled from large, stratigraphic cuts where no burials or other disturbances were observed. The only wild species of *Cucurbita* native to near the Vegas sites is *C. ecuadorensis,* which is endemic to southwestern Ecuador. It is not found in the driest parts of the Santa Elena Peninsula today where the Vegas sites are located. However, we cannot rule out the possibility that it may once have had a more widespread distribution. Nee (1990) and Andres (personal communication, 1996) believe that *C. ecuadorensis* was semidomesticated in prehistory because it has much larger fruits than other wild species of *Cucurbita* and the fruits often have a nonbitter flesh (see Chapter 3).

b

0 1 2 3

CM

PLATE 4.3   (*Continued*)

An analysis of phytolith size in modern and Vegas *Cucurbita* phytolith populations supports this contention, indicating that a domesticated form of squash was present at Site 80 by 9000 B.P. and also suggesting that another domesticated species of *Cucurbita* was probably introduced into Site 80 late in the eighth millennium B.P.

Table 4.2 shows that populations of present-day *C. ecuadorensis* have phytolith sizes intermediate between those of wild and domesticated species, a finding consistent with their semidomesticated status (Piperno and Holst, 1996b). Samples from nearly all the wild species of *Cucurbita* currently known, including the recently discovered *C. argyrosperma* ssp. *sororia* (?) from lowland Panama (Chapter 3), were analyzed. These have much smaller phytoliths than the domesticated squashes tested—*C. moschata* and *C. ficifolia*. A rind fragment of *C. moschata* from Vera Cruz, Mexico, donated by the New York Botanical Garden, has smaller phytoliths than those in the other domesticated samples.

The small size of our phytolith sample of *C. moschata* is due to the fact that Panamanian varieties of this species do not appear to produce the genus-specific "scalloped" phytolith from rinds (Bozarth, 1987; Piperno, 1993; Piperno and Holst, 1996b)[4] used in this analysis (Plates 4.4 and 4.5). A sample of this species from Caracas, Venezuela, and one from southwestern Ecuador also failed to yield this type of phytolith. Wild *Cucurbita* species, however, produce them in high amounts. Possible reasons for these differences are provided later in the chapter.

The reason for these size patterns seems to be simply that as seed and fruit size grow, the phytoliths, having more space to occupy in the rind tissue, grow with them. For many fruits studied, we were able to measure fruit and seed size and then compare them to phytolith size. As Figs. 4.7a. and 4.7b show, there is a strong positive correlation between phytolith length and fruit and seed size (Piperno and Holst, 1996b). Two curves are illustrated but others comparing, for example, phytolith width and thickness with fruit and seed size, show similarly strong correlations. It is also possible to estimate the size of archeological *Cucurbita* fruit and seeds directly from phytolith characteristics, which is accomplished by deriving a formula from a regression of the correlation coefficients shown in Fig. 4.7a. (Fig. 4.7c).

Cucurbita phytoliths from Vegas directly dated to between 9740 and 7170 B.P. show a clear trend for increasing size through time (Table 4.3). Those from the 9740 ± 60 B.P. context are characteristic of wild species. Their sizes and thicknesses suggest relatively little intervention with the genetics of the plant early in the tenth millennium B.P., when the fruits seem to have been about the size of a baseball (ca. 8 cm), typical of wild species today (Table 4.4). However, further analysis (currently under way) of pre-10,000 B.P. deposits from Vegas is needed to ascertain the type of squash that was exploited in terminal Pleistocene times.

A sample from excavation unit E8-9 110–120 cm in deep, undisturbed contexts yielded a ^{14}C phytolith determination of 9080 ± 60 B.P. The age is consistent

---

[4] *Cucurbita* phytoliths are distinguished from other species by their shape and surface decoration, the latter consisting of circular "scallops". They are found in the rinds of fruits.

TABLE 4.2   Phytolith Size in Modern *Cucurbita*

	$\bar{x}$ Length	Range	$\bar{x}$ Thickness	Range	n
Domesticated *Cucurbita*					
Cucurbita moschata					
Veracruz, Mexico	66	44–88	59	52–88	50
Manabí, Ecuador	93	60–124	70	40–100	50
Cucurbita ficifolia	114	92–148	98	80–120	50
	108	72–136	91	52–112	50
	114	88–132	95	68–116	50
	118	72–147	106	60–144	60
Semidomesticated[a]					
Cucurbita ecuadorensis	73	60–108	45	40–48	50
	86	56–116	56	40–60	50
	90	48–116	58	38–76	50
	92	52–120	60	48–76	50
	77	48–120	51	28–64	50
	84	48–116	48	28–68	50
	81	68–108	54	48–68	50
	83	60–120	58	48–76	50
	88	52–108	62	40–80	50
	75	60–96	48	40–64	50
	87	60–120	57	44–68	50
All	83		54		
Wild *Cucurbita*[b]					
Cucurbita argyrosperma ssp. *sororia* (?)	72	60–92	62	52–76	41
	74	48–100	59	32–80	21
	60	40–88	54	36–76	50
	56	36–88	49	36–64	50
	65	48–80	53	40–64	50
	58	40–68	49	42–60	50
	63	48–80	54	40–68	50
	55	40–80	46	36–60	50
	56	44–76	47	32–62	50
	56	48–64	—	—	34
	60	52–72	—	—	50
Cucurbita argyrosperma spp. *sororia*	74	52–92	62	52–76	100
Mexico	64	56–79	54	44–64	50
Cucurbita lundelliana					
Peten, Guatemala	71	56–100	57	40–68	50
Quintana Roo, Mexico	72	60–96	—	—	30
Cucurbita pepo ssp. *texana*	68	46–84	53	40–64	50
Cucurbita foetidissima	59	48–72	—	—	50
Cucurbita palmata	61	48–72	—	—	50

Note: Phytolith sizes are in micrometers.

[a] Eleven fruits from 11 individuals and populations.

[b] *Cucurbita argyrosperma* spp. *sororia* (?) includes at least three fruits each from three different populations and 10 different individuals in central Pacific Panama. The question mark reflects the uncertain taxonomic relationship of this plant to the domesticated species *C. argyrosperma* and *C. moschata*.

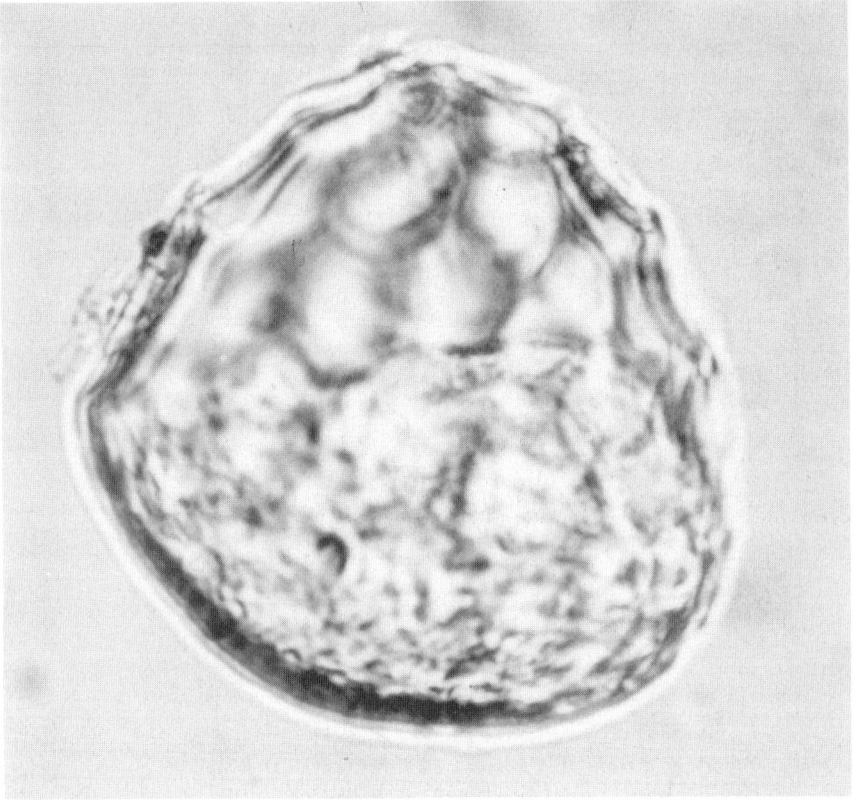

PLATE 4.4   Diagnostic *Cucurbita* phytolith from the rind of *C. ficifolia*. The phytolith is 98 micrometers long.

with the stratigraphic position of the sample in early Las Vegas–phase contexts. It contains phytoliths whose mean lengths have increased by 20% over 9740 B.P. contexts, and whose thicknesses are significantly outside the means and ranges for modern (semidomesticated) *C. ecuadorensis*. The sizes of the fruits and seeds from this squash were very likely larger than those in modern *C. ecuadorensis*. Estimations of fruit size via the phytoliths indicate more than a 50% increase from that of 9740 B.P. phytoliths. A genetically and morphologically altered form of squash about 12 cm long with many of the phytolith characteristics of the modern domesticate *C. moschata* from Ecuador appears to have been on the landscape at this time.

Whether this reflects development of *C. ecuadorensis* into forms not seen today because of the loss of human selection pressure on the plants, or introduction of a fully domesticated strain of squash from elsewhere, cannot be said at this time. It should be noted that this context also contains a tuber crop, *Calathea allouia,*

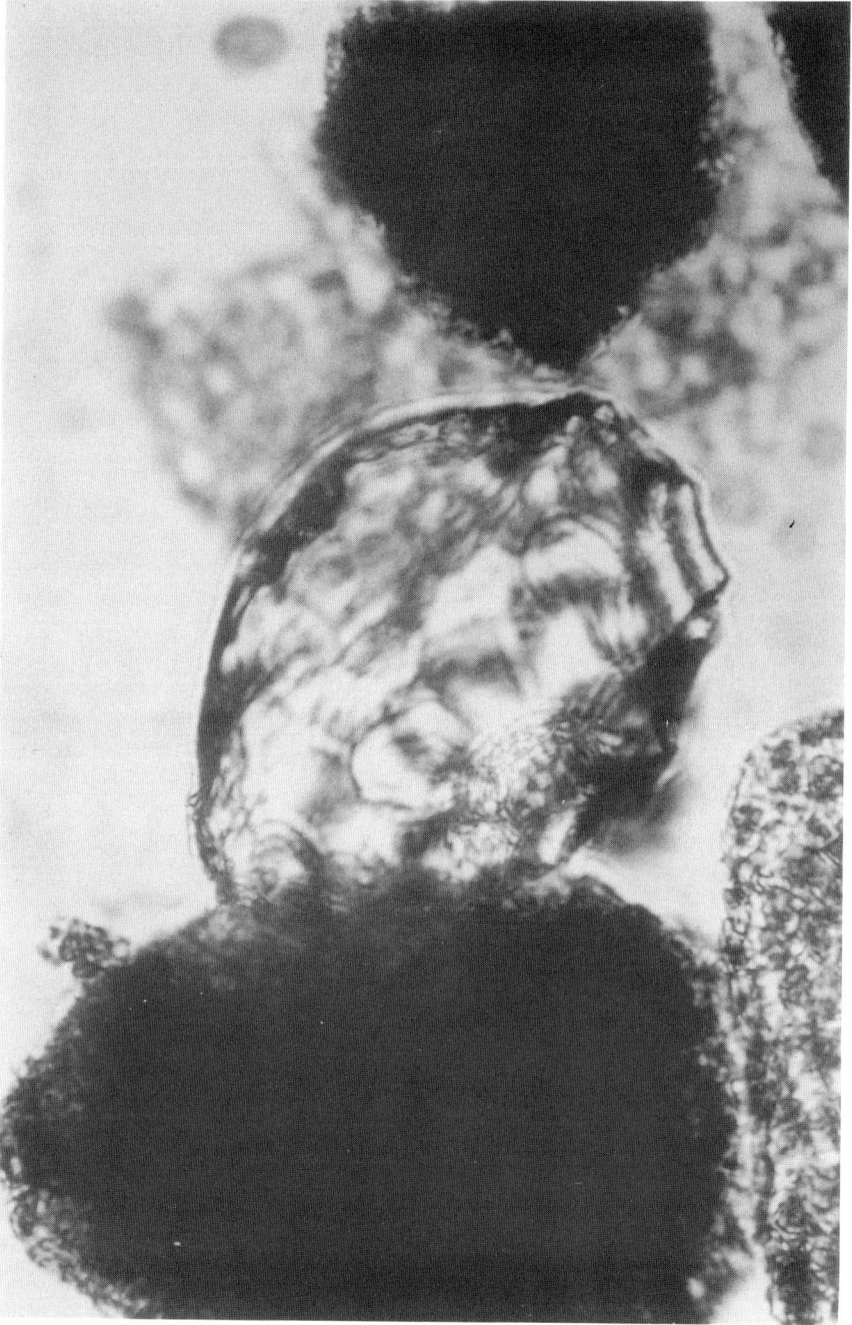

PLATE 4.5   Center, *Cucurbita* phytolith from Site 80, Unit E8-9 110–120 cm. Length of the phytolith is 104 micro-meters.

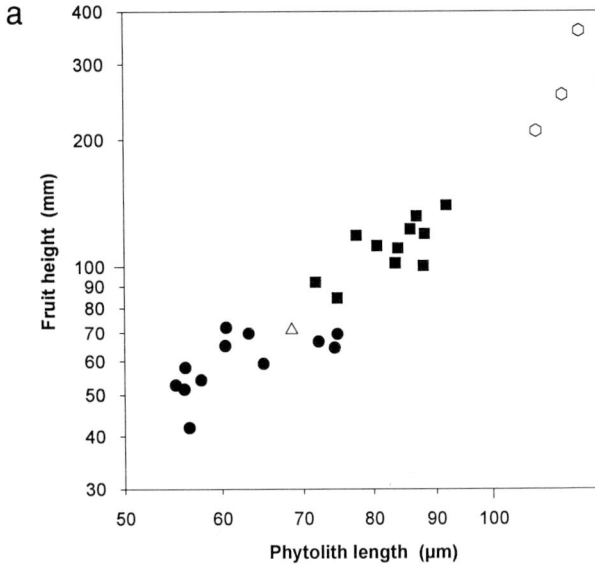

FIGURE 4.7   (a) Graph of the relationship between phytolith length and fruit height in various modern wild and domesticated *Cucurbita* species. Log–log scale; Pearson correlation coefficient (r) = 0.95; $p < 0.001$. (b) Phytolith length vs seed length. Pearson correlation coefficient (r) = 0.9; $p < 0.001$.(c) Regression analysis of phytolith length vs fruit height. A highly significant linear relationship exists between the two variables (F ratio = 210.881; R^2 = 0.894; $p < 0.001$). Closed circle, *C. argyrosperma* ssp. *sororia* (?); open circle, *C. foetidissima;* triangle, *C. pepo* ssp. *texana;* closed square, *C. ecuadorensis;* open square, *C. moschata;* hexagon, *C. ficifolia*.

that is almost certainly an introduction. If a form of *C. ecuadorensis* was present at 9000 B.P., it would also be unclear whether the plant was originally cultivated at the Vegas sites or at a moister location nearby because modern-day varieties of the species do not tolerate the dry conditions of this part of the peninsula.

Another phytolith sample containing numerous squash phytoliths from the large stratigraphic unit G10-11, between 130 and 140 cm beneath the surface, yielded a date of 7960 ± 60 B.P. The lengths and thicknesss of *Cucurbita* phytoliths are also increased substantially over 9740 B.P. contexts. Phytolith sizes indicate a fruit about 2 cm larger than that in the 9740 B.P. stratum, and they are consistent with those of modern *C. ecuadorensis*. *Cucurbita* phytolith sizes from the sediment sample dated to 9080 B.P. are substantially larger than those in the 7960 B.P. context, although both probably reflect genetic alteration from systematic planting. At face value, these differences may suggest that more than one type of squash was under cultivation at Site 80 during the ninth and eighth millennia B.P. The other possibility is that the 7960 B.P. phytolith date is too young, given its stratigraphic position near the bottom of the Las Vegas occupation of the site.

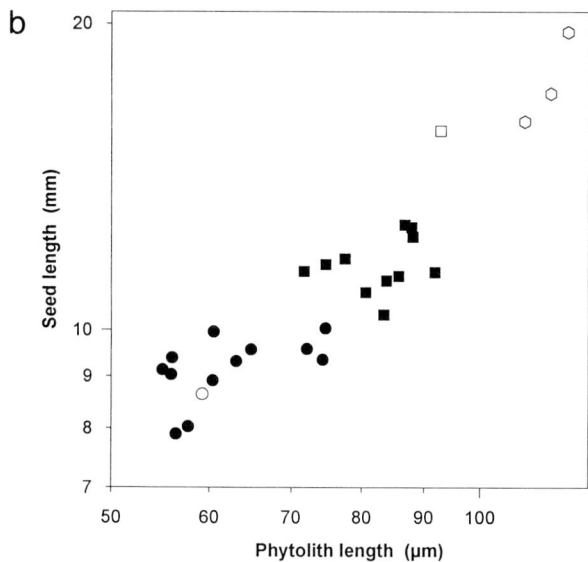

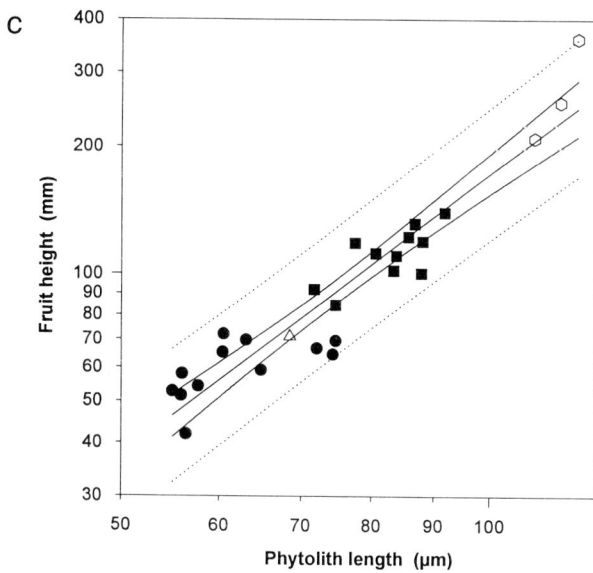

FIGURE 4.7   (*Continued*)

TABLE 4.3   *Cucurbita* Phytolith Size vs ^{14}C Phytolith Age at Vegas

^{14}C phytolith age (year B.P.)	$\bar{x}$ Length	Range	$\bar{x}$ Thickness	Range	n
5780 ± 60 Feature 1,112 cm	96	72–120	74	56–88	7
7170 ± 60 GH8-9, 105–110 cm	96	76–116	78	60–95	8
7960 ± 60 G10-11, 130–140 cm	77	56–108	63	40–76	48
9080 ± 60 E8-9, 110–120 cm	86	56–120	66	42–93	42
9740 ± 60 F8-9, 110–120 cm	71	48–96	55	36–68	47

Finally, in the phytolith assemblage directly dated to 7170 B.P., where maize first occurs, there is another increase in squash size, and phytolith characteristics now completely overlap modern domesticated species such as *C. moschata*. Fruits were approximately 16 cm long, almost double the size of the 9740 B.P. forms. Because no wild ancestors of modern, domesticated species grow anywhere near the Santa Elena Peninsula today, it is very likely that domesticated species were appearing with maize from elsewhere. Very large squash phytoliths also occur along with maize in a secondary burial, where they yielded a direct date of 5780 ± 60 B.P.

Bottle gourd phytoliths were recovered in small numbers in Vegas-phase contexts, beginning with the phytolith sample dated to 9080 B.P. Bottle gourd produces

TABLE 4.4   *Cucurbita* Phytolith ^{14}C Age and Fruit Height

Provenience (cm)	^{14}C Phytolith age (year B.P.)	Predicted fruit height (mm)[a]	95% Confidence interval for predicted fruit height (mm)	
			Lower	Upper
Feature 1 (112)	5780 ± 60	158	110	228
GH8-9 (105–110)	7170 ± 60	158	110	228
G10-11 (130–140)	7960 ± 60	100	70	143
E8-9 (110–120)	9080 ± 60	121	84	173
F8-9 (110–120)	9740 ± 60	81	57	116

[a] Using mean PL in the regression equation: $\log_{10} FH = 2.201 (\log_{10} PL) - 2.164$.

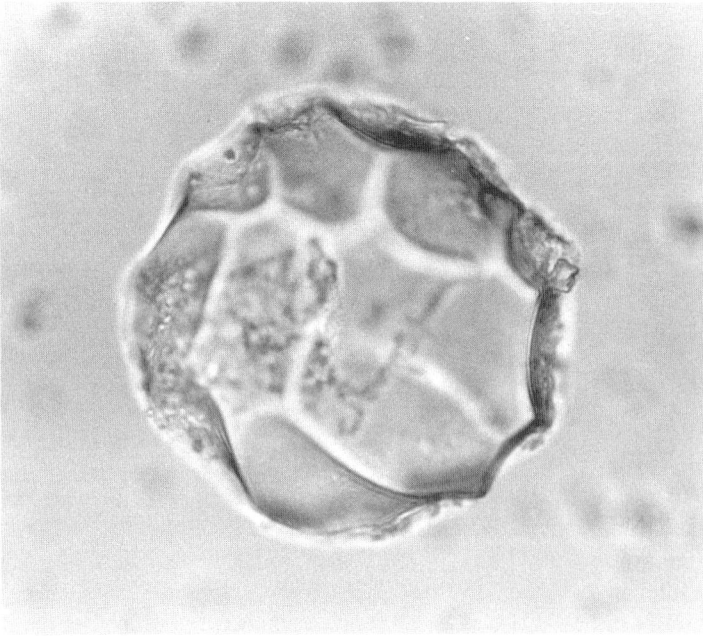

PLATE 4.6   Diagnostic phytolith from a modern bottle gourd from Colombia. Its length is 100 micrometers.

far fewer phytoliths than *Cucurbita* species, accounting for their limited distribution in archeological soils.[5]

In addition to the evidence for maize, squash, and bottle gourd, phytoliths identified as *Calathea allouia*[6] (leren) are present in the phytolith assemblages directly dated to 9080 B.P. and continue to be represented in later Vegas–phase assemblages. This domesticated root crop almost certainly represents an introduction because neither it nor its family, the Marantaceae, grow in the semiarid region today. That it is not present in earlier deposits from the site and that other kinds of phytoliths from the Marantaceae, an economically important family wherever it occurs naturally and a prolific phytolith producer, were not recovered from the sediments

[5] Bottle gourd phytoliths, which also derive from the rind of the fruit, can be distinguished from those of *Cucurbita* on the basis of differences in the shape of the scallops decorating the surface of the phytolith plus the three-dimensional characteristics of the phytolith (Plate 4.6). They appear to be species-specific.

[6] The identification of *C. allouia* is based on a comparison with a reference collection in Piperno's laboratory of seed phytoliths from more than 15 wild *Calathea* species, including the economically important *C. latifolia,* and more than 20 other species in the Marantaceae family. Only leren phytoliths have flat and undecorated upper bodies, in marked contrast to all the wild species studied from this family. The phytolith in question is confined to the seed of the plant.

also support the argument that it was not originally taken under cultivation at Site OGSE-80.

In summary, the Vegas phytolith evidence indicates that cultivation and domestication of certain seed and tuberous food plants took place during the early Holocene. To argue that the *Cucurbita* phytoliths randomly dispersed in the site sediments display the size trends through time previously described severely strains good logic. Similarly, if the maize phytoliths are contaminants from some undefined, more recent occupation, then they somehow moved down into late Las Vegas-phase levels but not into earlier deposits from the same excavation units directly below.

As Stothert (1985) notes, there is no ecological argument that would force the reevaluation of the conclusion that the Vegans were horticulturists. Vegas sites are located near small, seasonal rivers and streams of the semiarid peninsula where gardening still occurs today. The size and organization of the Vegas settlements suggest that the Vegas peoples were self-sufficient and egalitarian, particularly during the early Las Vegas tradition. Vegas peoples exploited a greater range of resources than later Valdivia culture peoples and did not undertake corporate projects requiring substantial efforts on the part of the community (Stothert, 1985). They were horticulturists but apparently not committed farmers. Currently, we are uncertain if the Vegas tradition represents an *in situ* development of food production or if the earlier squash remains are imports from horticultural occupations located in moister, tropical forest environments near the Vegas sites, i.e., elsewhere in the southwest Ecuador/northwest Peru area (see Fig. 3.18).

The founding of the cemetery at Site 80 late in the Vegas cultural tradition may reflect social changes that followed the development of horticulture in the area (Stothert, 1985) (Plate 4.7). Occasional grave goods, such as shell spoons, red ochre, and a polished stone axe, are present, and males and females have different burial orientations. Perhaps social cohesion and intercommunity organization among local groups became more important than it had been previously. Also, studies of the remains of more than 100 individuals from Vegas showed that they were healthy and did not suffer from any of the nutritional deficiencies commonly detected in prehistoric populations whose diets were mainly maize or other crop plants (Ubelaker, 1984, 1995). Pressures arising from intensification within the social sphere, overpopulation, and food shortages can probably be ruled out as factors explaining the Vegans' decision to cultivate plants.

An interesting feature regarding the burials from Vegas is that they bear rather striking similarities to those from the preceramic site of Cerro Mangote in Panama, whose midden dates from 7000 to 5000 B.P. In particular, some individuals from both sites were in identical types of secondary "bundle burials," whose form was very complex in relation to the placement of specific bones. This suggests that some kind of cultural contact between the two areas was occurring. These factors led Stothert to propose that early interactions among preceramic groups in northern South America and lower Central America were taking place, which would serve as a vehicle to transfer plants.

PLATE 4.7   An ossuary or circular massive burial from OGSE-80, partially excavated. One of the skeletons was dated to 6600 ± 150 B.P.

Our review of the hearths of the major crop plants in Chapter 3 suggested that *C. moschata,* the major species of squash grown today in the lowland tropics, may have had its origins in seasonally moist regions of southern Central America and Colombia. This is the species of domesticated squash most likely to be represented at Vegas during the eighth millennium B.P. and possibly earlier. *Calathea allouia* is thought to have been originally taken under cultivation in Brazil, Venezuela, and/ or the Guianas, although the evidence is currently unclear (Hawkes, 1989) and the potential zone of domestication of this species is quite large (see Fig. 3.19).

Southwestern Colombia

In the upper Cauca Valley near Popayán at an elevation of between 1700 and 1800 m (Fig. 4.5), Gnecco Valencia and Salgado López (1989), Gnecco Valencia (1994), and Gnecco Valencia and Mohammed (1994) have identified and excavated a number of early Holocene preceramic occupations. Although these sites are not in the lowlands, they are located near the edges of the Western and Central Cordilleras of the Andes in close proximity to low-lying tropical forest. The plant remains from one of them, San Isidro, are important in indicating a well-developed exploitation pattern of currently lowland elements of the tropical forest and possible cultivation of some its resources during the earliest Holocene.

San Isidro (Gnecco Valencia, 1994) is a single-component, open-air occupation with radiocarbon dates of 10,050 and 9539 B.P. Plant remains were recovered in fairly large quantity. Many of the identified taxa grow today in the lowland tropical forest. This indicates that a forest with no modern analog once existed in the area and/or that the remains represent imports from lower elevations. The lithics at the site included projectile points, bifaces, and edge ground cobbles. The site also yielded a polished adze.

More than 20 starch grains have been isolated from the grinding surface of an edge ground cobble from San Isidro and the pounding facet of this stone (Piperno and Holst, 1996a). The tool was thoroughly washed, the grains were from tiny pits and crevices in the surface (where they were protected from weathering and survived), and no soil particles were observed in the preparations. Therefore, we are confident that the grains represent plants that were processed with the tools. Thus far, starch grains from the tubers of *Maranta* cf. *arundinacea,* cf. *Xanthosoma/ Ipomoea* and/or *Manihot,* grasses, and legumes (genera currently unknown) have been identified. [The fissure is not visible on the *Xanthosoma/Ipomoea/Manihot* grains (see discussion of the Panamanian-site starch grains), preventing further discrimination.] It is not possible at this time to suggest either a wild or a domesticated status for the *Maranta* and other tuber grains, but it may be unlikely that they were distributed naturally in this region and at this elevation. It is clear that they were components of the diet and that they were being processed.

The carbonized plant remains from San Isidro include seeds and nuts from several species of palms, including the important lowland genus *Acrocomia.* Also present are *Persea* (avocado), *Caryocar* (a little discussed tree that is important in subsistence today), and *Virola* (a lowland forest genus, some of whose species are still used as a hallucinogenic substance by Colombian Indians). The subsistence strategy seems broad spectrum. Kernels of the *Acrocomia* palm were directly dated, yielding a determination of $10,030 \pm 60$ B.P. (β-93275) (C. Gnecco Valencia, personal communication, 1996). The fragments of *Persea* seeds recovered from San Isidro were medium to large (between 2 and 7 cm long), overlapping the sizes of those from modern, domesticated avocado. However, the separation of wild from early domesticated species of *Persea americana* is a difficult issue using just the seed. Seed length in modern wild individuals is thought to average about 4 cm (Sauer, 1993). The biggest difference between wild and domesticated avocados is the amount of edible fruit surrounding the seed (e.g., Sauer, 1993), a portion that would fossilize poorly. The avocado seeds from San Isidro will soon be evaluated by experts in *Persea* in order to further evaluate their status as possible early cultivars.

Pollen was fairly well preserved at San Isidro. The pollen record indicates a forested vegetation near the site. Early forest colonizers such as *Plantago* and the Compositae were present in significant amounts, suggesting some forest clearance. This finding, together with the presence of the adze, large avocado seeds, and the tubers, also suggests that incipient tree crop and tuber cultivation may have been taking place at the site.

Other preceramic sites near San Isidro (Gnecco Valencia and Salgado López, 1989) contain stone implements called "waisted hoes" that may have been used to till soil. These implements also date to the early Holocene and are widely distributed throughout the upper and middle Cauca Valley (they are discussed in detail below). It appears that early Holocene occupations are well represented in the Upper Cauca Valley area. The remains from San Isidro indicate well-developed adaptations to the tropical forest and inclusion of tubers as significant elements of the diet by early in the tenth millennium B.P.

Approximately 150 km north of Popayán, in the middle Cauca Valley near Calima in the western Cordillera of Colombia at elevations of approximately 1200 m, Herrera *et al.* (1992) and Salgado López (1989, 1995) have excavated several preceramic sites with radiocarbon dates between 9600 and 7300 B.P. The region spans both sides of a low pass that connects the Cauca Valley, on the eastern side of the Cordillera, with the low-lying Chocó rain forest and the Pacific coast on the western side of the mountains (Bray *et al.,* 1987) (Fig. 4.5). The natural vegetation of the area is sub-Andean forest, characterized by genera such as *Quercus, Myrica,* and *Alchornea.* Annual precipitation is between approximately 1200 and 1400 mm per annum. However, both semiarid vegetation and more humid forest can be found at short distances from the sites.

Two of the sites, Sauzalito and El Recreo, are small encampments at the edges of former quebradas. Their lithic inventories contain large tools with constricted bases and rounded, flat edges, called waisted hoes by the investigators (Plate 4.8).

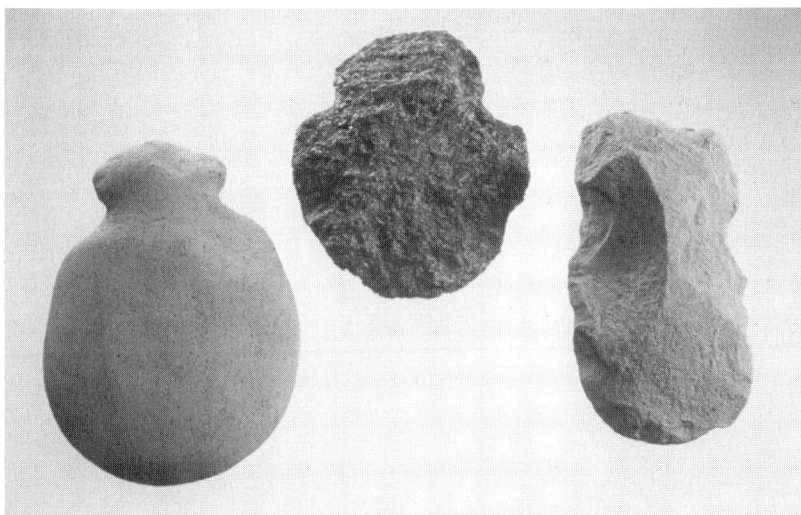

PLATE 4.8   Waisted hoes from Sauzalito and El Recreo. The one on the left is polished and dates to later in the early Holocene.

They certainly appear to be implements designed to till soil. The pedologist associated with the investigations believes that the sites' soil chemistries also indicate ground disturbance associated with tillage (Herrera *et al.*, 1992). Cobblestone grinders were also present in the tool kits.

The cultural periods containing the hoes and grinding stones are bracketed by dates of 9670 ± 150 and 9550 ± 110 B.P. (Sauzalito), and 7980 ± 120 B.P. and 7780 ± 140 B.P. (El Recreo). Sauzalito and El Recreo represent single-component occupations. Because the well-studied later occupations in the valley apparently were not using waisted hoes (Bray *et al.*, 1985; Cardale de Schrimpff *et al.*, 1992; Herrera *et al.*, 1982/1983), the early Holocene associations for the two sites appear to be secure. Interpretation of settlement and subsistence by the excavators is that the sites were seasonally occupied and positioned to exploit the bits of alluvium near the quebradas for plant husbandry. Of further interest is that the hoes from El Recreo, dating 1000–2000 years later than those from Sauzalito, are polished, indicating a transformation of the lithic technology as plant manipulation evolved through time (Plate 4.8).

What may actually have been cultivated at these Calima Valley sites is currently unclear. That the low and lower mid-elevation tropical forest flora were heavily involved in the economy is apparent from the plant remains. Considerable quantities of carbonized seeds and nuts were found in association with the hoes. These remains are still under study but include *Persea* (avocado) and palms. The avocado seeds are smaller than those typical of modern domesticated trees (R. Cooke, personal communication, 1996). Also present are abundant phytoliths of palms, bamboos, the Marantaceae, Annonaceae, Acanthaceae, and other lowland tropical forest plants (D. Piperno, unpublished data). Further analysis of the carbonized materials and starch grains from the lithics may reveal whether some kind of tree and tuber-based system was present.

This is likely because it is supported by pollen evidence from long sediment cores taken near Sauzalito and El Recreo, which reveal the appearance of large-sized cereal pollen grains identified as *Zea mays* at about 6680 B.P. in association with signs of small-scale forest clearing (Herrera *et al.*, 1992; Monsalve, 1985; Chapter 5). These data are in accord with the inference that people occupying Sauzalito and El Recreo at an earlier time practiced food production using plants less demanding on the environment, such as tuberous and tree species.

Other data relevant to the beginnings of food production and introduction of maize into Colombia come from analysis of lake and peat bog sections located at elevations of between 3000 and 3700 m in the high plain of Bogotá (Kuhry, 1988) (Fig. 4.5). Annual precipitation is low, ranging between 1000 and 1400 mm per annum. In a 12,500-year-old sedimentary record from a lake called Paramó de Peña Negra I (PPNI), maize pollen is continuously present in sections radiocarbon dated to between 8320 and 5210 B.P. Pollen indicators of forest disturbance and charcoal are also present at this time, when the climatic trend is toward warmer and wetter. Therefore, natural fires would seem unlikely. Immediately above the unit of the core with the 5210 B.P. radiocarbon determination is a thick ash level that effectively seals the earlier horizons with maize and beneath which little penetration of more recent pollen grains would seem possible.

At another site in the region, Paramó de Agua Blanca III (PAG III), maize appears in a level of the core immediately above a section radiocarbon dated to 6630 B.P. There is also an increase in pollen taxa indicative of human disturbance of the forest at this time. Kuhry (1988) considers that the maize at PPNI and PAG III probably traveled upward from lower elevations on the eastern slopes of the Magdalena Valley, from which ascending winds blow into the study sites. That people were well settled in the Magdalena Valley during the early Holocene is attested to by the number and wide distribution of sites being discovered (López Castaño 1993, 1995a,b).

In evaluating the presence of maize in these and the previously mentioned Calima Valley paleoecological sequences, one should note the following: First, maize is associated with other indicators of human forest disturbance. The levels of the disturbance indicated by the pollen and charcoal records are such that they can be accommodated with the presence of small and shifting settlements, as seem to be characteristic of the archeological sites El Recreo and Sauzalito. Second, if the argument is made that the earliest *Zea* pollen grains are intrusive, then it becomes difficult to understand why they did not intrude further down these long cores into even earlier deposits. Their appearance after 8300 B.P. and before 5200 B.P. is consistent with all the other microfossil evidence for the presence of *Zea* in lower Central America and northern South America.

The Colombian Amazon

The Araracuara region on the western edge of the Amazon Basin in Colombia is revealing a long history of human occupation dating to at least 9300 years ago (Cavelier *et al.,* 1995) (Fig. 4.5). Rainfall is high, about 3500 mm per annum, and the forests contain many of the same elements found throughout the moister areas of the greater Amazon lowland region, including numerous species of palms, bamboos, and other economically important plants.

The site of Peña Roja (Cavelier *et al.,* 1995), located on a terrace above the middle sections of the Río Caquetá, is a preceramic occupation with radiocarbon dates on midden charcoal of 9125 ± 250 and 9160 ± 90 B.P. There is also a sparse, ceramic-phase habitation of the site above the preceramic component that probably relates to after the time of Christ. The preceramic lithic assemblage contained milling stones and other plant-processing equipment as well as various stone artifacts similar to those recovered from the early Holocene period in Panama and Venezuela. Preceramic levels yielded carbonized fragments of nine species of palms and other unidentified tropical fruits (Morcote Ríos, 1994; I. Cavelier, personal communication, 1996) as well as phytoliths from *Cucurbita, C. allouia,* and bottle gourd (Piperno, 1997c) (Table 4.5). A carbonized seed of the palm *Oenocarpus*

TABLE 4.5   Cultural Chronology and Plant Remains Represented in Major Regions Discussed in the Text

Cultural chronology	Roots and tubers				Vegetables	Grains		Legumes	Industrial
	Calathea allouia (leren)	Xanthosoma/ Ipomoea/ Manihot spp.	Manihot esculenta (yuca)	Maranta arundinacea (arrowroot)	Cucurbita spp. (squash)	Zea mays (maize)	Chenopodium sp. (quinoa)	Arachis hypogaea (peanut)	Lagenaria siceraria (bottle gourd)
Southwestern Ecuador, Vegas phase (early preceramic) 10,000–7000 B.P.	**Phy**				**Phy**	**Phy**			**Phy**
Southwestern Colombia, Upper Cauca Valley (early preceramic), 10,000–9500 B.P.		SG		SG					
Calima Valley, Colombia (early preceramic), 9600–7800 B.P.									
Colombian Amazon, Middle Caquetá (early preceramic), 9300–8000 B.P.	**Phy**				**Phy**				**Phy**
Northwestern Peru, Zaña Valley (middle preceramic), 8400–6000 B.P., Las Pircas phase			M, SG		**M, P**		**M**	**M**	
Eastern Amazonia, C. de Pedra Pintada Paleoindian, ?10,800–10,000 B.P.									
Central Pacific Panama, Period IIa (early preceramic), 10,000–7000 B.P.	Phy			Phy	Phy				Phy

Note. M, macrobotanical; SG, starch grains; Phy, phytolith; P, Pollen. Bold print indicates that plant remains were directly dated by AMS.

[a] Seeds are the size of modern domesticates.

[b] Genera and species identified: Acrocomia.

[c] Genera and species identified: Astrocaryum aculeatum, Astrocaryum jauari, Astrocaryum sciophilum, Attalea sp., Mauritia flexuosa, Maximiliana maripa, Oenocarpus bacaba, Oenocarpus bataua, Oenocarpus mapora.

[d] Genera and species identified: Attalea microcarpa, Attalea spectabilis, Astrocaryum vulgare.

[e] Genera and species identified: Acrocomia mexicana.

yielded a radiocarbon determination of 9250 ± 140 B.P. The phytolith assemblage from the uppermost stratigraphic level containing *Cucurbita*, leren, and bottle gourd, which was associated with a charcoal date of 9125 ± 250 B.P., yielded a direct determination of 8090 ± 60 B.P. (Lab. No. UCR-3419; CAMS-27728) (Piperno, 1997c).

Only a few *Cucurbita* phytoliths were present in the preceramic deposits. Mean lengths of the phytoliths combined from the two preceramic levels containing

				Tree fruits						
Bunchosia sp.	Ilex sp.	Celtis sp. (hackberry)	Persea spp. (avocado)	Palmae (palms)	Caryocar	Virola	Sapotaceae (sapotes)	Hymenaea sp.	Bertholetia excelsa (Brazil nut)	Byrsonima crispa
			M^c	M^b	M	M				
			M	M, Phy						
				M, Phy						
M	M	M								
				M^d				M	M	M
				M, Phy		M				

them are 71 μm (range, 52–120; $n = 10$). The upper size limit of these phytoliths is significantly outside of those in modern, wild species. Also, the presence of a wild squash in this wet forest is unlikely, given that wild squashes have not been identified in the ever-wet tropics (Chapter 3). Hence, the remains very likely represent a cultivated species selected for this habitat.

Several preceramic levels beneath the dated phytolith sample yielded no squash or leren although other phytoliths were abundant. Also, ceramic levels of the site above the preceramic units were nearly lacking in Cucurbita, with only one such phytolith being observed. It appears that the cultivated plants were introduced into the site after it was initially occupied and by approximately 8100 B.P. It is interesting that pollen records from nearby Peña Roja indicate a drier than present period between 9000 and 8000 B.P. that would have created habitats more hospitable for squash (Cavelier et al., 1995). Calathea allouia phytoliths were not observed in the ceramic-phase levels. Because the C. allouia remains derive from the seeds of the plant, the absence in more recent deposits may suggest either that the plant had lost the ability to set seed under domestication pressure that focused on vegetative reproduction or that it was harvested before seed set.

The cultural remains from Peña Roja indicate that settlement of wet, evergreen forest in the western fringes of the Amazon Basin is ancient, and that two species of domesticated plants originally taken under cultivation in much drier areas had been dispersed into the wet Amazonian forest by 8000 B.P.

Zaña Valley, Northern Peru

Studies by Dillehay *et al.* (1989) and Rossen *et al.* (1996) in the upper Zaña Valley of northern Peru have revealed the probable presence of horticultural societies by 8000 years ago (Fig. 4.5). The sites lie at an elevation of 400–800 m and fall within Peru's closest juxtaposition of coast, sierra, and tropical forest. The environment at the time of site occupation appears to have been a deciduous tropical forest in close proximity to diverse, resource zones such as montane forest and thorny scrub habitats. Pollen records from the sites' soils contain such typical deciduous tropical forest trees as *Bombacopsis, Spondias,* and the Anacardiaceae and Moraceae, which shows that they were growing near the sites (J. Jones, personal communication, 1996).

More than 60 preceramic sites were discovered upon systematic survey during several field seasons in the 1980s. Settlements are dispersed single or multifamily units less than 1 ha in size along small streams in alluvial fans above the valley floor that flowed in the area before the historic period. Several of these sites along the Quebrada Las Pircas were intensively excavated, revealing individual small houses of approximately 2×3.5 m in dimension. Remains of the house structures contained intact floors, hearths, and postholes.

Macrobotanical plant remains were recovered from beneath ground stone slabs inside intact floors and from immediately outside the entranceways to the structures in possible toss areas. Most were desiccated, whereas a few were carbonized. Recovery of desiccated materials from an ancient site can be explained by their position underneath stones and grinding slabs, where they were protected from draining rainwater. The species list includes manioc (*Manihot esculenta*), peanuts (*Arachis hypogaea*), quinoa (*Chenopodium* sp. cf. *quinoa*), and squash (*Cucurbita* sp.), along with various fruits of trees and cacti (Table 4.5). As noted by Rossen *et al.* (1996), some of these plants, including manioc, peanuts, and quinoa, are far from their likely cradles of domestication. Quinoa was not discussed in Chapter 3 because it appears to be a central Andean domesticate.

Investigations by experts in the various plant taxa recovered revealed that peanuts were small and hirsute, and thus morphologically primitive (it is very difficult to find examples of hairy peanuts being grown by Peruvians today; Banks, 1990). The squash seeds found similarly could not be placed into a modern wild or domesticated species but were most like *C. ecuadorensis*. Remember that this species of squash was also present in a semidomesticated state at the Las Vegas preceramic site in nearby southwestern Ecuador. The archeological seeds from the Zaña Valley are larger than those of present-day *C. ecuadorensis,* a finding consistent with an early domesticate and with the presence of squash phytoliths larger than those in modern *C. ecuadorensis* at Vegas.

The chenopod remains also exhibited morphological differences from both wild and modern quinoa that were thought to be characteristic of an early domesticate. Fragments of the manioc tubers recovered were analyzed by Ugent (Rossen *et al.,* 1996) for starch grains, revealing the characteristic grains of the domesticate *M. esculenta.*

Associated charcoal from the house floors and other features, including that from beneath stone slabs where plant remains were found, revealed ages of between 7950 B.P. and 7640 B.P. As discussed by Rossen *et al.* (1996), AMS dates on the peanuts and squash belong to the later historic and modern eras, with the peanuts yielding a date consistent with a post-1950 deposition. However, present-day occupants of the valley cannot recall any agriculture being practiced or even any settlement in the vicinity of the sites in the modern era, by which time the streams were completely dried up. Geological evidence indicates that streams in the area have not been active since approximately AD 1000-1200 (T. Dillehay, personal communication, 1996). Examinations of late historic hacienda records in the local valley capital reveal no indication of historic settlement in the vicinity of the sites in question at the time indicated by the dates on the plant remains.

Phytoliths are not well preserved at the site, probably due to high pH, and cannot provide much information (D. Piperno, unpublished data). Among the few phytoliths identified was a species of wild *Calathea,* which might well have grown naturally in the streamside environments of the sites when they were better watered but does not occur near the sites today. This provides some support for the investigators' contention that the plant remains were not deposited during the recent past.

Pollen was not abundant but is present in a sufficient state of preservation and in enough quantity for many plant identifications to be made (J. Jones, personal communication, 1996). In addition to the trees mentioned previously, *Cucurbita* pollen was recovered from one of the same contexts that yielded squash seeds. Present in high frequencies are grass and Compositae pollen, a finding consistent with the creation of garden plots next to the sites.

Several other lines of evidence from the Zaña Valley support the interpretation of plant-based economies and horticulturists in the area in middle Preceramic times. Dillehay *et al.* (1989, 1997) found what appears to be a small-scale irrigation system associated with the preceramic occupations. It consists of furrows and short feeder ditches on alluvial flats and terraces in the upper quebradas near springs, which probably provided water to gardens outside adjacent occupational structures. Also, the settlement traits of the sites—small, dispersed stream-side houses and groups of houses—are like those of small-scale horticulturists today and do not fit the pattern of mobile foraging groups. The lithic assemblage contains numerous grinding stones and grinding bases probably used to process plants. Further, the lithics are all unifacial and include numerous types of flake forms that appear to have been used to harvest or process plants, based on microwear analysis. All the teeth of the recovered human remains exhibit heavy, grinding wear associated with mastication of plant foods.

Finally, Dillehay went back to the sites after the AMS dates were reported and placed test pits outside the house structures from the earlier and later preceramic periods. Soils were screened, floated, and searched for cultural and plant materials using the same methods as for the sites. No historic debris or plant remains, save a few grass fibres, were recovered (T. Dillehay, personal communication, 1996). Along with the excavators, we wonder how intrusive plants from historic and modern agricultural systems came to be deposited under, but not around, prehistoric grinding slabs and in, but not outside, ninth and eighth millennium B.P. house structures. We also wonder why the plants do not match any modern species of domesticate. We believe that the Zaña Valley sites represent settlements of an ancient horticultural society.

Currently, the earliest evidence for the use of exogenous domesticates seems to be the 7950 ± 180 B.P. determination on wood charcoal associated with the manioc, peanuts, and other plants discussed previously. The investigators believe that this early horticultural cultural complex lasted until approximately 7000 to 6000 years ago. At this time, coca (*Erythroxylon* sp.) and cotton (*Gossypium barbadense*) are also found at slightly later middle preceramic sites located along quebradas a few kilometers south of the Quebrada las Pircas localities (Rossen *et al.*, 1996; Dillehay *et al.*, 1997) (discussed in Chapter 5).

Along with the documentation of an intensive preceramic occupation of this lowland Peruvian valley by horticultural societies in a tropical forest setting, the middle preceramic record from the Zaña Valley highlights several important features relevant to the origins and dispersals of early food production. As at Vegas and Peña Roja, the first crop plants for which we have evidence in the Zaña Valley are food plants; coca and cotton may have been later additions to the horticultural economies in the region. These findings lend support to our belief that early food production was, indeed, a food-producing strategy and not simply an attempt to increase the supply of plants useful for prestige enhancement, storage, or other needs (e.g., Hayden, 1992).

Another significant finding is that preceramic sites were not located on the river floodplain. Rather, there is what Dillehay *et al.* (1989) call an "integrated and descending settlement pattern" in the area through time, with late preceramic sites located further downslope on the middle sections of streams. The positioning of sites near the river floodplain did not occur until the Initial period, several thousand years after the upper quebradas were first exploited. We find this sequence of settlement location through time to be repeated in central Pacific Panama (discussed later). It is clearly not in accord with the belief (e.g., Smith, 1995a,b) that the earliest food production and plant domestication in the New World took place in major river valleys.

Finally, some of the incipient domesticates evidenced at the sites, such as manioc, peanut, and quinoa, probably had been taken considerably outside their areas of original manipulation by 8000 B.P. Other cultural remains found at the sites, such as seashells and exotic lithics from higher elevations, also attest to exchange and/ or contact over considerable distances. The plant evidence from the Zaña Valley indicates that interactions among cultural entities over fairly large distances were already taking place during the early Holocene and that other, yet unstudied regions in South America have fascinating information on early plant manipulation to reveal.

On the other hand, the presence of a form of squash similar to *C. ecuadorensis* near the Vegas culture sites of southwestern Ecuador (discussed previously) at the time when squash phytolith size at Vegas indicates domestication suggests that locally available forms of wild squash were domesticated in this area of northwestern South America. Preceramic economies producing squash were apparently spread over the region by 8000 years ago.

Central Pacific Panama

During the past 10 years the drainage of the Rio Santa Maria in the central Pacific watershed of Panama has been a focus of archeological and paleobotanical study (Figs. 4.8 and 4.9) (see Cooke and Ranere, 1992a, for a summary). The 3500-km^2 area stretches from the Continental Divide, whose altitudes range from 1100 to 1600 m, to the mangrove-fringed zone of mudflats bordering the Pacific Ocean. Beginning shortly after the end of the last glacial epoch 11,000 years ago to before the onset of intensive human disturbance, lowland tropical forest graced virtually the entire region.

The region's annual precipitation ranges from 1000 mm along the coast to nearly 4000 mm near the Continental Divide. Except for areas near the divide, the rainfall is seasonally distributed, with virtually no rain falling during a 4- or 5-month period between December and April (Plates 4.9 and 4.10). The region possesses a broad coastal plain, which once supported a deciduous (dry) forest. Here, annual precipitation ranges from 1200 to approximately 1800 mm per annum. It is here that the forest was replaced by open vegetation during the late Pleistocene, discussed previously and in Chapter 2.

The Proyecto Santa Maria (PSM) was a multidisciplinary project initiated in 1982 by Anthony J. Ranere and Richard Cooke to study the evolution of settlement and subsistence in this region. Project goals were accomplished by systematic settlement survey following a stratified random sampling strategy, excavation of selected sites, and analysis of surface remains and excavated materials, including

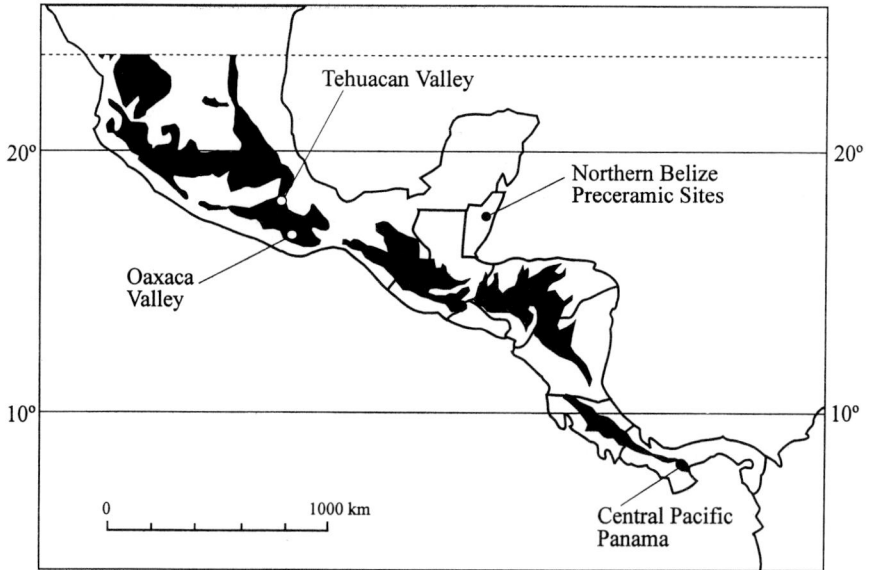

FIGURE 4.8   Map showing the location of the archeological sites in Middle and Central America discussed in the text.

plant remains (pollen, phytoliths, and seeds). In 1988 Piperno and Paul Colinvaux initiated paleoecological research to study the historical landscape and its relation to human settlement and subsistence through time. The archeological and paleoecological studies provide one of the most complete and detailed records available for early through middle Holocene human adaptations in a lowland seasonal forest.

The Archeological Record

Details and summaries of the archeological sequence may be found in Cooke and Ranere (1984, 1989, 1992a,b,c), Ranere (1992), Ranere and Cooke (1991, 1996), Ranere and Hansell (1978), Hansell (1987), and Weiland (1984). Summaries of the archeological botanical remains and the paleoecological records are found in Piperno (1988a, 1995a) and Piperno et al. (1991a,b, 1992).

The late Pleistocene paleobotanical and archeological records were described previously. They indicate that Paleoindians were both exploiting and modifying the tropical forest beginning approximately 11,000 B.P. Beginning approximately 10,000 B.P., a number of rock shelters situated in deciduous and other highly seasonal forests—in the foothill zones at elevations between 200 and 900 m (Corona, Los Santanas, and Carabalí), on the coastal plain (Aguadulce), and at the coast itself (Cueva de los Vampiros)—began to be more regularly occupied by people who continued to work stone bifacially (Fig. 4.9; Plate 4.11). The increased

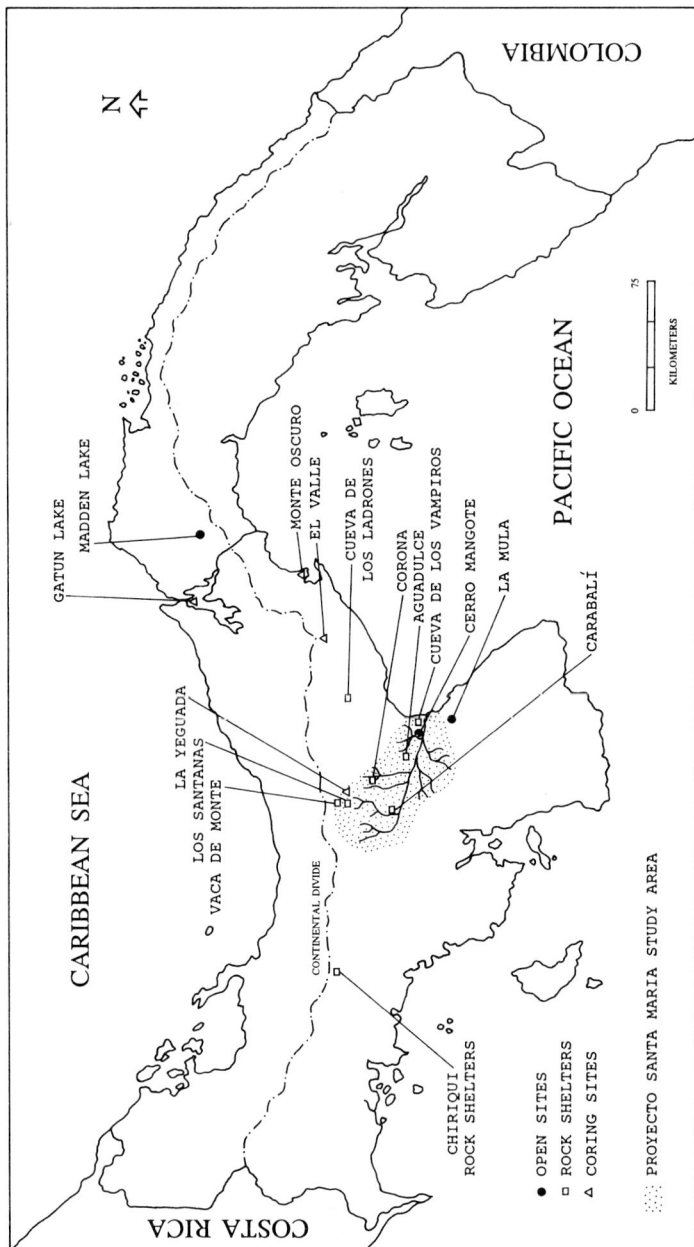

FIGURE 4.9 Map of Panama with the location of the archeological and paleoecological sites discussed in the text.

CARIBBEAN SEA

COSTA RICA

GATUN LAKE
MADDEN LAKE

MONTE OSCURO
EL VALLE
CUEVA DE
LOS LADRONES

CORONA
AGUADULCE
CUEVA DE LOS VAMPIROS

CERRO MANGOTE

LA MULA

CARABALÍ

LA YEGUADA
LOS SANTANAS
VACA DE MONTE

CONTINENTAL DIVIDE

CHIRIQUI
ROCK SHELTERS

PACIFIC OCEAN

COLOMBIA

N

• OPEN SITES
□ ROCK SHELTERS
△ CORING SITES

PROYECTO SANTA MARIA STUDY AREA

0   75
|___|___|___|___|___|
KILOMETERS

PLATE 4.11   (Foreground, right) the Corona rock shelter overlooking the foothill zone. In the distance to the right the coastal plain and ocean are visible.

utilization of these shelters during the early Holocene probably reflects a shift in focus from mobile resources (game) to more stable resources (plants).

Coastal resources were used from an early time in the region. At the Cueva de los Vampiros, currently located on the alvinas (mudflats) near the sea, estuarine mollusks and fish, including sea catfish and mullet, are present in a sealed stratum dated to 8600 B.P. An edge ground cobble/boulder milling stone complex and small, flat grinding stones are present at the rock shelter Carabalí by 8040 B.P.

In addition to the record of early Holocene occupation from these rock shelters, the systematic site survey carried out by the PSM recorded surface finds of more than 20 other sites that were occupied between 11,000 and 7000 B.P. They are small (<0.1 ha) and are often situated on flat spurs that overlook the minor alluvium of secondary water courses. Clusters of artifacts were sometimes detected that suggested groups of dwellings arranged into small hamlets. Precisely dating such kinds of small and open sites is difficult, but that they predate 7000 B.P. is certain because they possess the kinds of stone technologies (bifacial reduction strategies) that disappear from the rock shelter records by 7000 B.P.

It will be noticed that this settlement pattern in central Pacific Panama during the early Holocene—small house units and/or hamlet-like clusters situated along the alluvium of secondary rivers and streams—is like that recorded from the Zaña Valley, Peru, the Cauca Valley, Colombia, and the Vegas complex, Ecuador. We believe that it speaks to the same thing: small-scale horticulture practiced by house and hamlet units on small bits of alluvium away from the major river valley bottoms. This intepretation is supported by the paleobotanical records from nearby rock shelters and lakes, which indicate that a domesticated root (*Calathea*) and seed crop (*Cucurbita*) were used sometime before 7000 years ago.

Phytoliths

Phytoliths from the Cueva de los Vampiros point to the early Holocene use and possible cultivation of arrowroot (*Maranta arundinacea*). Seed silica bodies from this plant were recovered from a context radiocarbon dated to 8600 B.P. that was sealed underneath a hard stratum of culturally sterile soil (Cooke and Ranere, 1992a; Piperno, 1988a). The phytolith record from this site indicates a dry, open, and windswept environment, unlike habitats currently favored by arrowroot. Thus, it is possible that arrowroot was taken outside of its natural habitats, the moist areas beneath the forest canopy, and planted near the site. Arrowroot is also present in pre-7000 B.P. strata from the foothill (Corona and Carabalí) and coastal plain (Aguadulce) rock shelters. Even if it were not actually cultivated, its presence points to the use of indigenous tropical forest tuberous plants at an early date.

Arrowroot possesses a smallish tuber, which needs to be pounded and macerated in order for the starch grains to be released. It is likely that the early Holocene plant-processing kit was used to process arrowroot in addition to other kinds of tuberous plant parts. Before its discovery at the Cueva de los Vampiros, the use of arrowroot had not been documented in the Neotropical archaeobotanical record. At European contact, it was used as an antidote for arrow poison (Sturtevant, 1969). The plant appears to grow wild in Brazil and northern South America in addition to Panama (Hawkes, 1989), although how many different species of *Maranta* are under cultivation today in northern South America is unclear.

Direct evidence for the use of domesticated plants that probably antedates 7000 B.P. comes from the Aguadulce rock shelter (Ranere and Hansell, 1978). Here, phytoliths from the seeds of *Calathea allouia* and the rinds of bottle gourd are present in the bottom-most level containing cultural deposits, located just above sterile soil (D. Piperno, unpublished data). Leren phytoliths were not present in soils above this level. These phytoliths are morphologically identical to those recovered from Vegas and Peña Roja. Bifacial thinning debris was recovered in the bottom-most levels of two other excavation units at the site (A. J. Ranere, personal communication 1996), making it likely that the *C. allouia*/bottle gourd context dates from earlier than 7000 years ago.

The chronology of the cultural sequence at the Aguadulce shelter can be interpreted based on current evidence as follows. The earliest date from the site on nonphytolith material is 6180 ± 120 B.P. on shell from layer C4 (C, preceramic) of Block 3, the excavation block that produced the most informative phytolith record and the one discussed here (Cooke and Ranere, 1992a) (Fig. 4.10). Block 3 is underneath the current overhang. The soil deposits producing the phytolith

AGUADULCE SHELTER: EAST WALL OF BLOCKS 2 AND 3

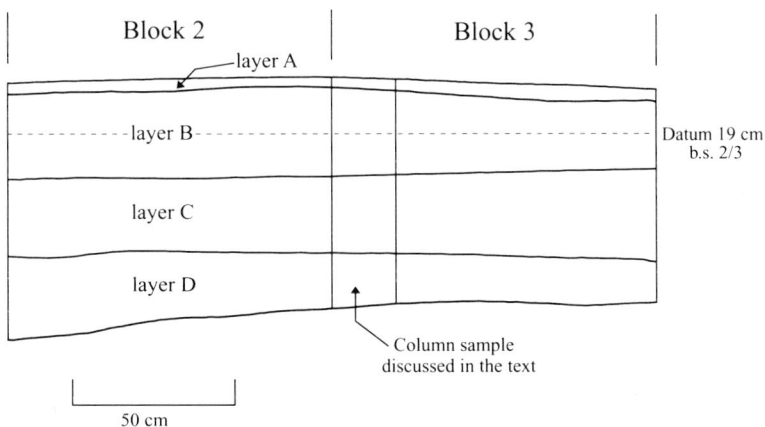

FIGURE 4.10   Profile of the east wall of Blocks 2 and 3 of the Aguadulce rock shelter. Layer B contains Monagrillo ceramics, layer C is preceramic, and layer D is sterile soil.

record are located near the back wall of the shelter and constitute a series of column samples taken from the cleaned profile wall and from the block as the excavations were in progress. A shell date from layer C5 of Block 3, at a depth of 35–45 cm below datum (b.d.), which contained the leren phytoliths, was determined to be 5840 ± 100 B.P. This date is too young, considering its stratigraphic placement at the bottom of the site and the presence of bifacial material in other deep levels of the site. The deposits in Block 3 were very compacted, increasing the possibility that this shell was derived from above.

Phytolith assemblages from three levels of the Block 3 column sample were directly dated to provide more information on site chronology (Table 4.6). Phytoliths from layer B1 at a depth of 2–7 cm above datum, containing Monagrillo ceramics (the earliest produced in Panama), yielded a determination of 4250 ± 60 B.P., consistent with the age of this pottery. Those from layer C2, at 21–26 cm bd and containing no ceramics, were dated to 6910 ± 60 B.P. A final sample from layer C4 at a depth of 28–33 cm bd, which contained the remains of what we believe to be a primitive variety of maize, yielded a date of 5560 ± 80 B.P.

This last sample contained few phytoliths other than the *Zea,* and produced the bare minimum needed for an AMS determination—0.130 mg final carbon (other samples had 10–20 times as much). That it yielded a date at all is a testament to modern science, but contamination with only 0.0200 mg of present-day carbon would cause a 2000-year error in a 7000-year-old specimen. Such a miniscule amount of carbon would be practically invisible to the naked eye. Although the

TABLE 4.6 Characteristics of *Zea* Phytoliths through Time at the Aguadulce Rock Shelter

Phytolith	% Decorated	Aguadulce rock shelter (Block 3 column sample)	% Decorated	^{14}C phytolith age (year B.P.)	Vegas	% Decorated	^{14}C phytolith age (year B.P.)
Balsas teosinte	73				Feature 1	20	5780 ± 60
	58	8–12 cm ad (layer A2)	13		105–110 GH8-9	50	7170 ± 60
	72	2–7 cm ad (layer B1)	20	4250 ± 60[a]			
		0–5 cm bd (layer B2)	15				
Maize		21–26 cm bd (layer C2)	33	6910 ± 60[b]			
Pepitilla, Mexico	40	28–33 cm bd (layer C4)	67	5560 ± 80[c]			
Reventador, Mexico	20						
Tabloncilla, Mexico	40						
Maiz Ancho, Mexico	14						
Bolita, Mexico	4						
Patillo, Ecuador	3						
Popcorn, central Peru	9						
Sweet corn, Peru	3						
Chillo, Ecuador Sierra	7						
Flour flint, central Peru	3						
Morochon flint, Ecuador	11						
Amarillo, Ecuador	11						
Cubano Blanco, Ecuador	5						
Purple dye corn, Ecuador	2						
Argentine popcorn	1						

Note. ad, above datum; bd, below datum.
[a] UCR-3418, CAMS-27727.
[b] UCR-3417, CAMS-27726.
[c] UCR-3462, CAMS-28743.

phytoliths were extensively pretreated in Piperno's laboratory to remove extraneous contaminants, a degree of control in the laboratory to which we possibly cannot aspire might be necessary to preserve the integrity of a sample of this size.

Therefore, the date is best taken as a minimal age marking the deposition of the phytoliths. Given its stratigraphic placement in the site, the lack of any observable disturbances, and, above all, its unique phytolith assemblage, the true date is probably sometime shortly before 6900 B.P. This would leave the leren phytoliths earlier, in accordance with their placement at the bottom of the site. Unfortunately, not enough soil remained to directly date the phytolith assemblages containing leren.

Also in pre-7000 B.P. levels at the Aguadulce shelter are other phytoliths from the Cucurbitaceae, including a few of the distinctive, spherical scalloped forms that are restricted to the genus *Cucurbita,* which were intensively studied at Vegas. However, the sample size is too small ($n = 2$) to enable a definitive identification of wild or domesticated, especially because a species of *Cucurbita* grows spontaneously near the site (see Chapter 3).

Another type of phytolith that is present in much smaller numbers than the scalloped forms in *Cucurbita,* and that can also be found in small numbers in another genus (*Calycophysum*) of the Cucurbitaceae, was recovered in asociation with *C. allouia* and bottle gourd (hereafter, this phytolith is called "Cucurbitaceae B") (D. Piperno and I. Holst, unpublished data). Those in the Aguadulce sediments are large (48–80 μm; n = 17), well outside the range of those in modern wild species of *Cucurbita* and *Calycophysum* (26–55 μm) and consistent with the sizes of the same phytoliths observed in modern *C. moschata* and at Vegas. Cucurbitaceae B phytoliths are much more common in the bottom unit at the site than in levels directly above, making downward intrusion seem unlikely. We believe that these forms are likely from a domesticated *Cucurbita.*

Another interesting factor relating to *Cucurbita* phytoliths is that rinds from modern Panamanian *C. moschata* fruits rarely produce the genus–specific scalloped forms that proved so useful for study at Vegas. Phytolith production by Ecuadorean varieties of *C. moschata* is also spotty. On the other hand, wild squash populations from Panama and elsewhere produce this phytolith in abundant numbers. It is possible that the large, scalloped phytoliths constituted a natural defense mechanism for wild fruits that was no longer needed when domestication occurred.

Reduction in levels of toxic chemicals made and used by plants for defensive purposes has accompanied the domestication of many food plants (Johns, 1990). If phytoliths act as a type of physical defense for plants and deter herbivory in tropical plants (Piperno, 1991), loss of or decreased numbers of phytoliths during the domestication process would be another logical consequence of human selection. What is pertinent for our purposes is the rarity of scalloped phytoliths at the Aguadulce shelter. Exploitation of wild *Cucurbita,* which still grows today not far from the site, would likely have left many more scalloped phytoliths, as in lower

levels at Vegas. Hence, rarity of these forms may be another signal of the domestication of *Cucurbita* before 7000 years ago.

In summary, archeological phytolith records indicate that at least two species
of domesticated plants, one a root and the other a seed crop, were used in central
Pacific Panama during the early Holocene period. As at Vegas and Peña Roja,
bottle gourd was also incorporated into horticultural systems in the lowland tropics
at an early date. We believe that arrowroot and leren belong to what is probably
a long list of starchy and oil-rich plants that were first taken under cultivation or
otherwise intensively manipulated, but whose importance as suppliers of calories
subsequently fell victim to the availability and increasing importance of maize,
manioc, and other crops that are more familiar to us today. In Panama today,
many campesinos do not grow arrowroot or leren, but speak of arrowroot as
having been somewhat more important in the recent past.

The microfossil records from the lake paleoecological sequences support arguments that an established tropical forest adaptation was in place during the earliest
Holocene, and that a small-scale horticultural system was developed before
7000 B.P. The disturbance horizon beginning at La Yeguada 11,000 years ago
(discussed previously) continues into the early Holocene and intensifies, with
particulate charcoal and *Heliconia* reaching their highest frequencies approximately
8600 years ago (Figs. 4.3 and 4.4). Burning and small-scale forest clearing must
have been widespread in the watershed.

Also, Fig. 4.4 shows that levels of burnt tree leaf phytoliths (% burnt arboreal),
which are from evergreen tree species, reach levels by shortly after 8600 B.P. that
are found today only in fields cleared for slash-and-burn cultivation. It makes
logical sense that fires set to burn areas cleared for planting would burn substantially
more numbers of arboreal phytoliths because the "slash" that is ignited by cultivators
includes much of the vegetative litter brought to the ground when trees and shrubs
are killed or felled. Burnt grass (Poaceae) phytoliths also increased substantially
8600 years ago. These indicators support the proposition that some active, if
relatively small-scale, clearing of the forest was taking place at that time for food production.

Macrobotanical Remains

Macrobotanical remains during the early Holocene are sparse. Even palms,
which must have been commonly utilized given their ubiquity in the phytolith
records, are poorly represented by charred remains. However, at two rock shelters,
Corona and Carabalí, carbonized palm fruits, including *Acrocomia mexicana* (formerly
A. vinifera, the coyol palm), the Sapotaceae, and other plants occur in deposits [14]C
dated from 8450 to 6500 B.P. (Cooke and Ranere, 1992a) (Table 4.5). Coyol is still
commonly eaten in Panama today, and its widespread occurrence in archeological
deposits suggests the same was true throughout prehistoric time.

Piperno (1989b) has suggested that the coyol palm is an introduction from Mexico because the tree is always found today around current and past settlements and never in established forest, as would be expected if it were native to Panama. Janzen (1983) has made much the same argument in Costa Rica, where coyol distribution parallels that in Panama. *Acrocomia mexicana* kernels are also ubiquitous in preceramic levels of rock shelters in Chiriqui province, Panama (Fig. 4.9), where they are dated by associated carbon to early in the seventh millennium B.P. (Ranere, 1980b,c). At one of these sites, Smith (1980) detected diachronic increases in *Acrocomia* fruit size between 6700 and 1600 B.P. that suggested human selection.

It is likely that the coyol palm was present in Panama by 7000–6000 years ago, a proposition we will soon test by directly dating some of the earliest kernels. Interestingly, coyol is reported from the Tehuacán Valley starting approximately 7000 years ago, where it was almost certainly cultivated because it does not grow wild in the valley and must be irrigated during the dry season. It appears that coyol is native to parts of Mexico that are more moist than the Tehuacán Valley and was under cultivation by 7000 years ago by which time it, along with maize, was dispersing south.

The early seventh millennium B.P. is a time being mentioned again and again in discussions of central Panamanian prehistory. It is evident that it marks a major watershed in the course of cultural evolution. First, settlement number and size appear to increase substantially. The number of sites dated to between 7000 and ca. 5000 B.P. vs earlier increases more than 15-fold, and the sites contain significantly more occupational debris. Second, bifacial thinning is no longer used as a core reduction strategy, and the edge-ground cobble/boulder milling stone complex becomes common, indicating a shift in, and intensification of, plant exploitation (Cooke and Ranere, 1992a; Ranere and Cooke, 1996). Despite these dramatic differences, other aspects of the lithic records indicate substantial cultural continuity between the pre- and post-7000 B.P. periods (Ranere and Cooke, 1996).

Also beginning 7000 years ago, coastal resources begin to be used intensively. At the site of Cerro Mangote, the earliest shell midden identified in the archeological record from Panama, faunal remains indicate that between 7000 and 5000 B.P., people focused on mangrove/estuarine shellfish, crabs, and fish (Cooke, 1992; Cooke and Ranere, 1989, 1997). Exploited mangrove fauna also included "terrestrial" animals such as the racoon. Cooke emphasizes that the important fact about Cerro Mangote's fish fauna is its "in-shoreness" (Cooke, 1992; Cooke and Ranere, 1997). There are no freshwater taxa. Rather, the dominant taxa are inshore forms that attain quite a large size and prefer turbid, shallow waters, especially mangrove edges and channels, and that can be easily caught with stationary traps. Fishing tools are absent at the site and boats probably were not used. Thus, this kind of "coastal" exploitation probably yielded a high return for labor invested. Its inception may well be a product of the stabilization of sea level and the presence of stable and highly productive estuarine resources. R. Cooke (personal communication,

1996) feels that beginning 7000 years ago, littoral faunal resources were probably more optimal than terrestrial ones.

Significantly, there are also signs that approximately 7000 years ago people were beginning to have difficulty procuring adequate amounts of terrestrial protein. Terrestrial mammals that are commonly taken by extant Neotropical forest hunters, such as monkeys, agoutis, coatis, and peccaries, are rare or absent from central Panamanian archeological assemblages after 7000 years ago. In contrast, small birds, reptiles, and amphibians become increasingly common in the middens (Cooke and Ranere, 1992c). The white-tailed deer is, however, present and commonly abundant in prehistoric sites of all ages.

Cooke and Ranere (1992a,b,c) think that both hunting pressure and human modification of the forests leading, finally, to habitat destruction were significant determinants of these faunal patterns. Domesticated dogs were present in central Panama by at least 7000–6000 B.P., and these animals probably made human hunters all the more efficient in their pursuit of terrestrial game (Cooke and Ranere, 1989, 1992c). Increasing human pressure on the terrestrial fauna may also have contributed to the heightened use of coastal resources. The white-tailed deer seems to thrive in areas filled with secondary forest, fallow, and field habitats, thus explaining its continued presence in environments heavily disturbed by humans.

Phytoliths, Pollen, and Starch Grains

Very significant changes in the paleobotanical records also take place at 7000 B.P. They indicate that (i) two new exogenous crop plants, maize and manioc, which were efficient caloric sources and, in the case of maize, much better suppliers of protein than the earlier cultivated root crops, were beginning to be incorporated into the subsistence system; and (ii) integration of more soil-demanding and nutritionally complete crops such as maize allowed populations to grow and also led to the intensification of the food-production system into full-fledged slash-and-burn agriculture.

Maize phytoliths and pollen are present in a rock shelter, the Cueva de los Ladrones, by or shortly after 7000 years ago (Piperno et al., 1985). Maize phytoliths are also present from near the beginning of the occupation of another rock shelter, Los Santanas, which dates to early in the seventh millennium B.P.

Starch grain studies carried out recently are providing valuable evidence of early plant use and cultivation in Panama (Piperno and Holst, 1997). As with the Colombian lithic specimens described above, the tools were washed, and the grains were recovered from tiny pits and crevices on their surfaces. At the Aguadulce shelter, a starch grain characteristic of maize was recovered from the pounding facet of a preceramic-phase edge ground cobble. This tool is from the soil layer that contained the primitive *Zea* phytoliths (described below), and is immediately above the layer containing leren and bottle gourd phytoliths, described above (Block 3, Layer C4) (Fig. 4.10). It dates to the late preceramic period (ca. 7000–6000 B.P.).

Another edge ground cobble from the same context yielded two starch grains from *Dioscorea* spp. and two starch grains identical in size and shape to those of modern, domesticated manioc. *Dioscorea trifida* can probably be ruled out from representation, suggesting that a native species of yam was being used in the local economy.

Five starch grains identical in size and shape to those from modern, domesticated manioc were also recovered from a grinding stone base at Aguadulce that occurred in the middle of the preceramic deposits from Block 2 (Plate 4.12; Fig. 4.10). This tool can also be assigned to the late preceramic period (7000–6000 B.P.). The manioc-like starch grains are compound, in contrast to the single grains found in many economic species such as palms, beans, various tree fruits, achira, and yams. They possess a type of central fissure and other morphological characteristics that distinguish them from compound grains found in grass seeds and other tubers, such as sweet potato, yautia (*Xanthosoma*), arrrowroot, *Calathea,* and other Marantaceae. They are indistinguishable from those of modern, domesticated manioc, and are unlike those in any other economic species that we can presently identify from Panama. They appear to be the remains of manioc.

Ten highly decorated phytoliths that appear to be from maize glumes (Piperno and Pearsall, 1993), and phytoliths like those isolated from modern arrowroot

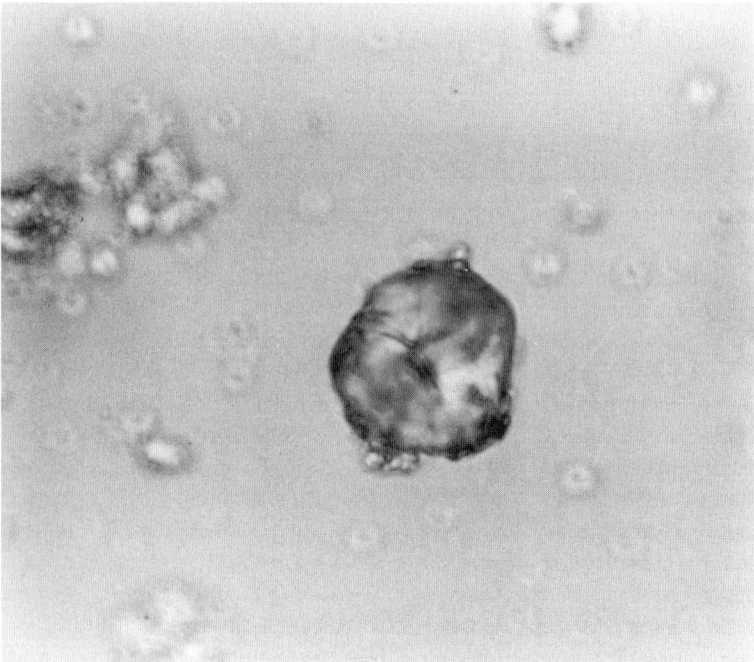

PLATE 4.12   A starch grain from manioc recovered from the surface of a grinding stone base at the Aguadulce rock shelter. It is 22 μm long.

tubers are also present on the grinding base, indicating that edge ground cobbles were used to process both tuber crops and small-seeded varieties of maize (below) (Piperno and Holst, 1997). Manioc-like starch grains and *Zea* glume phytoliths are the most common types of residues isolated thus far from the grinding base and edge ground cobbles, suggesting that these two species of plants were being consumed on a fairly regular basis.

Other starch grains present on the milling stone base include arrowroot and unidentified legumes. Some of them are similar to those in modern domesticated *Phaseolus* but are smaller in size.

Other elements of the phytolith records indicate that the cultivation system was undergoing change at this time. Arrowroot is in deposits from Cueva de los Ladrones and the Aguadulce shelter belonging to the 7000–5000 B.P. period but in smaller amounts than in pre-7000 B.P. deposits from Aguadulce (see Chapter 5). Leren is not observed in later deposits at the Aguadulce shelter, which again, however, may have as much to do with loss of the seed in response to the selection pressure being put on the tuber as with anything else.

If maize was being grown on the Santa Elena Peninsula in Ecuador by approximately 7000 years ago, it must have reached Panama sometime before in a fairly primitive state. When and in what form did it arrive in Panama? Phytoliths recently identified by Piperno from near the bottom of the Aguadulce rock shelter have shone new light on this problem. The deposits in question lie immediately above the deepest stratum from Block 3 that yielded the *C. allouia, Cucurbita,* and arrowroot phytoliths previously discussed (Provenience: Block 3, Layer C4, 28–33 cm bd; Fig. 4.10 and Table 4.6).

Phytoliths were found here that appear to have originated from the cobs of a primitive race of maize. These phytoliths did not occur in sediments from levels above or below. One of the types of phytoliths present occurs both in Balsas teosinte fruit cases and in the cobs of some of the most primitive modern Mexican races, such as Reventador and Maiz Ancho. However, those in the sediments are much thicker than those in modern maize races (Plate 4.13). Overall, the phytolith assemblage is much closer to that of Balsas teosinte in that it (i) includes three types of phytoliths found in the fruit cases of this plant but not in any modern race of maize ("IRP, short, and rectangular"; "elongate, spiney", and "elongate WCE") (phytolith shapes are explained in Piperno and Pearsall, 1993) (Plate 4.14) and (ii) contains many IRP forms and high percentages of "decorated" circular to acicular phytoliths, as does teosinte but not maize (Plate 4.15) (Table 4.6) (see Piperno and Pearsall, 1993). The phytolith assemblage also includes a type of phytolith ("blocky, rectanguloid") found only in primitive maize races.

It should be stressed that despite intensive coverage of phyoliths in the wild Neotropical grass flora, including *Tripsacum* spp. (Piperno and Pearsall, 1997), phytoliths typical of teosinte have not been isolated from any other plant. The fact that the Aguadulce phytolith assemblages are more similar to teosinte than they are to maize speaks to the age of the specimens, greater than 5560 years, a

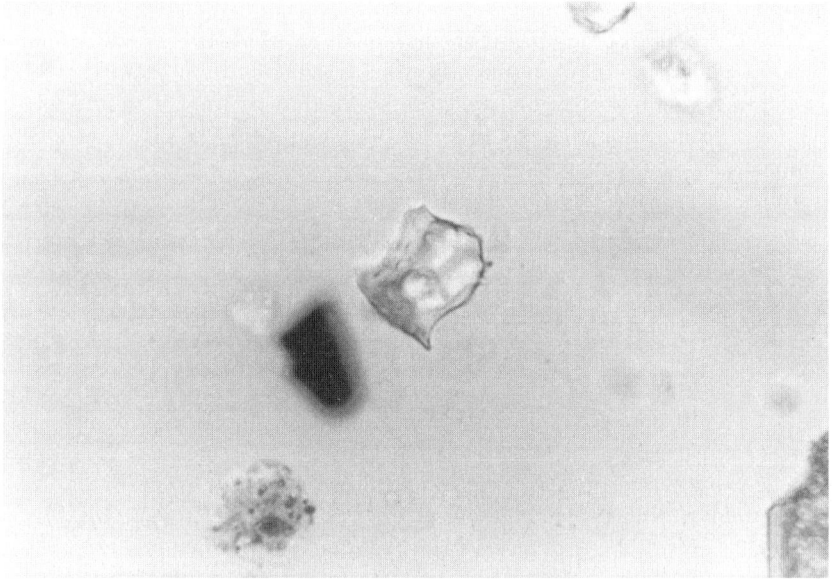

PLATE 4.13   An "oblong, half-decorated" phytolith from the Aguadulce shelter. Its longest dimension is 20 micrometers.

time for which no macroremain counterpart has yet been recovered from an archeological sample. Because sediments from directly below and above the primitive maize-bearing context do not possess the phytoliths in question, it is likely that maize was added to the subsistence cycle after the site was initially occupied and when tubers and squash were already being cultivated.

The implication from these data is that a morphologically primitive race of maize arrived in Panama and subsequently underwent further evolution there and/or in northern South America. What might this maize have looked like? To answer this question, we can examine the locus of silification in modern-day teosinte for the phytoliths recovered from the Aguadulce shelter. This has been revealed, by dissection and study of all the parts of the plant, to be the hard cupulate fruit case (Piperno and Pearsall, 1993).

The fruit case and cupule are almost certainly homologous, one having given rise under human selection to the other (Dorweiler et al., 1993). It seems to us that the presence of the fruit case bodies at the Aguadulce shelter may suggest that maize arrived with its kernels still at least partially enclosed by a fairly hard glume. The exposure of the kernel on the maize ear and development of soft glumes, controlled by the recently discovered tga1 locus (Dorweiler et al., 1993), must have represented a major advance in the use of maize as an important food plant. Given the implications of recent molecular data, which suggest that teosinte underwent a protracted period of cultivation before substantial morphological

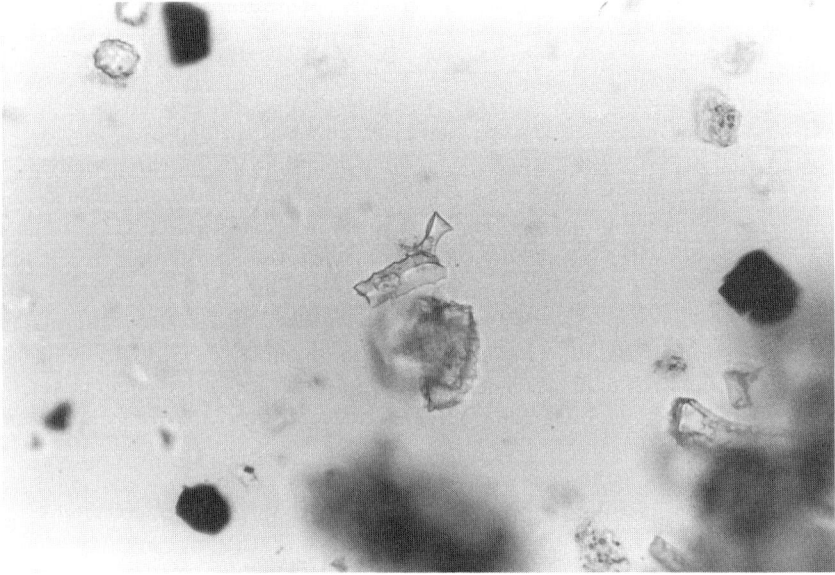

PLATE 4.14   (Center) an "elongate, WSE" phytolith from the Aguadulce shelter. These have only been observed in modern teosinte fruit cases. Its longest dimension is 30 micrometers.

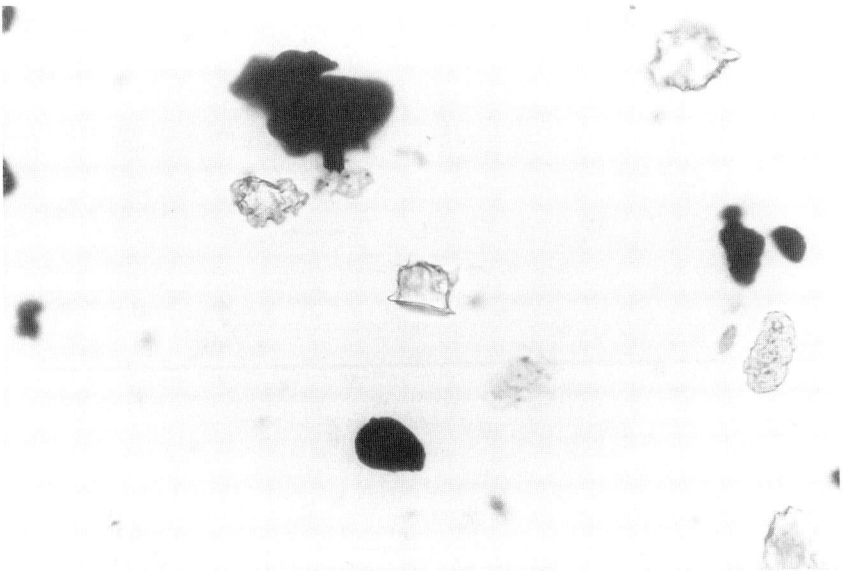

PLATE 4.15   (Center) a highly decorated circular phytolith on its side from the Aguadulce shelter. These occur in high numbers in modern teosinte fruit cases and in low numbers in modern maize cobs.

changes resulting in a cob occurred (Buckler *et al.,* 1995; Doebley, 1994a,b; Buckler and Holtsford, 1996), and these phytolith data, which indicate it dispersed at an early time out of its hearth, these advances need not necessarily have occurred in Mexico.

The maize-bearing Vegas sediments do not contain the primitive phytoliths recovered from the Aguadulce shelter, although here also the earliest *Zea* cob phytoliths differ from later ones at the site and from those in modern maize varieties (Table 4.6). Although it makes a certain amount of intuitive sense that maize may have embarked on a pronounced period of diversification and change after it left areas such as central Pacific Panama, which climatically and floristically is very much like the Rio Balsas watershed (e.g., Miranda, 1947; FAO, 1971), and reached the Andes and adjacent areas of high environmental diversity in Colombia, evidence from the Aguadulce shelter suggests that important changes in maize genetics and morphology may have been achieved by Panamanian cultivators shortly after they accepted the plant (Plate 4.16).

Table 4.6 shows that as one moves forward in time up the column sample from the Aguadulce shelter, the number of "decorated" circular to acicular cob phytoliths decreases. All modern-day maize races, including those thought to represent the

PLATE 4.16   An ear of modern Argentine popcorn. When maize arrived in Panama it may have been approximately this size, although the glumes were probably harder.

most primitive Mexican varieties, have much lower proportions of decorated phytoliths than occur in teosinte and the Aguadulce sediments with the most primitive maize. Conversely, proportions of decorated phytoliths in these sediments match those of Balsas teosinte.

Phytoliths directly dated to 6910 B.P. from the preceramic layer immediately above the primitive maize layer already have the inverted "decorated to undecorated" ratios characteristic of corn cobs, and the teosinte-type phytoliths present below this sample (elongate spiney; oblong, one-half decorated; and IRP, short, and rectangular) were not observed. Other IRP phytoliths continue to be observed but their frequencies are much reduced from the level containing the earliest *Zea*. Phytoliths from early ceramic levels of the site directly dated to 4250 B.P. further emphasize undecorated morphologies, as do many modern maize varieties. An evolution occurring in the maize cob seems to be evidenced at the Aguadulce shelter. The evidence suggesting that Panamanians may have changed primitive maize in substantial and important ways makes logical sense given the development and subsequent intensification of slash-and-burn agriculture in the region evidenced between 7000 and 4000 B.P. (see later and Chapter 5).

We will further evaluate these hypotheses in the future by studying phytoliths from a series of crosses between modern maize and teosinte provided by John Doebley. These should allow a closer examination of the sequence of changes in phytoliths as teosinte was transformed into maize. Also, as more early and mid-Holocene archeological sites are discovered in Central America and northern South America, we may be able to follow on a wider geographic scale the development and evolution of maize subsequent to the earliest changes of teosinte achieved by early Holocene cultivators in the Balsas River Valley.

Beginning approximately 7000 B.P., when the *Zea* at the Aguadulce shelter appears to have been undergoing experimentation and change and manioc likely arrived, the phytolith and pollen records from Lake La Yeguada demonstrate substantial changes associated with the intensification of the food production system in central Pacific Panama. For the first time, phytoliths of primary forest taxa decline substantially, while pollen grains of secondary forest trees greatly increase (Fig. 4.3; see Chapter 5; the pollen studies from La Yeguada were carried out by Mark Bush). Also, the frequencies of burnt grass phytoliths reach very high levels, further indicating that substantial areas of the watershed forest were now being cut and fired (Fig. 4.4). These factors, together with the continued high levels of microscopic charcoal in the cores, indicate the emergence of slash-and-burn agriculture in central Pacific Panama early in the seventh millennium B.P.

Grass pollen grains 100 μm in size that can definitely be identified as maize occur in the La Yeguada sediments immediately above a 3-cm core level ^{14}C dated to 5860 B.P. Grass pollen 82 μm in size occurs directly above a 5-cm core level ^{14}C dated to 6790 B.P. Because these seem to be outside the upper size range for *Tripsacum* spp., they are probably also from corn, although this cannot be said with complete certainty. However, an increase of *Zea* pollen size through time would fit with the changes in *Zea* being evidenced from the phytolith record.

Also, Piperno and Bush had identified lake sediments on the steep, southwestern shore of La Yeguada perched 5–7 m above the 1987 lake level that were exposed by the unusually low water levels of that much drier El Niño year (Bush *et al.,* 1992). These sediments were probably deposited during high-water stands during the early to mid-Holocene, when the climate may have been wetter than it is currently. Radiocarbon determinations on organic muds from depths of 6.00–6.30 m and 2.10–2.15 m   beneath the surface (bs) of the beds yielded ages of 12,200 ± 150 and 10,020 ± 150 B.P., respectively (Bush *et al.,* 1992). Results of phytolith analysis recently carried out on the sediments indicate that maize cross-shaped phytoliths are present in very high numbers at approximately 79–75 cm bs, when sediments become a gray clay with laminations of oxidized material (D. Piperno, unpublished data). Above and below this unit, the sediments are organic-rich lake muds.

The presence of high amounts of maize correlates to a time when the sediments do not appear to have been under deep and/or standing water throughout the year. Maize phytoliths were not recovered from below this section of the beds but are present above them. The implication from the nature of the sediment, and the abundance of cross-shaped phytoliths in them, is that maize was being planted directly on the rich soils at the lake's edge. Unfortunately, not enough of the sediment remains to be dated. However, its high clay content strongly suggests deposition between 7000 and 5000 B.P. because this is the only time during the Holocene that the sedimentation rate into the center of the lake slowed considerably, due to a reduction in stream flow from reduced annual precipitation, and lake levels were lowered (Piperno *et al.,* 1991b; Bush *et al.,* 1992).

In addition to the cross-shaped forms, phytoliths from the interstomatal cells of maize, which are much larger than those in any wild grass (including teosinte) but only infrequently produced in the leaf (D. Piperno, unpublished data), were found in the lakeshore beds. This finding is consistent with the other data indicating that maize was being cultivated at the lake's edge.

Cross-shaped phytoliths from sediments raised from the center of the lake had suggested maize presence beginning in the seventh millennium B.P., but like the pollen data it was not an unequivocal case, and we believed at that time that a conservative interpretation was warranted (Piperno *et al.,* 1991b). A single *Cucurbita* phytolith, which probably represents a cultivated species, was found in a level from the center of the lake that dates to approximately 6000 B.P.

To summarize, the trends in central Pacific Panama between 10,000 and 7000 years ago indicate substantial regional continuity, intensification of human modification of forests through burning and small-scale clearing, and the development of small-scale horticulture during the ninth millennium B.P. This system then intensified into slash-and-burn cultivation using a range of root and seed crops beginning approximately 7000 B.P.

It cannot be stated beyond doubt that the squash present in the Aguadulce shelter before maize arrived was the result of a local domestication of the plant, especially because the leren associated with it may have been an import. It seems that the exchange networks or population movements that were moving early cultigens south out of Mexico were also moving them north out of South America. Molecular studies in progress should reveal much about the relationship between the Panamanian populations of *C. argyrosperma* ssp. *sororia* (?) and *C. moschata*.

However, the occurrence of wild squash in central Pacific Panama, plus the persistence and intensification of the forest disturbance around La Yeguada between 11,000 and 8000 B.P., suggest to us that some native plants, including squash, arrowroot, and some trees, were taken under cultivation before maize and *Calathea allouia* arrived from elsewhere. It should also be noted that so little is known of the wild progenitor of *C. allouia* that an independent origin in Panama is not out of the question, especially because the native Panamanian flora is well stocked with *Calathea* species.

Panama shares in common with other areas where early food production emerged the important feature that a number of different resource zones were in close proximity to prehistoric populations. For example, in Panama coastal and, therefore, protein resources were not very far from even the most interior localities in the Pacific watershed of the central provinces, where some of the rock shelters discussed are located (Cooke and Ranere, 1992a). Mangrove and intertidal shells were found at Corona, one of the interior shelters, and directly dated to 6000 B.P. If hunting pressure between 10,000 and 7000 B.P. caused the available terrestrial fauna to decline, the nearness of the coast and estuaries provided essential proteins to people who were intensifying and expanding their food production systems.

OTHER AREAS WITH HUMAN OCCUPATION DATING TO THE EARLY HOLOCENE

Coastal Peru

A still poorly understood but important culture called Paijan existed on the northern coast of Peru during the early Holocene, between approximately 10,500 and 8700 B.P. (Chauchat, 1988; Hall, 1995) (Fig. 4.5). The campsites of these people are concentrated along the quebradas that course from west to east from the flank of the Andes to the ocean.

Organic materials left by the "Paijanese" people include the remains of land snails, small mammals, and birds. Fish bones are abundant. Marine mollusks are absent, unlike in later coastal preceramic occupations, and no large or medium-sized mammals are recorded. The excavators believe this evidence indicates an early orientation toward coastal resources, possibly in a time of moister climate than today because the land snails that were recovered occur only in places with higher water availability. The excavators also believe that the large bifaces recovered in quantity from the sites were used to spear fish, not large animals. The bifaces are thin and unbarbed at the base and would have been easily broken upon penetration of terrestrial game.

Numerous milling stones were also present in the deposits. What was processed is currently anyone's guess because plant remains have not been described from the sites. Chauchat (in Hall, 1995) suggests that seeds of the native, Peruvian mesquite (*Prosopis* sp.), which can still be found in the area today, may have been prepared with the grinding implements.

The issue of early Holocene climate along the northern Peruvian coast is being debated, with some investigators suggesting the presence of stable, warm tropical water and a substantially moister climate as far south as 10°S during both the early and middle Holocene (e.g., Richardson, 1995; Rollins et al., 1986; Sandweiss et al., 1996). These considerations are based on the presence of mollusk assemblages dominated by warm-water fauna in archeological sites dating to these periods (e.g., Reitz, 1994). Other researchers (e.g., DeVries and Wells, 1990) argue against climate change as the reason for the occurrence of warm-water faunas to the south of their current distribution, suggesting instead that local conditions (e.g., lagoon formation) provided suitable habitats.

The situation is clouded by disagreement among climatic modelers regarding the positioning and seasonal movement of the south Pacific anticyclone in the late Pleistocene and early Holocene and by a paucity of studies useful for reconstructing past vegetation and climate in northern Peru and southern Ecuador. However, some late Pléistocene faunal records [e.g., Talara tar pits, northwest Peru (Lemon and Churcher, 1961; Churcher, 1966; Campbell, 1982), and La Carolina locality, southwest Ecuador (Campbell, 1982)] do suggest that locally wetter conditions, such as ponds and swamps, characterized the outflow areas of some westward flowing rivers. Such environments could be supported by increased (or less seasonal) rain along the inland, Andean slopes (as argued by Simpson, 1975) or by melting glaciers, rather than by a more southward positioning of the air circulation system that brings seasonal rain to southwest Ecuador. A 25,000-year pollen record from a deep-sea core 300 km off coastal Ecuador (Heusser and Shackleton, 1994) documents the establishment of mangrove vegetation on the Ecuadorian coast, beginning at approximately 14,000 B.P. and peaking between 12,000 and 10,000 B.P., suggesting increasing run-off from rivers and streams at that time.

Chauchat (1988) argues the issue somewhat differently, at least for the period encompassed by sites of the Paijan tradition; Chauchat believes that 10,000 years ago the sea was still lower and at least 10 km further from the present shoreline near some of the Paijan sites. Under these more "inland" conditions, influence from the Andes would have been more significant, meaning more rain and runoff and a better watered environment, especially at streamside locations. During the early Holocene the glaciers were also melting, further increasing the discharge from the interior.

Although these conditions may have been true of the terminal Pleistocene and early Holocene in northern Peru, rising sea level would soon shrink the coastal plain and create environments more like those of the present day. Chauchat (1988) notes that the abundant fish remains present in the earliest Paijan sites indicate that substantial upwelling was occurring and that the sea was as bountiful as it is today. Substantial upwelling and bountiful seas probably mean cold waters.

If rain did fall along the upper Andean slopes year-round during the late Pleistocene (Simpson, 1975) rather than just during the South American summer, as is the case today, as conditions warmed and circulation began to take on seasonal patterning in the early Holocene, decreasing runoff would also have a severe impact on locally moist riverine and estuarine environments.

What is clearly important about the Paijan culture is that it may represent a change in adaptation from inland resources with a possible focus on large and medium-sized game to the resources of the sea. Although societies elsewhere at the same time were diversifying and intensifying their use of plants, the Paijan people turned their focus to the rich and productive fauna of the cold ocean. This distinctly coastal Peruvian adaptation would continue in some form deep into the Holocene and support the emergence of sedentary living and social complexity (a topic we discuss in Chapter 5).

The issues of whether warmer and wetter conditions extended 500 km further south along the western South American coast in the early Holocene, and, if so, what caused these conditions, are important ones that we feel cannot be resolved to satisfaction with the current data. If true, wild ancestors of a few crop plants found today near the current semiarid/humid margin of southwestern Ecuador, such as cotton, squash (*Cucurbita ecuadorensis*), and the jack bean, may have been distributed much farther south along the coast, thus providing the expanded possibilities for early Holocene experimentation and cultivation (e.g., Ugent *et al.,* 1982, 1984). Plant remains would be subject to the same kinds of destructive processes that affect those outside of the current desert environment and would be difficult to find.

However, wild ancestors of the numerous other domesticated plants discussed in Chapter 3 would almost certainly not have been naturally distributed in a moister-than-present northern Peruvian coastal zone. First, none of them are native to the western Andean slopes. Second, we feel it is unlikely that even under moister conditions, rainfall would have been high enough to support their growth. We are inclined to accept an explanation for early Holocene environments such as Chauchat's (1988) and Simpson's (1975)—that a more interior-like climate may have prevailed during the early Holocene because of reduced sea levels and increased runoff from the melting Andean glaciers and winter rains, creating a somewhat better watered habitat and lusher plant growth than exists today. We add that terrestrial paleoecological data would go a long way toward resolving the problem.

Venezuela

Settlement and subsistence data between 10,000 and 7000 B.P. from other areas of the lowland and mid-elevational Neotropics are few. Barse (1990, 1995) documents preceramic occupations in the Orinoco River valley of Venezuela that existed between 9200 and 7000 years ago. They are situated on alluvial terraces and near relict channels of the Orinoco River. At one of the sites, called Provincial, the preceramic component contained a hearth with a ground stone axe fragment, a hammerstone, and a pitted, nutting stone, all occurring beneath a thick, sterile horizon of eolian origin. Associated charcoal yielded an age of 9020 B.P. Carbonized nuts from two unidentified species of palm were common here but were the only macrobotanical plant materials reported. Charcoal recovered from an activity area near the hearth yielded a ^{14}C determination of 9210 B.P.

Barse (1990, 1995) has designated this earliest known occupation of the Orinoco Valley the "Ature" tradition. Another preceramic site in the region belonging to this tradition, called Culebra, contained flake tools and projectile points, the latter manufactured from exotic materials. At a third site, Pozo Azul Sur-2, a somewhat later occupation radiocarbon-dated to 7010 B.P. and placed in the Ature II phase is documented. The Ature cultural materials are consistent with well-developed early Holocene adaptations to the humid tropical lowlands together with some long-distance contacts manifested at sites from other regions discussed here. It is imperative that we get a better handle on plant exploitation such as sites like these because they are within one of the proposed hearth areas for manioc domestication (see Fig. 3.18).

The Interior of the Amazon Basin

We are loathe to relegate this huge area to the innocuous category "other area." However, in terms of human occupation and subsistence during the early Holocene, the interior of the Amazon Basin is very poorly known (Roosevelt, 1989, 1991). Although several large shell middens that may date to the early Holocene have been identified along the lower stretches of the Amazon River, the only securely dated and well-studied early Holocene sequences from the interior of the Brazilian Amazon seem to be from the sites of the Caverna de Pedra Pintada and Taperinha in the eastern Amazon Basin near Monte Alegre and Santarem (Roosevelt *et al.,* 1991, 1996; Fig. 4.5).

The 10,800–10,000 B.P. sequence of occupation at Pedra Pintada and its evidence for early exploitation of tropical forest and riverine resources was discussed at the beginning of the chapter. Human occupation appears to cease for a while at Pedra Pintada shortly after 10,000 B.P. Then, at 7600 B.P. the cave was reoccupied by people using ceramics. Specialized exploitation of the river fauna is also evident by this time. Roosevelt's team considers that this early ceramic occupation represents temporary visits from the main encampments near the Amazon River flood-plain, including from the site of Taperinha.

Taperinha (Roosevelt *et al.*, 1991) is located between Santarem and Monte Alegre on the south side of the Amazon River channel on the edge of an ancient river terrace (Fig. 4.5). It is approximately 20 km from the Caverna Pedra de Pintada. A dense midden was being formed at Taperinha between 8000 and 7000 years ago by ceramic-producing people who were concentrating on the extensive wild resources of the floodplain. The red-brown pottery, the earliest known in the New World, is grit tempered and unlike that of later time periods from South or Central America. Cooked in these pots were numerous wild foods from the river, including mussels, turtles, and fish.

The size and density of materials in the midden suggest a fairly sedentary occupation. Roosevelt and co-workers believe that the earliest economies at Taper-inha were based solely on wild resource procurement. We think that the case for the absence of food production is not yet proven, although, beginning during the middle Holocene, the alluvial plain of the middle and lower stretches of the Amazon River was probably one of the few tropical forest settings with the capability to support large and stable populations of foragers or complex hunting and gathering.

Pollen, phytoliths, and charcoal from two lakes that lie within 30 km of Pedra Pintada and Taperinha, called Comprida and Geral, also shed some light on the nature of early human adaptations in the eastern Amazon Basin. The lakes are located north of Monte Alegre in terra firme forest between 30 and 15 km north of the current main river channel (Bush *et al.*, 1997). As will be discussed in detail in Chapter 5, evidence for fires and human interference with the vegetation becomes apparent at one site, Geral, only shortly after 5800 B.P., after having been absent during the first 2000 years of this lake's depositional history.

At Comprida, charcoal and fire disturbance indicators are low to absent through-out the 10,000-year sequence of sediments. This is in marked contrast to the paleoecological records from Panama and Colombia, as well as Behling's (1996) records from Lake Curuca near Belem at the mouth of the Amazon River, which show burning and other forest modification early in the Holocene.

If humans were camping near lakes and setting forest fires, one would expect the presence of some charcoal in the paleoecological records. Fires generated consistently at distances up to tens of kilometers from the lake would also be expected to inject charcoal into the sediments (Clark, 1988). The near absence of charcoal, and phytoliths and pollen grains from early successional plants in the sediments from Geral and Comprida is strongly suggestive of an absence of system-atic human modification of the Amazonian forest near these lakes during the early Holocene.

The evidence that forms of human exploitation of the lower Amazonian forest involving persistent fire and land clearing may have occurred later than in other regions is consistent with the very poor character of the Amazonian terra firme soils, plants, and fauna. Possibly, these factors made usually attractive lakeside habitats less viable for a sustained and growing human presence. We suspect that when all is said and done, archeological and paleoecological data might reveal that from an early time, people focused more on the floodplain and riverine resources of the middle and lower Amazon River channel than on the dispersed and limited resources of the adjacent terra firme forest with its poor soils and small and circumscribed lake resource zones adjacent to it. This supposition draws some support from the archeological record at Pedra Pintada, where the earliest levels of the site contained freshwater fish, shellfish, tortoises, and aquatic turtles in addition to the remains of tree fruits.

During the early Holocene, river levels were lower and floodplains were less extensive than they are today, but these zones still probably offerred more concentrated and easily exploitable wild resources than did the dryland interior forest. Subsequent to the rise in sea level, the varzea eventually became as we know it, with its combination of abundant wild fauna and seemingly endless supply of good agricultural land. It is likely that the Amazon River had not risen sufficiently from its severe (100 m or more) late Pleistocene downcutting and did not accumulate enough sediment to maintain a varzea system until sea level had risen to near modern levels at approximately 8000–7000 B.P. (Van der Hammen and Absy, 1994). This chronology possibly explains that of the midden formation at Taperinha and the first evidence for specialized exploitation of river fauna at Pedra Pintada, which occurs at 7580 B.P. (Roosevelt *et al.,* 1996).

For now, we may consider that the paleoecological records are generating a set of hypotheses concerning human use of the Amazon Basin that will be tested with archeological settlement and subsistence data. More coring operations are under way to broaden the areal coverage of the Brazilian Amazon and determine whether regional differences in land use and modification existed.

Mesoamerica

Moving northward, the Belize Archaic Archeological Reconnaissance identified early Holocene cultural traditions belonging to the "Sand Hill" complex (MacNeish *et al.,* 1981; Zeitlin, 1984) (Fig. 4.8). The distribution and number of sites indicate that a wide range of environments, including the coast, were exploited. The subsequent "Orange Walk" complex, dated from approximately 8000 to 7000 B.P., contains milling stones and pestles, indicating an increasing use of seed and food plants by these people. Unfortunately, subsistence information is unavailable at this time.

It should be noted that the chronology of the previous sequence has been challenged by Kelly (1993). He believes, on the basis of later lithic analysis and excavation carried out in the area, that the projectile points used to establish the cultural periods and their associated time frames may largely belong to the late

Archaic period, when settlement densities appear to have substantially increased upon the introduction of slash-and-burn agriculture 5000 years ago (see Chapter 5).

There are very few data available on the early Archaic period from elsewhere in Mesoamerica, including lowland Mexico. Large shell middens on the Pacific coast of Guerrero near the cradle of maize may date to shortly after 5000 B.P. and appear to represent preceramic occupations. No plant remains are available for discussion (Adams, 1991).

We end with highland Mexico, and we do so symbolically because so many discussions of early plant use and domestication in the Americas have revolved around this area. We hope that by this point the reader is persuaded to consider that early and important developments leading down the road to agricultural systems may have been taking place elsewhere. The sequences of plant and other remains recovered from the Tehuacán and Oaxaca Valleys have been well published and discussed (e.g., Byers, 1967; Flannery, 1986c; Stark, 1986), including a recent publication by Pearsall (1995a), and we see no reason to repeat such a review here. Rather, we offer a possible explanation as to why an *in situ* development of food production and plant domestication may not have occurred in the central and southern Mexican highlands during the early Holocene and why exogenous domesticated plants, including maize, seem to have been grown in small quantities after they were accepted during most of the Holocene period.

The *Cucurbita pepo* squash present in early Archaic levels at Guilá Naquitz and avocado, chili peppers, coyol palms, hog plum, guava, maize, squashes, beans, and cotton increasingly evidenced at Tehuacán after 7000–5000 B.P. (MacNeish, 1967) were imports. With the likely exception of *C. pepo,* they were from the lower and wetter Pacific and Atlantic slopes (see Chapter 3). *Cucurbita mixta,* now called *C. argyrosperma,* found in small amounts in coprolite remains at 7000 B.P. and in larger amounts by 5000 B.P. at Tehuacán, may have been domesticated in the same region as maize, where its wild ancestor, *C. sororia,* is abundant in teosinte fields (T. Andres, personal communication, 1996). The two may have arrived together.

Cucurbita moschata, present at Tehuacán possibly as early as 7000–5000 B.P., was probably domesticated in southern Central America/northern South America. None of the many types of plants recovered from the Tehuacán and Oaxacan sites and native to these areas can be demonstrated to have been domesticated. It is possible that *Phaseolus* beans were planted at Guilá Naquitz, Oaxaca, although the evidence is equivocal. The case for cultivation of *Setaria* at Tehuacán made by Pearsall (1995a) has been revised in the face of an environmental reconstruction that shows that *Setaria* was likely available in large stands in the region (E. S. Buckler *et al.,* 1997). Neither shows evidence of morphological change associated with domestication. Why was this so?

To answer this question a short review of the late Pleistocene and early Holocene conditions of climate is necessitated. The paleoecological data presented in Chapter 2 indicate that much cooler and drier environments were characteristic of wide areas of the Neotropics during the late Pleistocene, including the central highlands of Mexico (Bradbury, 1989; Watts and Bradbury, 1982). Pollen studies at Lake Patzcuaro in Michoacan at 2040 m a.s.l. suggest a depression of vegetational zones by approximately 800–1000 m before 11,000–10,000 B.P. (Watts and Bradbury, 1982), an estimate in accordance with many studies carried out elsewhere. Such a depression would have brought elements of the oak–conifer forests occupying the higher stretches above the Tehuacán and Oaxaca valleys down to the elevations of the archeological cave sites, while probably leaving elements of the thorn–scrub–cactus vegetation of today in place but possibly in reduced quantities. The environment would have been more densely forested.

In environments with more and larger trees and less low browse, extinct and otherwise large game probably existed but not in the number seen in open landscapes. The considerable plant productivity of the present-day thorn–scrub–cactus vegetation may have been reduced. Medium and smaller-sized animals may still have been common, however.

On the basis of faunal remains identified in early Ajuereado (probably late Pleistocene) levels of Coxcatlan cave and their modern ecological preferences, Flannery (1967) reconstructed the Tehuacán Valley as an open temperate steppe before the beginning of the Holocene. We cannot gain much information regarding this question from the plant remains because in these earliest levels preservation is not good. However, as we will see, these differing interpretations relating to late Pleistocene environments may not much matter to the argument.

Late Pleistocene levels from Coxcatlan cave indicate that Paleoindians hunted horses, antelopes, and also many smaller types of game still found in the area today. They may have engaged in communal drives of jackrabbits (Flannery, 1967, 1986b). Less than 10% of the individual animals were larger game (horses and antelopes). Thus, from the earliest times the people at Tehuacán were catching large amounts of small game. Unfortunately, in these earliest levels at Coxcatlan Cave plant preservation is limited. However, mesquite and prickly pear are present, indicating these resources were used before the beginning of the Holocene (Smith, 1967).[6]

The early Holocene sequence of vegetation and climate in Tehuacán and Oaxaca is not well understood. Assuming that it was somewhat moister than today and considerably warmer and moister than during the late Pleistocene, the retreat of the mountain vegetation belts, or of a temperate steppe vegetation, resulted in the increase of oak woodlands and the productive thorn scrub/mesquite/cactus zones in the vicinity of the caves. Acorns are very common in the earliest macrobotanical records of Guilá Naquitz, whereas the various legume (including mesquite), fruit and succulent resources of the valley are also well represented. Mesquite becomes more common in the later Archaic deposits from the site (Smith, 1986). At Tehuacán, remains of the various resources from thorn scrub growth are well

---

[6] A recent reevaluation of the Tehuacán faunal remains and radiocarbon dates (Hardy, 1996) raises the question as to whether Coxcatlan Cave was occupied by people during the late Pleistocene. We note the controversy but defer on the issue until more information is available.

attested in the earliest Holocene strata with some significant vegetal preservation, El Riego, dated to ca. 8500–7000 B.P.

Given our return rate data for various habitats discussed in Chapter 2 and our general theoretical orientation, the reader can probably infer where we are headed with this argument. The early Holocene animal and vegetational resources of the highland Mexican valleys may have been as productive or only marginally less productive per unit of labor effort than the late Pleistocene resources, especially if large game were not exploited with any regularity. Also, and not insignificantly, many of the highly useful plants were perennials and, although providing a productive and stable resource base, would not yield quick returns if planted, assuming that one could keep them alive during the most arid parts of the year. Food-producing behavior was probably not nearly so advantageous as it was in other regions we have discussed.

Why small and mobile populations eating primarily wild resources continued to occupy the Tehuacán Valley well into the Holocene, accepting and growing small amounts of domesticated plants but continuing an essentially hunting and gathering way of life long after the termination of the Pleistocene and after productive systems of agriculture had developed in or spread into the Mesoamerican lowlands (Chapter 5), can, in part, be explained by the nature of the wild resource base they were given to work with during the earliest Holocene. The marginality of the regions for rainfall agriculture would have been another obvious drag on food production. Indeed, these regions would have been effective "refuge" areas for those early Holocene groups occupying more unfavorable habitats in the lowlands who did not experiment with food production.

We also note that the Tehuacán and Oaxacan records indicate that humans were drawing most of their calories and protein from a relatively small set of wild species available to them. Although efficiency of food procurement may not have favored early development or substantial incorporation of cultivated plants, it still shaped the subsistence strategies of these hunters and gatherers.

SUMMARY AND DISCUSSION

Many discussions of agricultural origins in the New World have focused on the dry highland zones, where archeology is more easily practiced and plant remains are well preserved. However, as more efforts are being made to study the humid and forested parts of the Neotropics, it is becoming increasingly clear that they were well occupied during the early Holocene. As discussed in this chapter, archeologists using site surveys have not had difficulty locating sites in regions where a seasonal tropical forest occurred. These regions include the middle Magdalena Valley and upper and middle Cauca Valleys in Colombia, the Zaña Valley, Peru, and central Pacific Panama. The inventories of material cultural, plant, and faunal remains retrieved from these sites are impressive. Even in wetter forests of the Colombian Amazon and the interior of the eastern Amazon Basin, sites dating to between 10,000 and 7600 B.P. and containing considerable cultural and plant debris, are present. Given these data, the "absence" of sites in other regions of the humid lowlands should be viewed as a call for archeological survey and excavation and should no longer be taken as proof that these areas were avoided by early humans.

We presented data in this chapter that we believe argues strongly for an independent development of food production, and early acceptance of cultivated plants, during the early Holocene in some of the regions listed previously. We argued in Chapters 1 and 2 that the question of food production origins in the Neotropics could benefit from NeoDarwinian theory and the application of hypotheses based on models from evolutionary ecology. The simplest version of cost/benefit optimal foraging models, the diet breadth model, was used to identify some of the incentives for, and constraints on, subsistence choices associated with late Pleistocene foraging and the transition to food production.

This model seemed particularly appropriate for three reasons: (i) The beginning of food production is inextricably linked to changing dietary breadth and subsequent specialization on certain resources leading to domestication, (ii) examining subsistence changes over long time periods with the diet breadth model does not require short-term or absolute precision in predicting what resources foragers will choose (a finding that between 11,000 and 10,000 years ago the return rate of every single resource in the diet of the average Neotropical forager was not above or equal to the average return rate of all resources eaten—one prediction of the optimal diet model—would not mean that energetic considerations were not considerably influencing subsistence choices and changes); and (iii) energetic return rates have been shown to be the single best predictor of foraging patterns among modern-day hunters and gatherers.

We believe that, particularly in the lowland Neotropics, energy may have been the most important selective constraint on human foragers during the early Holocene period. The dramatic decline in foraging return rates associated with the demise of the glacial-period resources, which we consider to have been the single most important factor driving subsistence changes, forced the need to inject more calories into the diet. This took place in a tropical forest ecosystem, one of whose hallmarks is its relative poverty of starch-rich foods and the high costs associated with collecting and extracting them from wild plants.

The common assumption that plant cultivation is more labor-intensive than plant collecting was shown not to be true in the tropical forest. Because the selection of a more efficient strategy of food procurement, cultivation, appears to have initiated food production in the Neotropics, the costs of foraging vs cultivating should be studied in other areas of the world. We suspect that cost/benefit ratios similar to those proposed for the Neotropics will also characterize other pristine areas of food production, a finding already made by Russell (1988) for the Near East and discussed in Chapter 6.

Before summarizing the Neotropical archeological records, it is worthwhile to review the basic predictions of subsistence and settlement changes under the diet breadth model (discussed in detail in Chapter 1). It is evident that it can account for many of the processes and changes commonly associated with food production origins in the Neotropics and perhaps elsewhere. The predictions are as follows. First, a decline in the abundance (encounter rate) of higher ranked resources on the landscape leads to the exploitation of lower ranked resources. Second, foragers will now choose a broader diet because it results in a higher return rate than could be achieved by more searching for food. Third, the reduction of search time permits greater investments in storage and food processing. Fourth, broader diets and decreased search time result in smaller foraging radii and, possibly, increases of residential stability.

The archeological and paleoecological records from the tropics seem to be in accordance with the basic predictions of the diet breadth model. Between 13,000 and 10,000 B.P., judging from the limited subsistence records and somewhat more informative lithic assemblages, human efforts were not yet focused on plant materials (although plants were certainly exploited) or on a variety of smaller game animals and marine fauna (although these were surely taken from time to time). Meat arguably provided the majority of calories in the diet. Lithic and faunal assemblages indicate that many groups were hunting large game animals, which, under conditions of optimal foraging, have the potential to create a relatively narrow dietary base and exclude many lower ranked plants from the dietary range. Subsistence was varied but cannot be characterized as broad spectrum, and resources occupying higher trophic levels in the food chain seem to have been emphasized.

A number of lines of evidence from archeological sites located in seasonal tropical forest indicate that after the termination of the Pleistocene at 10,000 B.P., people turned their attention to the plant resources of the forests and other lower ranked food items and increased the breadth of the diet under conditions of declining foraging return rates. This evidence includes (i) the presence for the first time of plant grinding implements and tools for working soil; (ii) the representation of varied plant remains from the tropical forest, including tubers, tree fruits, and herbaceous seed plants such as squashes; and (iii) persistent exploitation of marine fauna, such as fish and mollusks, occupying lower trophic levels. Following the diet breadth model, people would have started to cultivate some plants as soon as this practice increased the overall return rate (Kaplan and Hill, 1992).

Using paleoecological data to predict the threshhold at which this would occur, we find the time period between 10,000 and 9500 B.P. to be highly significant. This is when the closure of the landscape by forest occurred and when foraging return rates began to decline substantially. It would not have taken very long before farming was more efficient than foraging. The archeological record indicates that certain seed crops and tubers were taken under cultivation and domesticated in northern South America between 10,000 and 9000 B.P. Recent evidence from molecular biology, in addition to paleobotanical data from southern Central

America and northern South America, strongly suggest that maize must have also been an early Holocene cultivar.

The suites of plants that came to be characteristic of early food production systems in the Northern and Southern Hemispheres of the New World tropics may also be explained in terms of optimizing the efficiency of food procurement. Wild tubers generally give higher return rates per effort than wild grasses, possibly accounting for the early emphasis on tubers in southern Central and South America, where many suitable wild ancestors belonging to several different plant families were available. On the other hand, teosinte may have been virtually unique among wild Neotropical grasses in offerring acceptable energetic returns, explaining its prominence in Mesoamerica, which housed fewer good wild candidates for tuber cultivation.

If, as indicated by current evidence, seed crops are more costly to exploit than tubers (see Chapter 2), declining return rates and their intersection with the higher rates obtained from plant propagation would have resulted in the practice of tuber production before seed production, although we believe that the Mesoamerican and southern Central/South American developments were independently derived. The propagation of tubers may have been the earliest form of food production in the New World.

At first, tropical crops constituted a supplement, if a very important one, to the food base, which still contained a sizeable number and variety of wild products from the animal and plant worlds procured from outside of house gardens. The characteristics of tropical horticultural systems today, in which a prepared garden plot contains a variety of plants and as many "wild" as "domesticated" species, emphasizes the likelihood that early horticultural systems were of a similar nature. They would continue as such until the need arose to create larger scale food plots in the forest by slash-and-burn techniques in order to emphasize a smaller number of domesticated crops. The protein and fat resources from animal game found in the rivers, forests, coastlines, and gardens were more than adequate for the needs of the low-density and shifting settlements of the first tropical horticulturists. These foods also constituted essential resources after the demise of the big game at the end of the Pleistocene and the subsequent development of a biome dominated by protein-poor plants. They provided an important basis for early Holocene semisedentary and sedentary living and plant experimentation.

The early cultivation and domestication of an oil- and protein-rich seed source, *Cucurbita,* also helped to ensure adequate and stable supplies of these substances. It seems likely that the avocado was one of the earliest tree fruits to have been taken under cultivation. Containing one of the highest fat contents of any Neotropical fruit, it similarly provided a valuable source of food in the lipid-poor environments of the post-Pleistocene tropical forest.

Although our hypotheses seem to be supported by the available evidence, we note that archeological subsistence data during the crucial 11,000–10,000 B.P. period are rare. Also, the late Pleistocene climate and vegetation of several important regions, including the Mexican Rio Balsas watershed, await empirical documentation. Like all good hypotheses, the one we offer here may be fairly easily tested and refuted by an empirical finding that resource return rates probably did not drop substantially in a region after the close of the Pleistocene, but that food production developed independently in that region anyway. A finding that incipient food production involved an increase in the return to labor would also be at odds with our hypothesis. If, as seems likely now, a pre-Clovis occupation of the American tropics is established to the general satisfaction of the archeological community, we will have the interesting problem of which cultural tradition first propagated plants and we will have a longer sequence of precultivation subsistence to consider.

Finally, we believe that the development and initial spread of Neotropical horticulture and the emergence of pottery were not related. As Sauer (1936) argued, leren, yams, many other tubers, and the leaves and shoots of plants could have easily been prepared by baking in pits or roasting on coals and in ashes. Wooden vessels were also probably used for cooking. Other tubers and hard seeds such as maize required more laborious grinding before they could be cooked. All these techniques were available to early Holocene people. The development of a ceramic technology came later and, in many regions, was probably associated with the improvement and spread of certain crop plants such as beans and bitter manioc. We discuss this topic more fully in Chapter 5.

Risk Avoidance as an Explanatory Factor in the Emergence of Food Production

Few scholars, including us, would argue that optimal foraging can account for all the constraints, ecological and otherwise, that structure food choice. Other factors beside tendencies toward optimality in the rate of food procurement, such as risk avoidance, the division of labor, and territoriality, are thought to play important roles in foraging decisions (Jochim, 1988). It has been widely suggested that risk sensitivity may be a more significant constraint on human food choice than energetic considerations (e.g., Durham, 1981; Cashdan, 1982; Flannery, 1986a; Redding, 1988). Redding believes that humans optimize, but what they optimize is resource security. Risk reduction, characterized as making nature "more predictable" from year to year, has been cited as one of the primary reasons behind the initiation of food production in highland Mexico (Flannery, 1986a). Redding believes that risk factors were preeminent over efficiency concerns in causing food production behavior generally. Because this issue is a particularly important one in relation to early food production, we offer a discussion here in light of new evidence revealed by recent studies.

What are risk and risk-sensitive behaviors? Because severe, short-term variations in the food supply may occur and pose threats to human survival, humans may be said to be engaging in risk-reduction options when they try to reduce the short-term *variation* around the expected food returns (Kaplan and Hill, 1992; Winterhalder, 1986, 1990). At first glance, this puts the goal of foraging at odds with optimization because the latter is designed to achieve efficient long-term foraging rates. Humans can achieve short-term rates of food acquisition that have less variance in a number of different ways: by sharing food or information about potential food resources, storing resources, or altering the diet breadth to include resources with less variance. However, the latter option might well lower the long-term food acquisition rate, generally not a desirable strategy (Kaplan and Hill, 1992).

Winterhalder (1986, 1990) has provided important information on the question of how risk sensitivity structures food procurement strategies. Using a computer simulation, he studied the optimal resource selection by foragers under the assumption that the forager's goal is to minimize the risk of starvation or serious food shortages and not to maximize the rate of return. Winterhalder defined risk as the probability of failing to meet some minimum food requirement. His results showed that rate-maximizing and risk-minimizing diets are actually similar when the expected return of food does not either substantially exceed or fall below the minimum requirement. This finding is another important line of evidence indicating that attention to efficiency in foraging is a very beneficial strategy for humans, and that diet breadth models can be used to study directional changes in food choice over time.

Also, contrary to numerous expectations in the literature, Winterhalder (1986, 1990) found that expanding the diet breadth (adapting more generalized subsistence strategies) under conditions of expected food need did lower variance but at such a cost to the average foraging efficiency that it increased rather than reduced the chance of a shortfall. Under conditions of impending food shortages, a risk-minimizing diet that is beneficial will *narrow* the food base. It is only under conditions of food surpluses that an attempt to minimize risk that broadens the diet should occur (Winterhalder, 1986, 1990). It is important to repeat that risk-generated responses that altered the diet breadth at the expense of energetic efficiency were predicted to occur only when the expected food intake was much higher or lower than the minimum needed to prevent a shortfall.

In light of these considerations derived from Winterhalder's (1986, 1990) simulations, and the fact that early Holocene Neotropical foragers were not altering their diets following the demise of the Pleistocene animal and plant resources in environments of food abundance, it is ever more likely that the expansion of diet breadth at the end of the Pleistocene was substantially a response to the changing availability and densities of resources under conditions of optimal foraging.

Another important aspect of risk revealed by Winterhalder's (1986, 1990) studies is that even when foragers minimized risk optimally, they did not always completely avoid it. Winterhalder's simulations indicated that food sharing was more effective than changes in dietary breadth in reducing risk. Benefits accrued from sharing required the participation of only six or fewer individuals and reduced variation in consumption by 60%. Winterhalder (1986, 1990) and Hill *et al.* (1987), among other evolutionary ecologists, believe that food sharing was the primary device used by humans to miminize short-term variance in food procurement, and that food sharing, therefore, is an ancient strategy.

There is substantial empirical evidence from studies of modern tropical hunters and gatherers and horticulturists that food sharing is the most important device used by humans to counter risk. For example, resources that have the highest variance in acquisition (peccaries and tapirs) were always pursued by Aché hunters. These resources were also the most likely to be shared by the Aché (Kaplan and Hill, 1992). Similarly, among native horticultural groups in the American tropics today, highly sought-after resources subject to high variance in yield, such as large game and fish, are the objects of extensive household exchange (Hames, 1990; Baksh and Johnson, 1990). Conversely, low-variance resources—those obtained from horticulture—were infrequently or never shared, as expected if sharing was designed to reduce variance in consumption (Hames, 1990).

Another important point to consider in discussions of risk and tropical horticultural origins is that traits such as semisedentary and scattered settlements with household autonomy and a diverse selection of crop plants grown in dispersed fields, which are characteristic of modern and past tropical food producers, are, in and of themselves, potent risk aversives, even if they were not intended to serve in that role. In summary, these empirical and simulation data indicate that human sensitivity to risk is probably not a confounding factor in considerations of foraging strategies and food production origins in the humid tropics that are based on optimality models. It seems that risk sensitivity may often have few effects on the basic food choices of some foragers and simple farmers.

From Small-Scale Horticulture to the Formative Period: The Development of Agriculture

*All anthropologists who have studied the problem of the emergence of
civilization are well agreed that civilization cannot appear until a truly
productive agricultural system has been developed.*

Donald W. Lathrap (1975)

INTRODUCTION

In this chapter, we discuss the development of the earliest forms of small-scale
horticulture into truly effective agricultural systems and the emergence of societies
dependent on agriculture for subsistence. The evidence we summarize leads us to
believe that the first expressions of productive agriculture in the New World
emerged in the humid tropical lowlands, where they took the form of shifting or
slash-and-burn cultivation.

As with Chapter 4, we begin in South America and make our way north,
ending with Mesoamerica. The northwestern parts of South America, comprising
southwest Ecuador, the Cauca Valley (Colombia), northwest Peru, and the Amazonian regions of Colombia and Ecuador, generally contain more and earlier evidence
for the intensification of food production systems than do regions east and north,
such as the interior of the Amazon Basin, Venezuela, and the Caribbean littoral of
northern Colombia. This situation is, in part, due to lack of systematic archeological
survey/excavation, botanical recovery, and paleoecological research in many regions, some of which, such as the southern portions of the Guianas and their

borders with northern Brazil, probably saw primacy in the domestication of certain plants. It is also probably a result of a later adoption and intensification of food production in some regions, as we discuss for the interior of the Amazon Basin.

Accordingly, we divide our discussion into six main geographic areas: northwestern South America, the desert coast of Peru (where food production occurs later in time and probably is not indigenous, but where excellent preservation leads to an abundance of data), the interior of the Brazilian Amazon, other areas of South America not included in the preceding three areas, southern Central America, and Mesoamerica. As in Chapter 4, cultural chronologies and summaries of the sequences of crop plant appearances are provided for areas where the evidence is more substantial. We do not discuss at any length post-3000–2000 B.P. society in the lowland tropics for two reasons. In many regions of Central and South America, it appears that the complexes of crop plants that supported indigenous populations at the European contact were already in place by this time. The emergence of chiefdoms and complex societies in the Neotropics, and the artificial modification of landscapes into ridged and mounded fields (e.g., Bray, 1995; Parsons, 1969; Roosevelt, 1987) that took place in some regions largely at this time are subjects outside of the scope of this book.

NORTHWESTERN SOUTH AMERICA

Coastal Ecuador: The Beginnings of the Formative

Phytolith evidence from the type site of the preceramic Vegas culture indicated that the cultivation and domestication of *Cucurbita* (squash) and *Calathea allouia* (leren) took place prior to 8000 B.P. and that maize was introduced into the area by or shortly after 7000 B.P. Bottle gourd (*Lagenaria siceraria*) was also cultivated at this time. Sites of the Vegas culture appeared to be confined to the driest areas of the Santa Elena Peninsula (Fig. 5.1) and were positioned, in part, to exploit small areas of alluvium along seasonal watercourses for horticultural activity.

There is a hiatus of approximately 1500 years in the archeological record after the end of the Las Vegas occupation and the appearance of the Valdivia culture, the earliest ceramic complex of the region, at 5500 B.P. (Table 5.1). The youngest date from the midden of the Vegas type site, OGSE-80 (Site 80), is 7150 B.P. (on shell, which, when corrected for the reservoir effect, would probably be approximately 300–400 years younger). The cemetery founded during the last part of the site's occupation contained skeletons that yielded dates as young as 6600 B.P. Phytoliths from a large, undated secondary burial at Site 80 produced a radiocarbon determination of 5780 B.P., suggesting that people continued to bury their dead at Site 80 while not living there. Shell from the surface of another Vegas-phase site near Site 80 produced a date of 5830 B.P., suggesting that the peninsula may have been occupied, albeit sparsely, after OGSE-80 was abandoned as a living site.

Stothert (1985) believes that the occupation hiatus in the region is more apparent than real, and that although the driest portions of the Santa Elena Peninsula may not have been occupied later than approximately 6800 B.P., populations shifted somewhat eastward where rainfall was higher and began to focus on the alluvial zones of more substantial rivers where food production systems could be intensified more easily (Fig. 5.1).[1] This scenario is in agreement with the cultivar record from Vegas Site 80, where maize appears to have been introduced very near the end of the occupation of the site. Squash, bottle gourd, and leren could grow very well on the small bits of alluvium near Site 80 and by "pot irrigation" on soils removed from streams of the peninsula but a system that included maize would do better on larger expanses of alluvium. Stothert (1985) also notes that there is evidence for a cultural connection between Vegas and Valdivia. For example, shell spoons were made in an identical fashion, stone flaking technologies are similar, and burial patterns are identical.

Evidence for possible continuity between Vegas and Valdivia peoples also comes from a site in the region called Altomayo, located approximately 15 km southeast of Vegas Site 80 and 4.6 km north of the current shore (Fig. 5.1). Here, Damp and Vargas (1995) identified a phase transitional from Vegas to Valdivia called Chuculunduy that they estimated to date between 6600 and 5500 B.P. Altomayo contains worked clay pieces and ceramic artifacts, including figurine pieces, that may represent an experimental stage in ceramic production. One figurine piece is remarkably similar to that from Valdivia I contexts elsewhere. Remains of a house were also identified in the midden.

Altomayo also contained shell, bone remains, and macrobotanical specimens that are yet to be identified. Although subsistence information from Altomayo is sparse, this site has the potential to reveal many aspects of the Vegas to Valdivia transition.

Other investigators believe that the origins of the Valdivia culture are to be found eastward in the Guayas Basin, where higher rainfall results in moist tropical forest vegetation and Valdivia sites were also found (e.g., Lathrap *et al.,* 1975; Raymond *et al.,* 1980). Raymond (1993) sees little cultural connection between Vegas and Valdivia. Valdivia would then represent a tropical forest tradition that spread west into the coastal plain of southwest Ecuador as populations grew and expanded. The Guayas Basin may have considerable importance in Ecuadorian prehistory. Its moister climate and denser vegetation have made site location and study more difficult than on the drier southwest coast.

---

[1] Changes in climate (e.g., Sarma, 1974) have also been invoked to explain the 7000–5500 B.P. patterns in the cultural sequence. However, the available botanical data from the Vegas and Valdivia sites lead us, as well as Stothert (1985), to conclude that the basic climatic patterns were the same as those today and have characterized the area since the end of the Pleistocene (Pearsall, 1979).

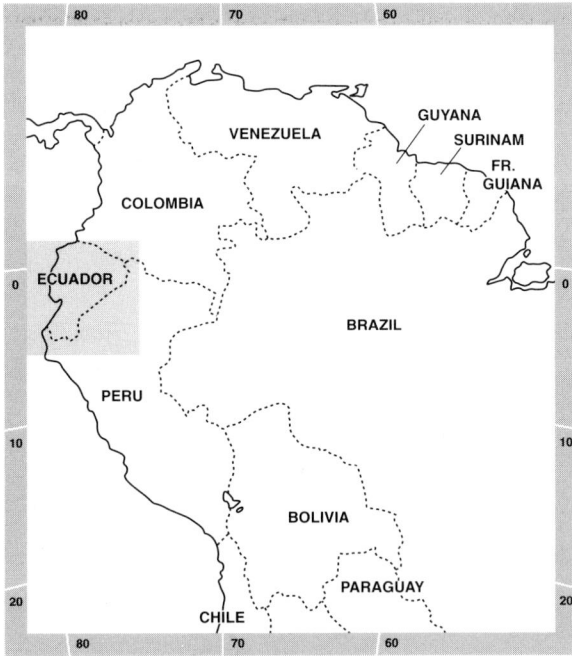

FIGURE 5.1   Map with the location of coastal Ecuadorian sites.

Whatever the case, it has become clear that Valdivia peoples practiced multicrop agriculture along the alluvium of substantial watercourses and lived in the first sedentary farming villages in the New World that have been identified. Thus, it is likely that the Valdivia culture is the first expression of Formative society in the New World. That is, Valdivia culture is characterized by permanent settlements, some large, with truly productive agricultural systems and with indications of social complexity in the form of site hierarchy and ceremonial construction (Lathrap *et al.*, 1977; Lathrap *et al.*, 1975).

Sites of the Valdivia culture are located in southwestern Ecuador from the Guayas Basin west to the confluence of the Rio Verde and Rio Zapotal, called the Chanduy Valley, north into Manabí province, and south into coastal El Oro province (Fig. 5.1). Numerous radiocarbon determinations date the culture from approximately 5500 to 3500 B.P. (Lathrap *et al.*, 1975; Damp and Vargas, 1995; Raymond, 1988). More than 150 sites have been located, and they were formerly thought to be a cultural complex centered around marine resources (Meggers *et al.*, 1965). However, it is now clear from several systematic settlement surveys that the sites are positioned to exploit prime agricultural land near rivers, streams,

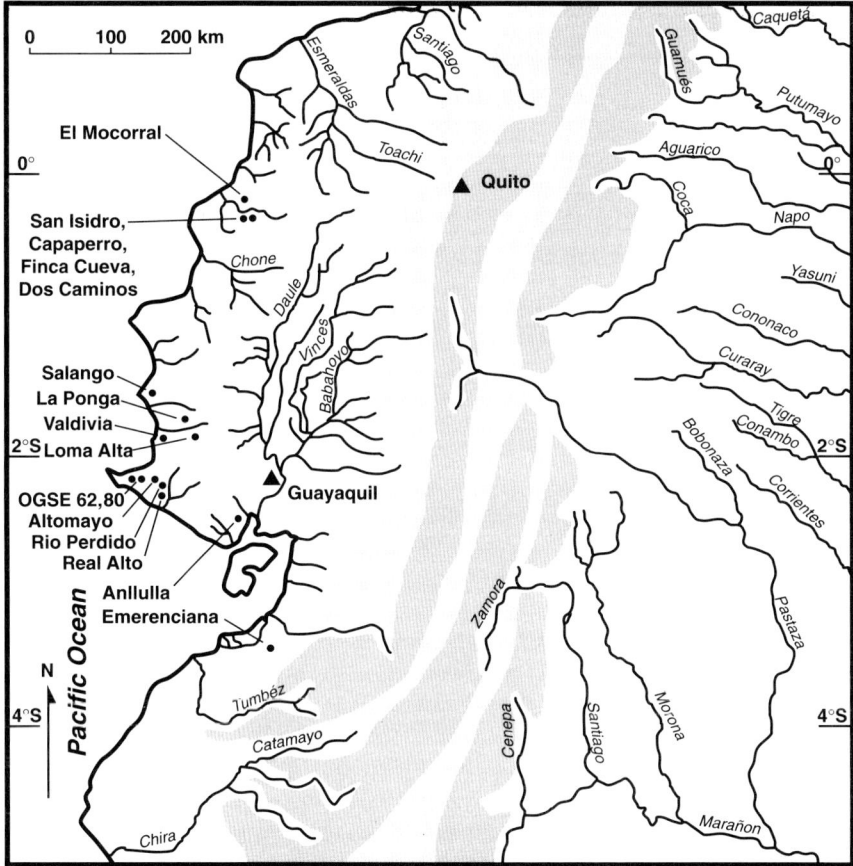

FIGURE 5.1   (*Continued*) Detail of coastal Ecuadorian sites (shaded area of map facing).

and lakes (Raymond, 1988; Schwarz and Raymond, 1996; Staller, 1992, 1994; Zeidler, 1986, 1991; Zeidler and Pearsall, 1994).

For ease of discussion, Valdivia can be divided into four periods: early (Valdivia 1–2, 5500–4400 B.P.), middle (Valdivia 3–5, 4400–3800 B.P.), late (Valdivia 6–7, 3800–3600 B.P.), and terminal (Valdivia 8, 3600–3500 B.P.) [periods are after Schwarz and Raymond (1996) and Zeidler and Pearsall (1994); phase numbers correspond to Hill's (1972–1974) ceramic chronology; refer to Zeidler *et al.* (1997) for calibrated dates for sites in the Jama Valley, Manabí]. Sites of early, middle, and late Valdivia affiliations are currently known only from southwest Guayas and southern Manabí provinces (e.g., sites such as Real Alto, Loma Alta, and Valdivia) (Fig. 5.1). By contrast, cultural sequences in northern Manabí (e.g., San Isidro site), the Guayas estuary region (e.g., Anllullo site), and coastal El Oro (e.g., La

TABLE 5.1   Cultural Chronology and Plant Remains in Major Regions of
Northwest South America

Cultural chronology	Roots and tubers				Vegetables	Grains		
	Calathea allouia (leren)	Maranta arundinacea (arrowroot)	Canna edulis (achira)	Manihot esculenta (yuca)	Cucurbita spp. (squash)	Zea mays (maize)	Chenopodium (quinoa)	Canavalia (jack bean)
Zaña Valley, Peru, 6400–5000 B.P. (middle preceramic), Tierra Blanca Phase				M	M		M	
Coastal Ecuador Valdivia (early Formative), 5500–3500 B.P.	[a]	Phy	Phy		Phy[b]	Phy		M
Machalilla (middle Formative), 3200–2800 B.P.	[a]	Phy	Phy			Phy, M		M
Chorrera (late Formative), 3000–2500 B.P.	[a]	Phy	Phy		Phy[b]	Phy, M		
Middle Caquetá Colombian Amazon (Preceramic), ?–4700 B.P.						P		
Preceramic, 4700–2000 B.P.				P		P		
Calima Valley Colombia, Ilama (early Formative), 3000–2000 B.P.		Phy			Phy	Phy, M		

Note. M, macrobotanical; Phy, phytolith; P, pollen.
[a] Phytolith preparations not yet checked for this plant.
[b] Cucurbitaceae (*Cucurbita* or *Lagenaria*) in this case.

Emerenciana site) begin with terminal Valdivia. This suggests that the period just prior to 3500 B.P. was a time of population expansion into previously unoccupied regions of coastal Ecuador, perhaps in response to a growing need for new agricultural land (Pearsall, 1997).

Our knowledge of early Valdivia plant use comes primarily from the Loma Alta and Real Alto sites. Early Valdivia deposits at Loma Alta were excavated by

Cultural chronology	Legumes		Industrial		Stimulants		Tree fruits		
	Arachis hypogaea (peanut)	*Phaseolus vulgaris* (common bean)	*Lagenaria siceraria* (bottle gourd)	*Gossypium barbadense* (cotton)	*Erythroxylum* (coca)	Palmae (palms)	*Annona* sp. (soursop)	*Sapotaceae* (sapotes)
Zaña Valley, Peru, 6400–5000 B.P. (middle preceramic), Tierra Blanca Phase	M			M	M			
Coastal Ecuador Valdivia (early Formative), 5500–3500 B.P.			M	M		M, Phy	M	M
Machalilla (middle Formative), 3200–2800 B.P.								
Chorrera (late Formative), 3000–2500 B.P.		M				Phy		
Middle Caquetá Colombian Amazon (Preceramic), ?–4700 B.P.								
Preceramic, 4700–2000 B.P.								
Calima Valley Colombia, Ilama (early Formative), 3000–2000 B.P.		M				M, Phy		

Raymond in 1980 and 1982 as part of a survey and site testing program in the Valdivia River valley. Loma Alta is located on a relict Pleistocene river terrace 12 km from the coast. Occupied beginning approximately 5000 B.P., it is the largest of the early Valdivia sites in the valley (Raymond, 1988). Macroremain and phytolith data are available from the site (Pearsall, 1988, 1998).

Overall, few charred food remains were recovered. This is likely the result of the age of the site, the flotation technology employed, and the small volume of soil processed (Pearsall, 1995b). The Sapotaceae and Palmae were among the taxa contributing tree fruit and dense cotyledon fragments recovered. One almost entire *Scirpus/Cyperus* root was found. A second unknown root was also present. Fragments of *Canavalia* beans were also recovered (Table 5.1).

Phytolith analysis was carried out both to evaluate the presence of maize and to determine what other plants were utilized at the site. Maize phytoliths were observed in two of eight samples examined. Palm, *Maranta, Canna,* and Cucurbita-ceae phytoliths were also observed (Table 5.1). Because of high soil pH, phytoliths were not well preserved.

The Real Alto site, located on the lower course of the Rio Verde in the Chanduy region of southwest Guayas province, was excavated by Lathrap and associates in 1974 and 1975 (Lathrap *et al.,* 1977). Smaller scale excavations were later conducted by Damp (1984a, 1988) in the early Valdivia village. During early Valdivia times, Real Alto was a small village (ca. 1 ha in Valdivia 1 and 2.25 ha in Valdivia 2), located in the northwestern sector of the larger Valdivia 3 village. The beginning of ceremonial mound construction dates to Valdivia 2 (Marcos, 1978; Zeidler, 1991).

Charred maroremains were not very abundant in early Valdivia samples for the reasons discussed previously. However, fragments of tree fruit rinds and meats, seed endosperm fragments, and root/tuber pieces were recovered (Pearsall, 1979). *Canavalia* beans, sedge, and cotton were also present (Damp and Pearsall, 1994; Damp *et al.,* 1981). Cotton was a commonly used plant at the early Valdivia village. Whether it was domesticated or wild cannot be determined from the remains, however.

Large cross-shaped phytoliths were recovered from 3 of 4 samples analyzed from early Valdivia contexts. These were identified based on size as likely to have come from maize (Pearsall, 1979). Reanalysis using size plus cross-shaped three-dimensional characteristics, and a discriminant function to separate maize and wild grasses, confirmed that maize was present in one Valdivia 2 sample and was likely to have been present in one Valdivia 1 sample (Pearsall and Piperno, 1990). Rescanning these samples also led to the identification of *Canna,* the Cucurbitaceae, and *Maranta* in early Valdivia contexts at Real Alto.

Although recovery technique may be the primary factor contributing to the lack of charred maize remains in Real Alto flotation samples, it is also likely that maize was uncommon in early Valdivia times, and that several types of tree fruits, root/tuber foods, wild and cultivated legumes, and wild/weedy annuals (not discussed here) contributed to the plant portion of the diet (Pearsall, 1979).

Our knowledge of plant use in middle and late Valdivia times is limited to Real Alto (Pearsall, 1979). The Valdivia 3 village at Real Alto was a large (12.4 ha) elliptical settlement, with a central ceremonial precinct (discussed later). This is the period of maximum village size and residential population (Zeidler, 1991; Marcos, 1978). After Valdivia 3, population declined at the Real Alto ceremonial center, and hamlets appeared along the river. Real Alto was abandoned after late Valdivia.

Botanical data are similar in many ways to the data from the early Valdivia village. Tree fruit rinds and meats, seed endosperm fragments, and root/tuber pieces were present in many samples, suggesting that a mixed subsistence strategy continued. Maize phytoliths were more common, and achira (*Canna*) and Cucurbitaceae phytoliths were present in a few contexts. Charred *Canavalia* was present.

As discussed previously, the terminal Valdivia period marks an important change in early Formative settlement in coastal Ecuador. For the first time, sites are found outside the driest part of southwest Ecuador. Botanical data are available from three terminal Valdivia sites; San Isidro, Capaperro, and Anllullu (Fig. 5.1).

Zeidler and Pearsall and colleagues have worked at the San Isidro site (M3D2-001) in the middle Jama River valley, northern Manabí province, since 1982. From 1988 to 1991, a valleywide survey and site testing program was carried out; results of the early stages of this research are published (Zeidler and Pearsall, 1994). Final analyses of survey and excavation data, including botanical materials, are nearing completion. San Isidro is a regional ceremonial center located approximately 25 km inland on a major left bank tributary of the Jama. The site is 40 ha in area with stratified archeological and natural deposits ranging between 3 and 6 m in depth. Occupation at the site, and ceremonial mound construction, began with terminal Valdivia (Piquigua phase, 3600–3500 B.P.) and continued through the Chorrera period (Tabuchila phase, 2900–2300 B.P.).

Because terminal Valdivia deposits are deeply buried, relatively few contexts dating to the period have been exposed. Few charred materials are present probably because of the depth of the deposits, the clay-like nature of the soil, and sample size. Data are from Pearsall (1994a,b, 1996) and from unpublished lab documents. The full array of food plants identified in sites in the Jama Valley are present beginning with the terminal Valdivia occupation at San Isidro. Maize, *Maranta, Canna,* squash or gourd, palm, and sedge are identified in the phytolith samples (maize in four of six samples tested); charred gourd rind was also recovered.

The Capaperro site is located adjacent to San Isidro and was excavated in 1989 and 1990 as part of the Jama Valley project. Flotation samples come from 10 terminal Valdivia contexts. Few charred materials were present, and recovery of food remains was poor, being limited to tree fruit and dense cotyledon fragments and small pieces of porous endosperm tissue. Analysis of phytolith samples is under way at Pearsall's lab.

The Anllulla shell mound, located on a small estero leading into the Estero Salado south of Guayaquil (Fig. 5.1), was tested by Lubensky in 1973. This multiple-component site has a thick Valdivia occupation (from 140 to 280 cm below surface (bs) in the test pit) characterized throughout by Valdivia D ceramics (E. Lubensky, personal communication, 1995). A ^{14}C date of 3560 B.P. at 180 cm confirms the deposit as terminal Valdivia. Maize kernels were recovered from a depth of 210 and 220 cm bs. Root/tuber material occurred at 220 cm and below, and tree fruit fragments occurred at 320 cm.

To summarize, a number of crop plants have been evidenced from macroremains and phytoliths (Table 5.1). Maize is present in the phytolith assemblages from earliest to latest Valdivia contexts. Also present from earliest Valdivia times are squash, achira, cotton, and *Canavalia* beans. The record is silent on some important crops, including manioc. Because manioc is in central Pacific Panama by this time, it is likely that sweet manioc cultivation was taking place at early Valdivia sites. Other unidentified macrobotanical materials include fruit and nut-like seeds and tubers.

A well-developed agricultural system with a focus on rich bottom lands is apparent, although one of us (D. Pearsall) is hesitant to say that domesticated crops produced most of the calories consumed at Valdivia sites because preservation of macrobotanical data, which would permit an estimation of the relative abundances of wild and cultivated foods, is poor at sites studied. Piperno believes this is the case because the crops already evidenced grown on fertile valley bottoms would have provided a rich supply of food. Certainly, an array of domesticated, tended, and wild tree food resources was used by Valdivia peoples. They include palm, soursop, hackberry, legumes, and the Sapotaceae family, among others (Table 5.1).

Other plants evidenced in the form of carvings or ceramic decorations include carved bowls that mimic the shape of half of a bottle gourd and the figurine of a man with the quid of coca in his cheek, both dating from Valdivia 3 (4400–4000 B.P.). The on-site plant use varies considerably by time period and region. When taken together with the repertoire of definitive crop plants, there is little doubt that many components of the plant assemblage are derived from the tropical forest.

Faunal data from Valdivia sites reveal that coastal resources continued to play roles in subsistence. Fish bones, mangrove oysters, and other shellfish are present at many sites (Byrd, 1976, 1996; Stahl, 1998). However, seafood is often less abundant than terrestrial game. Specialized fishing equipment such as fishhooks carved from mother-of-pearl shell is present for the first time, and some of the fish species may have been caught by fishing in the open ocean (Lathrap *et al.*, 1975). The most common type of terrestrial animal is the white-tailed deer, which thrives under conditions of vegetational disturbance and, especially, the presence of crop fields.

Analysis of isotope ratios from skeletons dating to the Valdivia 1–3 periods (nine individuals; eight from Loma Alta and one from Real Alto) (Van der Merwe *et al.*, 1993) suggest a terrestrial diet of primarily C_3 plants with no significant dependence on maize and negligible input from marine foods. Delta ^{13}C values are somewhat depleted (-18 to -20) with slightly enriched delta ^{15}N from marine and estuarine resources in the diet. We note that the collagen in these individuals was studied. Apatite (the carbonate fraction of the bone) is now known to be more sensitive for detecting and measuring maize consumption because collagen disproportionally represents dietary protein, potentially underestimating C_4 plant input, whereas apatite accurately represents the whole diet (Norr, 1995). Thus, the contribution of maize in the Valdivia diet is still to be determined. It does seem, however, that maize did not dominate the diet as it would later in time, and that much of the dietary energy during Valdivia was supplied by C_3 plants. Unfortunately, only one individual from Valdivia 3 contexts at Real Alto yielded collagen of sufficiently high quality to be studied. Thus, variability, if any, in maize consumption from Valdivia 1 to Valdivia 3 times is difficult to assess.

Skeletal data from Real Alto suggest low to intermediate levels of stress on health (Ubelaker, 1998). Caries rates are on the high side, a finding consistent with the proposition that a sticky starch was being heavily consumed. Given the bone isotope data, this starch probably came from combinations of manioc, other root crops such as achira, and maize.

As Schwarz and Raymond (1996) note, the locations of early Valdivia villages in the Chanduy and Valdivia valleys remained in stable locations for 1000 years. This pattern is also typical of societies dependent on productive agriculture and indicates a much less transitory system than usually associated with swidden cultivation. Pearsall (1983) found little evidence from the archeological carbonized wood specimens from Real Alto that the interfluve trees were being cut through time, a finding in accord with the strong preference of Valdivia people for fertile river bottomlands.

We view early Valdivia culture to be prima facie evidence that lowland tropical agriculture based on a variety of tuber, seed, and tree crops, and deriving a substantial number of its calories from tubers, can support sedentary living, growing populations, and increasing social complexity. As Stothert (1992) argues, the coastal Ecuadorian case illustrates that there are multiple foundations of, and routes to, civilization in the Andes.

The organization of the community at Valdivia sites also points to a Formative way of life with increasing cultural complexity through time. The early Valdivia village at Real Alto, the most completely excavated Valdivia site, was laid out in the shape of a "U" and approximately 1 ha in size (Damp, 1984a,b; Raymond, 1993). Individual "houses" were elliptical pole and thatch constructions approximately 5 × 4 m in dimension that surrounded a vacant, central space. This space was aligned along a north–south axis. Raymond (1993) and others have noted that this pattern is very similar to that of the settlements of ethnographically known tropical forest villages, and that today the bounded, central village space is where ceremonial and ritual activities participated in by all the community take place.

Most artifacts occurred around the house structures and included spindle whorls for weaving cotton, numerous manos and metates, pottery, stone and ceramic figurines, and distinctive ground stone axes with ears that are still used by tropical forest peoples today to clear forest for agriculture (Damp, 1984a,b; Lathrap et al., 1975). It is likely that a "heddle loom" was used for weaving cloth instead of simple twining, and this would represent the earliest such advance in textile industry in the New World (Lathrap et al., 1975). Items used for personal adornment become much more common compared to earlier periods in the area. They included necklaces, labrets, and ear spools. Seashells were commonly used as beads.

The florescence of the Valdivia culture took place in Valdivia 3 (4400–4000 B.P.) (Lathrap *et al.*, 1975, 1977). Real Alto covered an area of 300 × 200 m and may have had a population of between 1500 and 3000 people (Plates 5.1, 5.2). Houses were large, sturdy structures, 10 × 8 m, and may have served as residences for extended families (Zeidler, 1984). House structures surrounded a linear plaza on which were two ceremonial mounds that were remodeled and enlarged many times. Clusters of deliberately broken manos and metates suggest rituals related to agricultural production, and a tomb of a high-status individual was lined with manos and grinding slabs (Marcos *et al.*, 1976; Marcos, 1978).

PLATE 5.1   View of the Valdivia 3 surface of the Fiesta House Mound, Real Alto site. The linear arrangement of depressions or troughs are of unknown function. Courtesy of J. A. Zeidler.

PLATE 5.2   View of a cached mass of used manos and metates from the Valdivia 2 phase at Real Alto, interpreted as a ritual cache of ground stone related to harvest. Courtesy of J. A. Zeidler.

Raymond (1993) notes that the low overall consumption of maize indicated by the bone isotope record would make sense if maize was primarily a feast food important in the ceremonial life of the community. By the end of Valdivia 3 at 4000 B.P., Real Alto covered an area of 12 ha and was a large residential center.

By the end of the late Valdivia period at ca. 3400 B.P., the Chanduy and Valdivia river valleys were packed with settlements. Highest settlement density may have occurred in southern Manabí province, where rainfall is higher, the valley bottoms are more extensive, and agriculture was probably more productive than that in the drier part of southwest Guayas (Lathrap *et al.*, 1975). Real Alto had become an administrative and ceremonial center, regulating "satellite" settlements dispersed along the rivers of the Chanduy Valley, and it may have controlled 600 acres of prime riverine agricultural land (Marcos *et al.*, 1976). Such factors suggest some degree of stratification of the society. The Valdivia culture was beginning to spread widely outside of southwest Ecuador, into northern Manabí and El Oro provinces (Zeidler and Pearsall, 1994; Staller, 1992).

As noted by Lathrap *et al.* (1975) and recently by Drennan (1995), southwestern Ecuador during the late sixth and fifth millennia B.P. stands out from other regions in a number of important ways. The manifestations of the Formative Valdivia period—sedentary and nucleated villages with efficient farming systems, ceremonial centers, and well-made ceramics—antedate similar developments in highland Mexico or in the arid mountains and coast of the central Andes by 1500–2000 years and may have inspired these latter developments.

Schwarz and Raymond (1996) comment that it is also informative to consider the local consequences of the Valdivia tradition. They note that ceremonial activity appeared to end in southwest Ecuador by the termination of the Valdivia period, and later Formative populations in the area did not participate in increasingly elaborate ceremonial life and political complexity but rather returned to a settlement and social pattern reminiscent of tropical forest—and early Valdivia—culture. The early Formative peoples of southwestern Ecuador apparently chose ultimately not to elaborate their social and economic systems and unify them over large areas.

This was not the case everywhere on the coast, however. As mentioned earlier, Valdivia culture apparently spread out of southwest coastal Ecuador at the very end of the period. San Isidro, founded at this time in northern Manabí province, and La Emerenciana, in coastal El Oro, each show evidence of continued ceremonial activity in the form of earthen mound construction (Zeidler and Pearsall, 1994; Staller, 1992, 1994). The central platform mound of San Isidro, for example, grew from a minimum size of 30–40 m in diameter during terminal Valdivia to the 100-m diameter mound that dominated the 40-ha site by the end of Muchique 3 (AD 1250) (Zeidler, 1994). Regional differences in the course of later cultural developments in coastal Ecuador raise interesting questions about the factors leading to the emergence of complex chiefdoms and other high-level societies that are outside the scope of this book.

The Middle and Later Formative

The cultures that followed Valdivia are named Machalilla and Chorrera or Engoroy. Occupations belonging to the Machalilla period are radiocarbon dated to between 3000 and 2700 B.P. The Chorrera cultures lasted from ca. 2700 to 2000 B.P. (Table 5.1). The subsequent cultural phases of the regional development period, which we will not discuss here, had their roots around the time of Christ and extended for approximately 500–700 years into the post-Christian era.

The Machalilla, or middle Formative period, is not nearly as well documented as the Valdivia. Ceramics show overall stylistic continuity with late Valdivia and site location continues to be focused on the bottomland. Several sites in the Valdivia Valley are much larger than late Valdivia occupations and two are larger than any Valdivia settlement (Schwarz and Raymond, 1996). This may reflect a reversal of the Valdivia trend toward smaller, dispersed settlements with a few large, ceremonial centers and full-time ceremonial specialists and the reestablishment of large, nucleated settlements. During the early Machalilla period, population appears to have substantially increased compared to Valdivia times.

Only two sets of botanical data are available for Machalilla—from the Río Perdido and La Ponga sites (Table 5.1). Río Perdida is a small site located less than 1 km east and north of Real Alto (Lippi, 1983). It is one of approximately 26 Machalilla hamlets (average size is 0.46 ha) recorded during pedestrian survey in the valley (Ziedler 1986). Because of the small number of proveniences analyzed (three for charred remains and seven for phytoliths), it is difficult to say more than that maize was used, as was achira, and that dense, porous, and rind materials occurred (Pearsall, 1979). No *Canavalia* remains were identified.

La Ponga is located 15 km from the coast in the Valdivia River valley in southern Manabí province (Lippi, 1983; Lippi *et al.,* 1984). Only the analysis of maize remains has been reported. Maize was recovered in all flotation samples, including those associated with the earliest Machalilla strata at the site. Two varieties of small-kerneled maize may have been present.

Bone isotope data are available from only one coastal Machalilla-period site. They indicate a marked enrichment in the δ ^{13}C values, but given the associated nitrogen values, the investigators note that the data may largely mean marine foods were heavily relied on (Van der Merwe *et al.,* 1993). Fewer crop plants have been identified for the Machalilla than for the Valdivia period (Table 5.1), a product, no doubt, of the fact that the middle Formative has been much less intensively studied.

The late Formative or Chorrera period in southwest Ecuador basically saw the continuation of settlement patterns demonstrated by the Machalilla cultures (Schwarz and Raymond, 1996). Bone isotope values indicate that maize was consumed more at some sites than it had been during preceding periods but possibly not at others (van der Merwe *et al.,* 1993). For example, the mean δ^{13}C for Chorrera-period skeletons from Loma Alta (13 individuals) (van der Merwe *et al.,* 1993) falls just outside the range for populations dependent on C_4 terrestrial resources. C_3 plant foods and marine foods are still in the diet. However, the isotope data from the Salango site indicate less use of maize. We note again that isotope studies of the apatite portions of the bones would clarify whether the dietary calories were coming from C_3 or C_4 plants. The few available health indicators for this period (La Libertad site; Ubelaker, 1988) show a generally healthy population—healthier than the subsequent Guangala-phase populations.

The best botanical data for the late Formative come from four Chorrera sites (Tabuchila phase) in the Jama Valley (Pearsall, 1996). The sites are San Isidro, the central place of the valley; Finca Cueva and Dos Caminos, two multicomponent sites located nearby; and El Mocorral, a multicomponent site located in the Narrows, a stretch of the Jama river with rugged topography and no active alluvium. The previous three sites were tested by Engwall in 1991 and 1994 as part of Zeidler and Pearsall's research in the region. Charred maize is nearly ubiquitous at one site (Dos Caminos), and maize phytoliths are present in most samples analyzed, regardless of the presence of charred maize. Two root crops, arrowroot and achira, and the common bean are present, as are various tree fruits. Further evidence for subsistence comes from phytolith studies of an alluvial cut in the Jama Valley (Veintimilla, n.d.; Pearsall, 1996), which reveals significant forest clearing by Chorrera times. Open-area animal taxa are also common at sites (Pearsall and Stahl, 1996).

Late in the period new developments emerged. Settlements were founded in zones not occupied before, in more "marginal" uplands considerable distances from riverine bottomlands (Schwarz and Raymond, 1996; Pearsall and Zeidler, 1994). Apparently, growing populations had filled up tl•e most favorable agricultural areas and were forced to seek other living and planting spaces. In the Jama case, this shift began in Chorrera and accelerated thereafter.

In short, the available data indicate that maize was making a substantial contribution to the diet by Chorrera times, that root crops were still important in subsistence, and that growing populations required new lands. By the time of Christ, more settlements were occupying the inland areas of the Ecuadorian coastal plain, and extensive areas of the Guayas Basin to the east of the area that we have previously discussed had been converted from swampland into raised agricultural fields (e.g., the Peñon del Rio complex of fields), in which maize appears to have been extensively grown (Pearsall, 1987).

The Ecuadorian Amazon

> *Since stone and pottery may not have been used at all, or but little, archeologic sites may be lost, unless recognized by an anomalous vegetation.*
>
> Carl O. Sauer (1958)

Lake Ayauch[i] is situated at the base of the Andes in the southeastern part of Ecuadorian Amazonia at an elevation of 500 m above sea level (Fig. 5.2). Phytolith and pollen studies of its 7000 year-old record indicate that starting approximately 5300 B.P. the watershed forest was being cut for plots of maize (Bush et al., 1989; Piperno, 1990b). The section of the core where both maize phytoliths and pollen begin to be continuously recorded is 10 cm below a 2-cm section ^{14}C dated to 4510 B.P. and 20 cm above a 2-cm section ^{14}C dated to 7010 B.P. Thus, the date of 5300 B.P. is arrived at through interpolation, not extrapolation, and because sedimentation rates in the core are slow the maize almost certainly antedates 5000 B.P.

The first evidence for maize is associated with a rise in *Cecropia,* a secondary forest tree, and herbs and with the disappearance of the Bombacaceae, arboreal taxa that are prominent in mature forest. Charcoal is present in high quantities, also indicating that woody taxa were being systematically cut and burned. However, at this time and later in the sequence, when cultivation practices appear to intensify, arboreal taxa are not reduced to very low levels, nor do the Poaceae and other herbaceous pollen and phytoliths achieve the levels reached in other sequences in which maize agriculture is implicated in forest clearing at this time, e.g., at Abejas

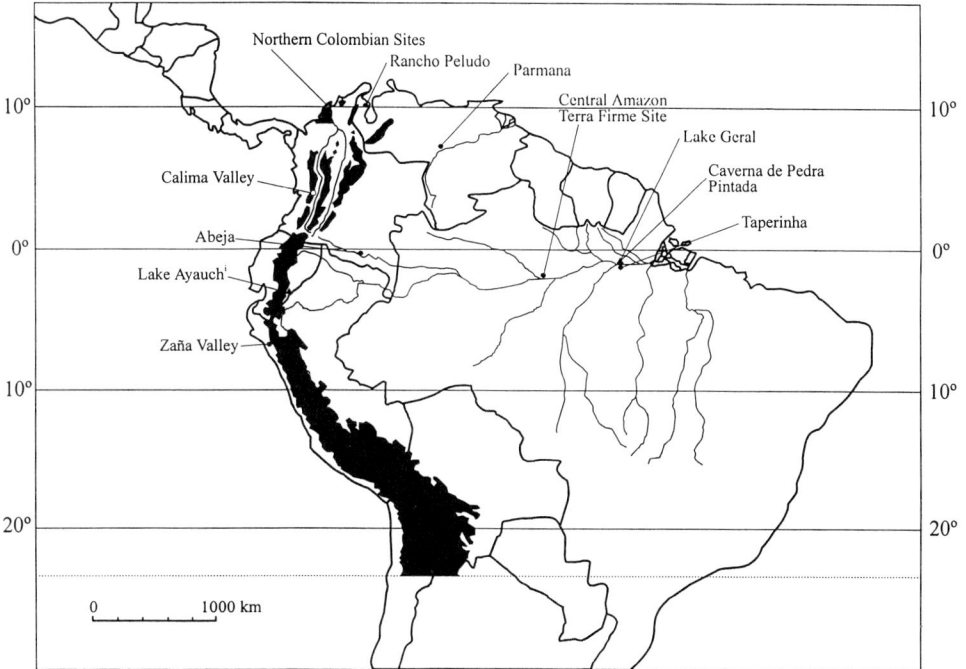

FIGURE 5.2   Map with the location of the principal sites in South America discussed in the text.

and in Panamanian and Belize sequences (discussed later; see Figs. 5.3 and 5.14 for summaries of the sequences from Ayauch[i] and other lakes discussed). This probably indicates that agricultural intensities and population densities were not as high in this particular part of the Amazon as in other areas where lake cores have been analyzed, a factor that might relate to soil fertility.

Also, maize yields are probably more dependent than those of manioc and other root crops on the efficiency of the swidden burn, which converts the nutrients found in the vegetation into ash and enriches the soils, as well as on the general fertility of the soil (Harris, 1971). Hence, the moister climate in this part of South America compared with other, drier areas studied may have diminished maize productivity and overall population number. All these factors may have resulted in a less destructive form of slash-and-burn cultivation than seen in other areas.

Beginning at 2400 b.p. frequencies of maize phytoliths and pollen greatly increase, other Poaceae phytoliths become more common, and levels of some arboreal taxa fall. An agricultural intensification is suggested for this region of the Amazon at this time, when agricultural and social complexity was accelerating throughout the Neotropical lowlands. Again, there do not appear to be the massive levels of deforestation that many other regions outside of the Amazon experienced. Many

YEARS B.P.	ECUADOR Lake Ayauch[i]	AMAZONIAN COLOMBIA Rio Caquetá	COLOMBIA Hacienda Lusitania	EASTERN AMAZON Lake Geral
	Agriculture abandoned	Settlement abandonment		Forest Regrowth
			Renewal of	----?500---
	---- 800 ----	Cotton, cacao,	intensive	
1000		chile pepper,	agriculture	Agriculture
		avocado		intensifies
	Agriculture	---- 1565 ----		but arboreal
	intensifies.		-- Undated --	pollen
2000	Zea increases			frequencies
				never
	---- 2500 ----		Forest	substantially
			Regrowth	decline
3000		SLASH	Undated	
		AND		-- 3350 ZP--
	SLASH	BURN		
	AND		SLASH	
4000	BURN		AND	First forest
			BURN	disturbance.
				Probable
		---- 4700 MP--		S+B
5000		ZP and forest		
		disturbances	---- 5150 ----	
	5300 ZP, Phy	before 5000 B.P.		--- 5760 ---
6000			ZP	
	Mature forest	Base Undated	Undated	Semi-evergreen
	with fire			forest. No
	disturbance			disturbance
7000				recorded
	---- 7100 ----			--- 7760 ---
	BASE			BASE
8000			Mixture of	
			Andean and	
			Sub-Andean	
9000			forest	
10000				
11000			BASE >39,000 B.P.	

FIGURE 5.3   Summary of pollen and phytolith data from the paleoecological sites in South America discussed in the text. P, pollen: Phy, phytolith: Z and M, first appearance of *Zea* and manioc in pollen (P) or phytolith (Phy) records: S+B, slash-and-burn agriculture.

arboreal taxa continue to be well represented during the agricultural maximum. Because the date of 2400 B.P. also witnessed the first appearance of carbonized remains of maize along the Orinoco River (Roosevelt, 1980), it is likely that maize agriculture was becoming more intensive and dispersing throughout the Amazon Basin around this time.

A number of other important points of information were revealed by the Ayauch[i] studies. Charcoal was abundant in pre-5300 B.P. sediments. Although little other evidence of disturbance could be found, Piperno (1990b) believes that the charcoal probably represents human burning of the forest because of its abundance and the fact that the climate was moist.

The Ayauch[i] record also probably represents an example of riparian agriculture in the Amazon Basin (Piperno, 1990b). During the agricultural maximum, percentages of maize phytoliths and pollen reach levels seldom found outside of maize fields in modern phytolith and pollen spectra (Piperno, 1990b). From the sedimentological record, Bush and Colinvaux (1988) identified a dry period between 4200 and 3150 B.P. during which lake levels were lower. Diatom evidence also indicates that lake levels began to fall at 4400 B.P. and did not begin to rise again until approximately 2065 B.P. (M. Redinger, personal communication, 1996). This drier interval would have provided suitable fertile areas of at least seasonally exposed sediment in the crater itself, on which maize could be planted. It is likely, then, that both slash-and-burn cultivation in the watershed forest and "wetland" agriculture using the margins of lakes were being practiced in this part of the Amazon Basin several thousand years before the current era.

Cauca Valley, Colombia

Early Holocene sites, such as Sauzalito, El Recreo, and El Pital from the middle Cauca (Calima) Valley discussed in Chapter 4, seemed to manifest the typical settlement patterns for early tropical food producers. They were small, possibly seasonally occupied, and positioned near small parcels of alluvium that could serve as backyard gardens. The latest ^{14}C determination for the occupation of these sites was 7300 B.P. Between this time and a few centuries before the time of Christ, there are few archeological sites that have been studied. The best evidence for occupation and subsistence in the region comes from pollen and phytoliths from lake and swamp cores (Table 5.1; Fig. 5.2 and 5.3).

Approximately 12 km south of El Recreo and Sauzalito lies a large, swampy valley called El Dorado, which until shortly before the time of Christ held a sizeable lake. Two long sediment cores were taken from this valley and analyzed for pollen by Monsalve (1985). One of the core sequences, called the Hacienda Lusitania (HL), revealed the presence of maize pollen 15 cm below a 5-cm-thick peaty soil ^{14}C dated to 5150 $\pm$ 180 B.P. (Fig. 5.3). The dated horizon marks the significant decline of arboreal elements in the core, such as oak and *Hedyosmum,* with a concurrent marked increase in the weedy Compositae family. The appearance of maize and other changes in the pollen record indicate the presence of a well-developed agricultural system with substantial forest modification by late in the sixth millennium B.P.

At the Hacienda Lusitania, maize pollen increases markedly 10 cm above the dated peat and grass pollen also increases, pointing to the intensification of agriculture. The core subsequently shows a decline of the agricultural system and some regeneration of the forest, followed by another increase in agricultural activity. These parts of the sequence are undated.

The pollen records from Hacienda Lusitania are corroborated by a similar core sequence from a part of the valley 2 km from HL. At the Hacienda El Dorado, maize is continually present starting in a level directly dated by radiocarbon to 6680 ± 230 B.P. (Bray et al., 1987). Shortly after, maize pollen appears in very high frequencies. Obviously, we would like to see more segments of these cores dated, particularly the upper sections that follow the course of agricultural development. However, they are remarkably consistent in pointing to the early incursion of maize into the region and consequent shift to more intensive agriculture.

The people who engaged in the earliest maize growing and forest clearing have yet to be identified archeologically. The next archeological manifestations of cultural activity in the Calima Valley after the early preceramic El Recreo and Sauzalito occupations are dated to approximately 3000 B.P. They belong to the Ilama culture, which appears to represent sedentary occupations based on the cultivation of maize, beans, and squash (Bray et al., 1985, 1988; Cardale de Schrimpff et al., 1992) (Table 5.1).

More cemeteries than habitation sites have been defined for the Ilama people, who buried their dead on large cemeteries on hilltops. Currently, the only habitation site identified and excavated for this culture is El Topacio, located on the Río Calima. Carbonized maize, common beans, and palm kernels, along with phytoliths from maize, squash, arrowroot, and palms, were retrieved (Bray et al., 1988; Cardale de Schrimpff et al., 1992; Kaplan and Smith, 1988). Smith believes that the maize belongs to a precursor of the Chapalote/Nal Tel/Pollo complex.

The hill slopes appear to have been the foci of cultivation for the Ilama peoples because most of the valley bottom would have been too inundated to have been planted without water control mechanisms. The hill slopes were fertile, being volcanically derived alfisols and inceptisols (Bray et al., 1987). Although evidence for settlement pattern and subsistence is still fragmentary, the Ilama culture appears to belong to a Formative way of life. Whether Ilama is the earliest expression of the Formative in the middle Cauca area cannot be said at this time.

Later in the prehistoric period in the Calima Valley, the lake in the El Dorado Valley dried into a swamp and the valley floor became suitable for agriculture. Drainage systems dating to shortly after the time of Christ during the Yotoco period were constructed and eventually raised fields were also built. Bray et al. (1985) believe that the main purpose of the ditching was to keep the water table as low as possible. Phytolith and pollen evidence make it clear that maize was one of the most important crops grown in these systems (Piperno, 1985b).

Colombian Amazon

The site of Peña Roja, located along the middle stretches of the Caquetá River in the Araracuara region of the Colombian Amazon, provided evidence for small-

scale cultivation of leren, bottle gourd, and probably squash, along with the exploitation of a diversity of tropical forest plants during the ninth and eighth millennia
B.P. (see Chapter 4; Cavelier *et al.,* 1995). In the immediate vicinity of Peña Roja,
sites occupied at a later time have been identified and studied. It is likely that they
represent cultural continuity from the early Holocene Peña Roja settlement (Table
5.1; Fig. 5.2). The sites are situated on the Araracuara structural hill or plateau, a
major geological formation that rises behind the Caquetá River. These occupations
document the presence of both maize and manioc in the region by at least the
fifth millennium B.P. and point to the development of considerable social complexity within a wet, lowland tropical forest habitat (Mora *et al.,* 1991).

One of the sites that most clearly shows this development is called Abeja. It
lies 12 m below the top of the plateau and 140 m above the river. The cultural
sequence at Abeja was studied through an intensive series of augoring and excavation that defined living areas with individual house structures, along with probable
loci of former gardens and fields.

There were three main occupations of the site. The earliest for which radiocarbon dates have been obtained (4695 ± 40 and 4330 ± 45 B.P. from a 5-cm section
of an organic soil horizon of a major core that the investigators believe penetrated
former field areas) is called Tubaboniba. Both maize and manioc pollen are present
in the core approximately 10 cm and more (the maize) below the dated soil horizon
(Figs. 5.3 and 5.4). Associated with the cultivar pollen are high levels of charcoal,
grass pollen, and other forest disturbance pollen types, indicating that substantial
sections of the forests were being cleared for crop planting.

There is an earlier undated occupation of the site that also contains considerable
amounts of maize but not manioc pollen and occurs 35 cm below the 4695 B.P.
level of this 1-m-long core sequence (Fig. 5.4). Charcoal is present in smaller
quantities than it was at ca. 5000 B.P., indicating that a smaller scale form of forest
modification was being practiced than during the fifth millennium B.P. Because
the soils of this period are separated from those of 4695 B.P. by a zone with a
very different pollen spectrum, indicating a drier climate, and loss of maize, the
provenience of the early maize seems secure. The investigators of Abeja consider
that this earliest cultural activity at the site represents an early maize agricultural
occupation in western Amazonia that predates 5000 B.P., and whose absolute
chronology needs to be determined (I. Cavelier, personal communication, 1996).
Considering the fluctations of climate indicated by the pollen record and the loss
and subsequent return of maize pollen, it seems that adjustments were being made
to the food production system over time in response to changing environmental
conditions.

The final occupation of Abeja is called Méidote. Six radiocarbon dates place it
between 1565 and 775 B.P. By the end of the Méidote phase, the site, including
cultivation plots and houses, covered 6 ha. Substantial numbers of people must
have been involved because they produced a brown anthropic soil 40 cm thick
in a time span of approximately 800 years. Mora *et al.* (1991) concluded that Abeja

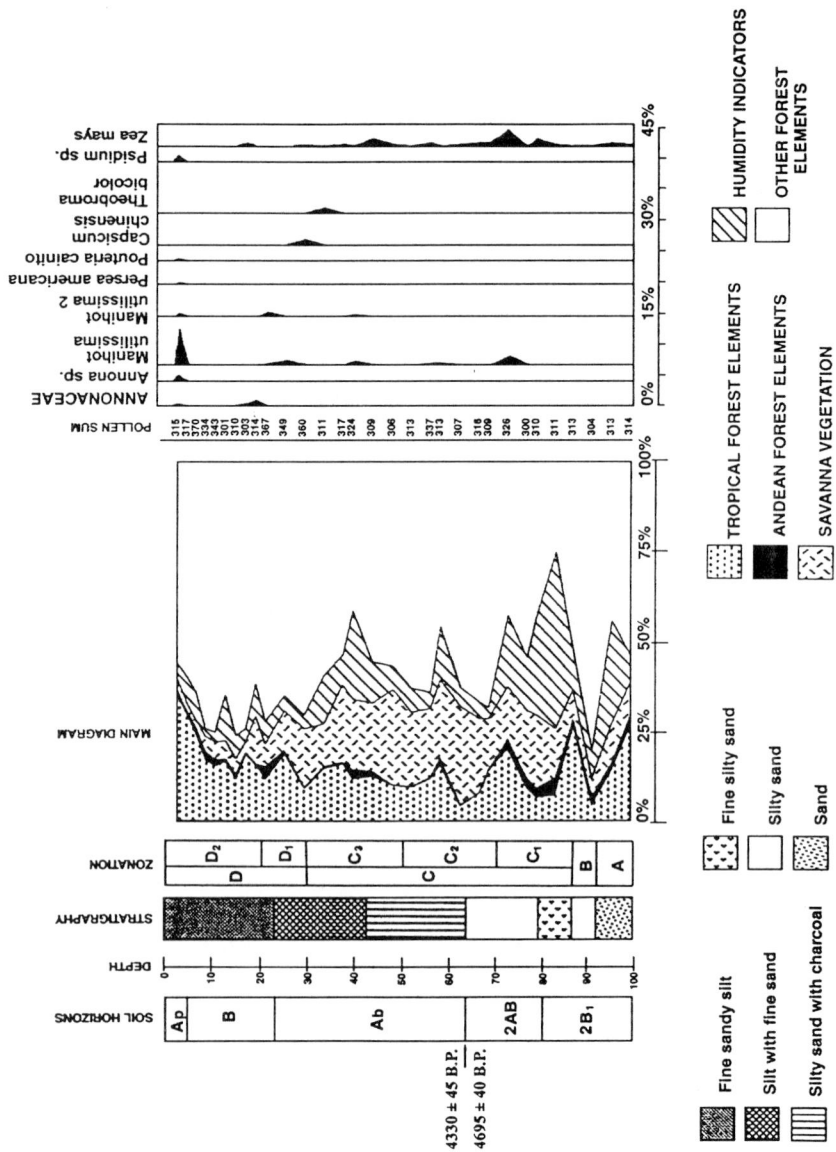

FIGURE 5.4   Pollen profile from the Abeja site showing the major vegetational changes at the site and the presence of cultigens through time.

was a permanent village that did not shift its location during the Méidote phase. The Méidote phase saw the first pollen evidence for other crops, such as cotton, cacao, chile peppers, and avocado, that may have been later components of these agricultural systems or, perhaps more likely, were less important and visible earlier in time. Forest disturbance frequencies in the pollen cores remained high, indicating that the area was widely cleared for settlement and agriculture.

The termination of the anthropic soil and cultural materials indicates settlement abandonment approximately 775 B.P. Frequencies of forest pollen return to dominate the pollen record as the forests regenerated. The continued, spotty presence of maize, manioc, and fruit trees suggests that, like indigenous Amazonians today (e.g., Denevan *et al.,* 1984; Balée and Gély, 1989), people returned to use the abandoned fields and regenerating fallows. Settlement abandonment at Abeja coincides with an influx of population at another settlement located on the plateau only 3 km away called Aeropuerto. Here, the sudden appearance of algae in the pollen record at 1200 B.P. indicated that people were adding alluvial silt and organic debris to soils to increase their fertility. Thus, techniques of soil improvement were not unique to the Mayan area. It is interesting that a diversification of cultivars is seen at this site and Abeja at this time.

Mora *et al.* (1991) believe that factors such as the transportation of large amounts of silt from the valley bottom to the site, high population densities, the presence of probable satellite communities near main population centers, and sophisticated ceramics and other cultural materials point to the development of centralized leadership and complex societies in the Caquetá region shortly after the time of Christ. That these developments occurred within a wet, lowland tropical forest indicates that such environments are not inimical to the fostering of cultural complexity.

Zaña Valley, Peru

Sites dating to the early eighth millennium B.P. in the Zaña Valley of Peru yielded macrobotanical remains of squash, manioc, peanuts, and quinoa, whose morphologies indicated an early stage of domestication but whose ages could not be unequivocally determined. Somewhat later middle preceramic sites in the Zaña Valley have been investigated by Dillehay *et al.* (1989, 1997) and Rossen *et al.* (1996). Habitation during this period, called the Tierra Blanca phase, dates from approximately 6400 B.P. until 5000 B.P. Like during the earlier period, sites were situated above the river floodplain on terraces in lateral quebradas near the headwaters of streams, beside small but fertile areas of alluvial soils. However, some important differences between these and earlier occupations are demonstrated by site complexity and organization as well as by the inventory of domesticated plants.

Growth in the food production system is suggested by the recovery for the first time of cotton and coca, whereas chenopods and peanuts continue to be evidenced. Also present during this time period are special-purpose, nonresidential sites used for public ceremonies and other functions. One of these, the Cementario de Nanchoc, was intensively excavated by Dillehay *et al.* (1989, 1997). It is 2 ha in size and contains two earthen mounds up to 1.3 m in height. The construction and use of the first artificial surface is dated to 7700 B.P., when the site probably contained only a low knoll marked by aligned stones. Activities at the site appeared to have ended approximately 5000 B.P.

One zone of the Cemetario de Nanchoc seems to have served as a special activity area for lime production, as indicated by the remains of multiple hearth features and unburned chunks of lime found there. The lime was probably chewed with coca leaves to help release the active components in the coca. Thus, this practice, still important today in indigenous ritual activity and daily life, appears to be very ancient. Lime is also currently used as a nutrient additive to quinoa and various tubers, both probably grown at the habitation sites nearby.

The likelihood that the Cementerio de Nanchoc was a public site where lime processing was perhaps organized at the community level is indicated by the rare remains of domestic refuse such as postholes, floor stains, and storage pits. Dillehay *et al.* (1997) suggest that household autonomy probably characteristic of the earlier, Las Pircas-phase occupations (discussed in Chapter 4) declined as communal activity increased in the area. Although it cannot be suggested who may have controlled the production of lime at the site, it was probably redistributed to the Zaña Valley neighborhood at large and may have been a product for exchange. Hence, as in coastal Ecuador, a significant elaboration of the social sphere oriented around increasing group identity and cohesiveness can be detected in the Zaña Valley as horticulture developed.

The evolution of subsistence and settlement in the area can also be followed by the trajectory of settlement location through time. No preceramic sites have yet been located on the river flood plain. Later preceramic sites than those discussed here are situated farther downslope on the middle sections of streams. The Initial period sites, with larger residential structures, pottery, and stone and earthern mounds that served administrative purposes, are positioned near the river floodplain. At these later sites positioned closer to the valley floor, the diversity and density of artifacts, particularly grinding stones, also increase. As discussed in Chapter 4, this pattern is *contra* the notion held by some investigators of American agricultural origins that the earliest farmers lived in sedentary and substantial villages situated along major floodplains.

Summary

An impressive feature of the records previously discussed is the substantial continuity and intensification of the food production system demonstrated through time in several regions (Table 5.1, Fig. 5.3). Coastal Ecuador, the Cauca Valley, and the middle Caquetá regions of Colombia, and the Zaña Valley, Peru, which had all seen an early Holocene development or introduction of food production, subsequently saw new crops added during the middle Holocene. An increasing labor investment in the food production system is indicated by the development of slash-and-burn agriculture (Cauca Valley, middle Caquetá areas, Jama Valley, Ecuador) and utilization of more substantial areas of alluvium along larger rivers (southwest Ecuador).

As indicated by the middle Caquetá and Lake Ayauch[i] data, the intensification of food production and development of slash-and-burn agriculture was not confined to lower rainfall areas supporting a seasonal tropical forest with a long and marked dry season. That maize is present during the sixth millennium B.P. in these areas is consistent with the likelihood that it initially diffused into northern South America through the drier inter-Andean valleys and lower elevations to the east of the eastern Cordillera. Maize seems to have become adapted early on to wet, evergreen forests, where it is associated with the appearance of slash-and-burn cultivation and subsequent intensification of agricultural systems. The adaptability of maize has always been considered one of its preeminent characteristics as a food plant.

THE COASTAL PERUVIAN SEQUENCE

We now look at the sequence of cultivated plants and some of their associated cultural inventories on the arid coast of Peru, spanning the period from 7600 to 2300 B.P. (middle preceramic through early Horizon). The cultivated and domesticated plants documented at the sites are in almost all cases sure imports from elsewhere; many examples came from the humid forested lowlands. Some are not well evidenced or documented at all, by either macro- or microbotanical remains, at sites nearer to their areas of origin. Therefore, following the course of these plant introductions into this arid landscape can provide insight into what may have been happening elsewhere. We hope to soon refine the chronology and taxonomy of the sequence of crop plant introductions by restudying some plants, such as the squash remains, and directly dating some of the earliest plants from Paloma and Chilca 1.

It will be shown that there appears to be a delay between the appearance of certain crop plants on the Peruvian coast and their origins/introductions into the more humid, forested areas of South and Central America. Why this may be necessitates a consideration of the distinctive ecology of the coastal desert with its immensely productive marine zone (refer to the discussion in Chapter 2) and, possibly, of the El Niño phenomenon.

The El Niño–Southern Oscillation

The El Niño–Southern Oscillation, or ENSO "event" (referred to here as El Niño) is an interaction of atmospheric and oceanographic conditions that alters atmospheric pressure, wind, rainfall, and ocean currents and sea level in the tropical and subtropical latitudes of the Pacific (Glynn, 1988). El Niño events occur on average every 4 years. Basically, warm, nutrient-poor surface waters from the equatorial latitudes move south along the coasts of Ecuador and Peru, bringing torrential rains and disrupting the cold-water biota of the Peruvian waters. Depending on the severity of the event, massive ecosystem disruption can occur. The 1982–1983 event is an example of an unusually severe El Niño.

The annual migration of the equatorial trough of low pressure, called the ITCZ, to the south in January (see Chapter 2) brings seasonal rains and warmer ocean surface to the western coast of South America, typically to around the border of Peru and Ecuador (5°S latitude). An El Niño is essentially an intensification of this pattern: Warm water extends much deeper and moves south into central Peru (stronger flow of the north equatorial countercurrent and weakened flow of the Peru current). Rains are heavier, bringing flooding and sedimentation. Nutrient concentrations in off-coast waters decline because of weakened upwelling; upwelling does not stop, but the water is from a deepened and impoverished thermocline (Enfield, 1992). Collapses in populations of anchoves, sardines, and guano birds are well documented, but influences move throughout the food chain. Tropical taxa move with the warm water, replacing cold-water taxa. El Niño events vary greatly in intensity and duration: Typically they are of 12 months duration, beginning in January with an incursion of warm water down the coast, and peaking in April–June, with a second, weaker warming in November and December (Glynn, 1988).

The effects of El Niño are not limited to the western coast of South America, however. Aceituno (1988), in a widely cited study of the modern functioning and effects of El Niño/Southern Oscillation, describes three pronounced climate anomalies of the negative southern oscillation (El Niño) phase in South America: (i) excessive rainfall along the Ecuador–Peru littoral, (ii) dry conditions in northeast Brazil and parts of the Amazon Basin, and (iii) a weak tendency for relatively wet conditions in the Paraná River basin. Marengo et al. (1993) also document that dry conditions in northern Amazonia are associated with El Niño conditions, whereas abundant rain in that region corresponds to anomalously cold surface water in the eastern Pacific (i.e., the La Niña or positive southern oscillation phase). Low rainfall in northern and eastern Amazonia is caused by altered wind, air pressure, and surface water temperatures in the tropical north Atlantic (Aceituno, 1988).

Diminished rainfall also characterizes parts of the northern Andes, the Caribbean, and Central America during El Niño years (Leigh et al., 1990; Philander, 1990). The southward displacement of the near-equatorial trough in the eastern Pacific that brings heavy rains to the western South American littoral may be a factor in the reduced rainfall in the northern low latitudes (Aceituno, 1988). Panama, southern Andean Peru, and Bolivia experience drought conditions in El Niño years (Leigh et al., 1990; Philander 1990), whereas in the southern high latitudes (central Chile and Patagonia), there are regional differences in the relationship between ENSO and precipitation (Villalba, 1994).

Thus, the recent literature on El Niño, especially the studies stimulated by the severe 1982–1983 El Niño, have established that El Niño/Southern Oscillation effects climate not only throughout the tropics but also into the temperate latitudes. It is not a regional phenomenon affecting only coastal Peru; its effects are not "blocked" by the Andes. Variability and abundance of rain in northern and eastern Amazonia are affected by El Niño. The amplification of climate variability—the "excess" variability as Nicholls (1989) terms it—caused by El Niño potentially has some impact on human adaptation throughout the New World tropics and would profoundly affect adaptation on the desert coast of Peru. However, how long has this process been in place?

Antiquity of El Niño

A variety of data, including historic records, pollen cores, deep-sea cores, beach ridge data, flood data, and lake levels, can be used to address the previous question. Although the phenomenon probably existed in some form during previous glacial and interglacial periods of the Quaternary, as demonstrated in cores documenting upwelling during the past 430,000 years (Oberhänsli et al., 1990), the point that concerns us is when during the past 10,000 years it began to have an impact. There is disagreement on this point. Historical data document El Niño back to AD 622 (Quinn, 1992; Quinn et al., 1987; Quinn and Neal, 1992). Historic records can also be matched to well-dated recent proxy records (pollen cores, ice cores, beach ridge data, and lake levels) to establish the "signature" of El Niño events in the geological record. These signatures can then be used to document prehistoric events. Very strong El Ninos, such as the 1982–1983 event, are most likely to leave recognizable signs in the proxy records.

The archeological and geological records have proven to be informative in studying the age of El Niño. Using geoarcheological evidence from the north-central coast of Peru, Rollins et al. (1986) and Sandweiss (1986; Sandweiss et al., 1996) suggest a 5000 B.P. date for the birth of El Niño. Furthermore, as discussed in Chapter 4, they place this event in the context of a major reorganization of the east Pacific water structure, proposing that the boundary between the warm Panamic province and the cold Peruvian province was located 500 km south of its current location (i.e., at approximately 10°S rather than 5°S latitude) prior to that date. They base this on a change from warm-water to cold-water molluscan fauna at archeological sites at this latitude at 5000 B.P.

In the same area, the oldest of nine beach ridges, argued to represent sediments from major El Niño events, also dates to this time. Similar beach ridges are reported in the Chirca Valley (5°S); these are also dated after 5000 B.P., when sea level stabilized along the coast (Richardson, 1983). Overbank flood data from the Casma Valley (9°30'S) document five large flood events for the period from 7500 to 3000 B.P. that are probably correlated with El Niño (Wells, 1987, 1990).

Beach ridge data from the central Brazilian coast also show evidence for the operation of El Niño back to 5100 B.P. (Martin *et al.*, 1993). In this case, beach ridges are caused by a reversal of longshore sand transport during an El Niño. Unfortunately, there are no data prior to 7000 B.P., when modern sea level was achieved. From 7000 to 5100 B.P, rising sea level (a submergence episode) prevented beach ridge formation. There is clear evidence during an emergence episode, 5100–3900 B.P., for seven reversals (El Niño-like episodes). The next period of emergence, 3600–2800 B.P., shows no reversals—El Niño-like episodes were not present—whereas the most recent, 2500 B.P. to the present, documents three. Flucuations in sea level limit the usefulness of beach ridge data, especially for the early Holocene.

Fortunately, other proxy data, such as water levels in Lake Titicaca and pollen and lake levels in the Amazon, can be used to fill in some of the gaps (Martin *et al.*, 1993). From 7000 to 3900 B.P., water levels in Lake Titicaca fluctuated around a position much lower than is current found. The mean does not change, suggesting not a permanently dry situation but a succession of droughts throughout this period (i.e., El Niño-like conditions).

This correlates with the beach ridge data for 5100–3900 B.P. and further suggests that El Niño-like conditions occurred back to 7000 B.P. A pollen and lake level study from eastern Amazonia, an area that would be drier during an El Niño event, shows forest regression from 7000 to 4000 B.P. that is interpreted not as permanent dryness but as a series of dry and wet periods (Martin *et al.*, 1993). Again, this suggests El Niño-like conditions.

From 3900 to 3000 B.P., Lake Titicaca shows a steady rise in lake level (i.e., no El Niños). This correlates with the beach ridge data for 3600–2800 B.P. and suggests that the second submergence period also lacked El Niño conditions. The Lake Titicaca data show four or five droughts after 3000 B.P., again correlating with and extending the beach ridge data. Martin *et al.* (1993) believe the duration of these El Niño-like conditions was several decades, and that they were numerous before 3900–3600 B.P. and rare after 2800–2500 B.P. These relatively short-term events nonetheless may have contributed to the enhanced Holocene seasonality of the South American tropics.

Although the studies discussed previously make a good case for El Niño, or something similar, back to 7500–5000 B.P., DeVries (1987) argues that identifying El Niño events based on data for flooding demonstrates only increased rainfall and not the duration of the event. Such increases in rain could be the result of climate change. There are also alternative explanations for beach ridge formation; for example, tectonic, eustatic, and climatic changes (DeVries, 1987).

What about the warm-water molluscs in middens along shores currently in the Peruvian, i.e., cold-water faunal, province? DeVries (1987) notes that most warm-water species are those that favor the quiet waters of lagoons or embayments and that species from other warm-water habits are lacking, and he suggests that the faunal data are too coarsely resolved to distinguish between periodic events and longer term change. Sandwiess *et al.* (1996) respond that not all the taxa present could have survived in embayments.

The bulk of the evidence leads us to accept that El Niño was operating by approximately 7000–5000 B.P. As Martin *et al.* (1993) indicate, the duration of El Niño or El Niño-like conditions may have been prolonged in the past, lasting for decades or longer instead of a year or two. During periods when seasonality contrasts were lower in the Southern Hemisphere tropics than they are today, such as in the early Holocene, it is possible that El Niño events did not occur or were not the same as they are today. The increasing seasonality of climates during the middle Holocene may mark the emergence of El Niño as a dominant feature of climate in the tropics. Interestingly, this time period is when crop plants began to be introduced into the Peruvian coastal desert.

Maritime and Plant Subsistence along the Peruvian Coast

In Chapter 4 we discussed how a shift to a focus on maritime resources from a probable former terrestrial animal-based economy seems to have begun during the early Holocene on the northern coast, as evidenced by the subsistence remains at the Paiján sites. Other sites dating to the early Holocene are known, but subsistence data are few. Beginning approximately 7600 B.P., there are good databases available for examining the evolution of maritime and plant subsistence strategies through time (Fig. 5.5 and Table 5.2). The sites discussed in the following sections were chosen because in most cases subsistence reconstruction was a focus of the research design and systematic botanical recovery procedures were employed.

Middle Preceramic

The two earliest sites with good representation of plant and animal remains, Paloma (7700–4800 B.P.) and much of the occupation of Chilca 1 (5600–5000 B.P.), were occupied before the cotton preceramic, a convenient horizon marked by the appearance of abundant cotton remains in coastal sites. Table 5.2 summarizes the occurrence of cultivated plants at these and other coastal Peruvian sites. The earliest cultivated plants to appear on the coast, at the Paloma site, are bottle gourd, squash, some kind of *Phaseolus* bean (probably lima bean), and guava (a tree fruit) (Benfer, 1982, 1984, 1990; Dering and Weir, 1979, 1981; Weir and Dering, 1986). None of these appear to be dietary staples. All are probably imports. A local tuber, *Begonia,* may also have been grown. Four other cultivated taxa appear at the Chilca 1 site: cotton,[2] *Canna edulis* (achira), *Pachyrrhizus* (jicama), and *Canavalia* bean (Jones, 1988; Weir *et al.,* 1980; Engel, 1988). Lima bean is also present. Thus, there are a total of eight cultivated taxa for the middle preceramic. Lima beans at

[2] Cotton occurs at Chilca 1 in tomb contexts.

FIGURE 5.5   Map with the location of sites from Peru discussed in the text.

Chilca 1 directly dated by AMS yielded an age consistent with other age assessments for the site (L. Kaplan, personal communication, 1996).

Both sites had good preservation, and good recovery techniques were employed. The low richness of cultivated plants at the sites and the fact that wild plant foods are much more abundant than domesticated ones, probably reflect low use of cultivated taxa rather than preservation or recovery biases. At both sites a variety of remains from the sea—shellfish, fish, and mammals—were common. Marine resources were undoubtedly a mainstay of diets and over the course of occupation at Paloma subsistence became increasingly maritime, with sea mammals and anchovies becoming more common and terrestrial mammals less frequent in the midden (Benfer, 1990). Strontium data show a decline in plant consumption through time.

Population rose at this sedentary village, and then declined (Vradenburg *et al.*, 1997) as population outstripped the resources of the lomas (fog oasis) location and people moved to the adjacent river valley. Judging from skeletal studies, people were healthy (Benfer, 1984, 1990). Paloma, then, provides convincing evidence that marine-based subsistence can support sedentary life and may thus serve as a foundation for the later emergence of complexity.

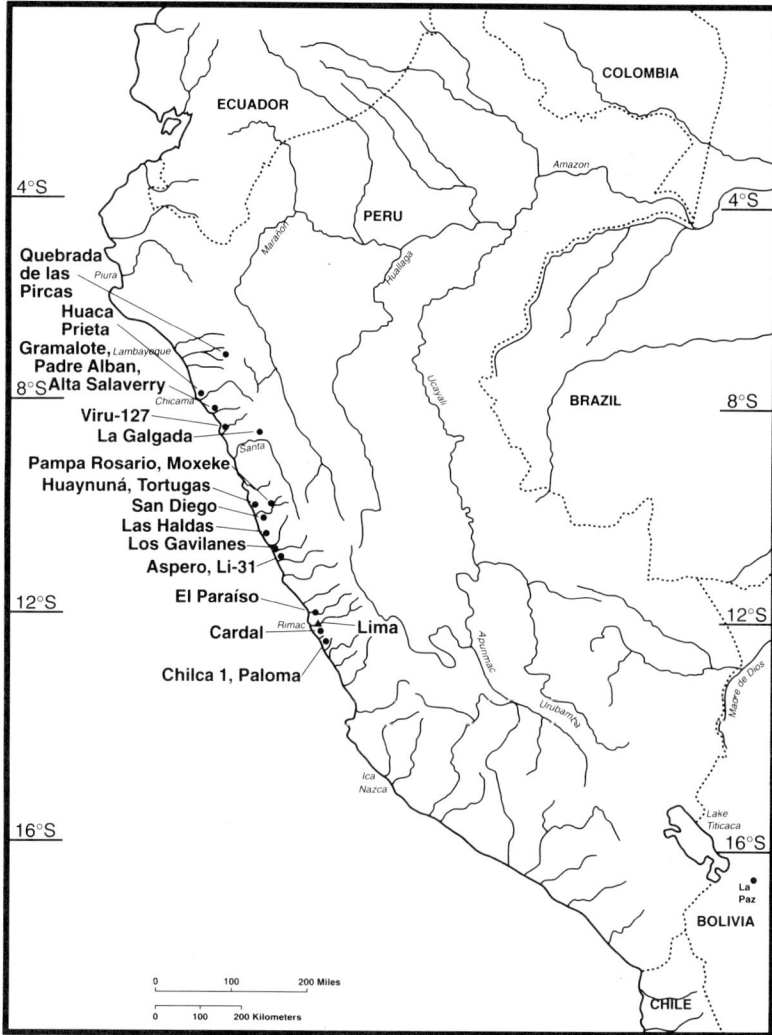

FIGURE 5.5   (*Continued*) Detail of Peruvian sites (shaded area of map facing).

Cotton Preceramic

The cotton preceramic sees the first appearance of monumental architecture on the Peruvian coast. We are fortunate to have several sequences with systematically recovered floral and faunal data within the context of valleywide settlement studies, as well as a number of single-site studies, that allow us to examine subsistence change during the late or cotton preceramic (4600–3500 B.P.) and after on the central and north coast of Peru.

TABLE 5.2   Cultural Chronology and Plant Remains in Coastal Peru[a]

	Roots and tubers					Vegetables	Grains			Tree fruits	
Cultural chronology	Canna edulis (achira)	Manihot esculenta (yuca)	Ipomoea batatas (sweet potato)	Solanum tuberosum (potato)	Pachyrrhizus (jicama)	Cucurbita spp. (squash)	Zea mays (maize)	Persea americana (avocado)	Psidium guayaba (guava)	Annona (soursop)	Lucuma bifera (lucuma)
Coastal Peru middle Preceramic, 7700– 5000 B.P.	P[b]				P	P			P		
Cotton Preceramic, 4600– 2800 B.P.	67	11	22	11	22	89		44	78	11	56
Initial period, 4000– 3500 B.P.	17	33	33	50	17	100	50	67	50		100
Early Horizon period, 2800– 2200 B.P.	60	80				100	100	100	60		80

[a] Refer to text for sites contributing remains. Percentage of sites with a given crop is indicated for all but the earliest time period.

[b] P, present.

Late preceramic sites with botanical remains considered here include Los Gavi-lanes[3] (Bonavia, 1982; Popper, 1982), La Galgada (Grieder, 1988; Smith, 1988), Aspero (Feldman, 1980), Huaca Prieta (Bird et al., 1985), El Paraiso (Quilter et al., 1991; D. Pearsall, lab documents), Las Haldas, Huaynuná (Pozorski and Pozorski, 1987), and Padre Alban and Alto Salaverry (Pozorski, 1983).

The north-coast Casma Valley sequence (Pozorski and Pozorski, 1987) appears to be fairly representative. Early preceramic sites of the Paijan and Mongoncillo lithic traditions (considered to postdate Paijan) are found only at the valley mouth (1 site; of the Paijan tradition) and in a lomas area south of the river (38 sites; 1 Paijan and the rest Mongocillo) (Malpass, 1983). No subsistence remains are available from this survey, but sites all appear seasonal—small occupations oriented toward exploitation of the lomas (and in one case the varied resources of the valley mouth/estuarine environment). In the late preceramic, settlement has shifted, with 2 substantial sites (1 with mounds) appearing on the coast. Both Huaynuná and Las Haldas are oriented toward good areas for fishing and shellfish collection; neither

[3] We do not include the maize found at Los Gavilanes in the tally of domesticated plants at this site because direct ^{14}C dates on the remains (ranging from 200 to 800 B.P.) do not support preceramic status. Bonavia (1982) rejects the maize dates, arguing that no special precautions against contamination were taken in handling the remains while they were studied prior to being tested.

Cultural chronology	Inga feuillei (pacae)	Bunchosia armeniaca (ciruela)	Legumes				Industrial		Condiments	Number domesticates
			Phaseolus lunatus (lima bean)	Canavalia (jack bean)	Phaseolus vulgaris (common bean)	Arachis hypogaea (peanut)	Lagenaria siceraria (bottle gourd)	Gossypium barbadense (cotton)	Capsicum (chile peppers)	
Coastal Peru middle Preceramic, 7700– 5000 B.P.			P	P			P	P		8
Cotton Preceramic, 4600– 2800 B.P.	56	33	56	22	56	11	100	100	78	19
Initial period, 4000– 3500 B.P.	67	67	50	33	67	100	100	100	100	19
Early Horizon period, 2800– 2200 B.P.	60	60	40	80	60	100	80	80	80	16

is in the river valley (Las Haldas is at the edge of the lomas area mentioned previously, approximately 20 km south of the river).

As might be expected from site locations, all faunal materials from Huaynuná are marine in origin (mostly shellfish from rocky substrates), whereas materials from the preceramic levels at Las Haldas combine abundant shellfish with use of land snails from the lomas. Cultivated plants are present at both sites. Pozorski and Pozorski (1987) emphasize the abundance of "industrial" plants, such as cotton and gourd, that are useful for marine exploitation (all other plant taxa are very rare, rare, or moderate in occurrence, but three tubers are present at Huaynuná).

These late preceramic sites document that a change in orientation toward marine resources, from an earlier pattern of lomas exploitation, took place. Other late preceramic sites illustrating a pattern of marine settlement orientation and faunal exploitation with plant cultivation include Huaca Prieta in the Chicama area (Bird et al., 1985) and Padre Aban and Alto Salaverry in the Moche Valley (Pozorski, 1983).

Where were crops being grown? Most likely in fields in the valleys; there is no "rainfall" cultivation on the coast of Peru. Coastal rivers run mainly between January and May (when it rains in the sierra), with cultivation beginning with the onset of water flow. Simple water control features to spread water out and use of areas with high water table are likely mechanisms for cultivation in the lower reaches of the valley. Limited cultivation may have also been possible using runoff from the lomas (Benfer et al., 1987).

There is no evidence for permanent canal structures at this period. In the Casma case, survey has found no late preceramic sites in the Casma Valley itself (Malpass, 1983); perhaps there were (seasonal?) settlements near fields that have been lost over the centuries of intensive use of the lower valley.

Although most central and north-coast valleys show the pattern outlined for the Casma—late preceramic sites, often substantial, oriented to take advantage of marine and coastal wetland resources—major sites are also known from valley settings. El Paraíso de Chuquitanta is an impressive example (Quilter *et al.*, 1991). This late cotton preceramic site, occupying 50 ha and including eight or nine stone buildings, is located near the mouth of the Chillón Valley, adjacent to the river. Eleven crops were present, which is a large increase compared to the number of crops at middle preceramic sites, and cotton and tree fruits were abundant. Gourd was present but rare. Fish and molluscs were also common. Land mammals (deer, fox, and rodents) and birds were present but relatively rare, leading Quilter *et al.* (1991) to propose that fish were the major animal food at the site and that subsistence was relatively broad, incorporating a variety of cultivated and wild plant foods. Site location adjacent to agricultural land also perhaps signaled the importance of controlling production of cotton for textile manufacture.

Aspero, located in the Supe Valley (Feldman, 1980), is another example of a late preceramic site with monumental architecture located near arable land. In this case, location on the north edge of the lower Supe Valley would have given populations ready access to the sea and coastal wetlands as well as to land suitable for small-scale cultivation (lower valley lands are prone to salination problems).

To summarize, the cotton preceramic sees the first appearance of monumental architecture on the Peruvian coast. In association with this development, richness of cultivated plants present at sites jumps, with most sites with good preservation containing 9–13 crops, and there was a total of 19 kinds of crops along the coast. Ten new plants were added during this period. These include cotton, chile peppers, 5 tree fruits (avocado, *Lucuma bifera*, *Inga feuillei*, *Bunchosia armeniaca*, and *Annona* sp.), the common and lima bean, and 2 tubers (the white and sweet potato). Yuca (*Manihot escualenta*), present earlier in the Zaña Valley, also appears on the coast in the Cotton Preceramic. Although the tree fruits (except avocado) are possibly local, all the other species are introduced, many from the sierra or eastern lowlands (Pearsall, 1992; see Chapter 3).

As the percentage presence data in Table 5.2 suggest, however, no one food plant dominates (note the mixture of seed crops, root crops, and perennials). Cotton, however, is widespread and abundant. Furthermore, site location, with some exceptions such as El Paraiso, is to take advantage of rich shellfish and fishing areas. Sites are typically not oriented toward productive agricultural lands, although such lands were undoubtedly used. Thus, although the abundance of cultivated plant taxa present suggests to us an increase in reliance on plants in general, and on domesticated plants in particular, during the late preceramic, site location argues for a primary maritime focus for subsistence. In a number of cases, utility plants, such as bottle gourds and cotton, are the most abundant crop remains.

The Initial and Early Horizon Periods

A settlement shift to locations adjacent to productive land in the lower and middle valleys occurs throughout the central and north coast during the subsequent Initial period (4000–2800 B.P.), marked by the appearance of ceramics. This is not to say the coastal zone is abandoned. On the contrary, sites (e.g., Tortugas in the Casma Valley and Gramalote in the Moche Valley) are established in productive shellfish and fishing areas along the arid shore, but these sites now appear to be involved in exchange of marine foods for agricultural products with inland valley sites (Pozorski and Pozorski, 1979; 1987; Pozorski, 1983).

Initial period sites with botanical remains considered here include La Galgada (Grieder, 1988; Smith, 1988), Pampa de las Llamas Moxeke, Las Haldas, and Tortugas (Pozorski and Pozorski, 1987), Gramalote (Pozorski, 1983), and Cardal (Umlauf, 1988, 1993). One important new domesticate (maize) appears and *Annona* is not present, leaving the number of cultivated plant taxa present at 19 (Table 5.2). Maize, although present at three Initial period sites, is far from abundant. There is one cob at La Galgada, where it is not considered a local product (Smith, 1988; Grieder *et al.,* 1988), two cobs at Gramalote (Pozorski and Pozorski, 1983; T. Pozorski, personal communication, 1996), and maize phytoliths in four contexts at Cardal (in addition to one cupule fragment; Umlauf, 1993).

The real growth of major sites with ceremonial architecture is in the river valleys, including locations where canal irrigation could be practiced. There are a number of exceptions such as Las Haldas, the site out of the Casma Valley, where a temple was built during the Initial period and occupation continued into the early Horizon. It may be an outlying or supporting site of Sechin Alto, and perhaps it was indirectly supported by irrigation agriculture via an exchange system (T. Pozorski, personal communication, 1996).

At major inland sites, such as Pampa de las Llamas-Moxeke in the Casma Valley and Caballo Muerto in the Moche Valley, although marine resources still appear to dominate the animal component of diets, there is evidence for an increase in use of terrestrial animals. At Caballo Muerto, for example, marine and terrestrial animal foods appear to contribute equally to subsistence, as indicated by food value calculations (Pozorski, 1983). The Cardal site, one of four late Initial period ceremonial centers in the lower Lurín Valley, is located 15 km up the Valley, providing easy access to both agricultural land and lomas vegetation (Burger and Salazar-Burger, 1991). However, most animal protein is maritime in origin (lomas land snails and some small game also occur).

Some data on health and nutrition are available. Skeletal data from the Cardal site (Vradenburg, 1992) indicate that general community health was poor. Infectious disease, identified as endemic syphilis, was prevalent and contributed to enhanced childhood morbidity and mortality and to adult morbidity. Caries data indicate a rate of infection characteristic of a mixed horticultural subsistence base. There is some indication that elite individuals consumed more cariogenic foods.

Vradenburg, (1992) suggests that poor health, with possible reduced fecundity, may have contributed to the ultimate abandonment of the site [i.e., rather than warfare or conquest from the outside, as suggested for the abandonment of Initial period sites in the Casma region (Pozorski and Pozorski, 1987)].

These trends, i.e., growth in site numbers in the valleys and sites located adjacent to productive land, continue into the subsequent early Horizon period (defined by the appearance in some areas of Chavin-style pottery, 2800–2200 B.P. or later), when there is also clear evidence for irrigation. Early Horizon sites with botanical remains considered here include Las Haldas, San Diego, Pampa Rosario (Pozorski and Pozorski, 1987), Li-31 (Feldman, 1980), and Viru-127 (Ericson et al., 1989).

There is some indication of a drop in the richness of crop plants used. Fewer kinds of root crops appear at any one site; achira and manioc are the only ones documented. Maize occurs at all sites. Use of tree fruits and legumes continues. The reduction in the number of root crops suggests a narrowing of the subsistence base. Maize is still not a mainstay of the diet, however. Isotopic studies of human bone from Early Horizon sites in the Viru Valley suggest maize may have made up 10–20% of the energy input (Ericson et al., 1989).

In the Casma Valley, animal foods continue to be dominated by marine taxa, more so at sites close to the coast and less so at sites further inland (Pozorski and Pozorski, 1987). For example, virtually all fauna is marine at the San Diego site (5.5 km from sea) (and is narrower in focus than at earlier sites), whereas both marine and terrestrial animals, including camelids, fox/dog, guinea pigs, and land snails, occur at the Pampa Rosario site (16 km inland). Site V-127, located in the lower Viru Valley, has faunal remains suggesting marine dominance (Ericson et al., 1989).

In the Supe Valley, the early Horizon Li-31 site, located near Aspero, combines a wealth of plant remains (approximately five times the abundance as at the late preceramic site) with marine fauna use, including shellfish, fish, and sea lion (Feldman, 1980)

Discussion

The shifts in settlement and subsistence outlined previously are complex. The interpretation that the emergence of social and political complexity in the late preceramic was based on a maritime, not an agricultural, resource base (i.e., the maritime hypothesis, Moseley, 1975, 1992) has especially received much attention. We essentially agree with this proposition. A shift to increased use of maritime resources is demonstrated in the Paloma sequence and supported by the location and faunal assemblages of many late preceramic sites. It is also consistent with the relative natural abundances of marine and terrestrial resources on the arid Peruvian coast. We also agree with Pozorski and Pozorski (1990, 1991) that the existence of complex sites *without* ceramics apparently simultaneously with complex sites *with* ceramics in some valleys signals the need for further research, and especially dating, of the critical preceramic–ceramic transition along the central and north coast.

In our view, the available botanical and faunal data from late preceramic coastal sites are not robust enough to reconstruct caloric input of the various plant and animal foods, as attempted by Raymond (1981), to argue against the maritime hypothesis. However, Raymond (1981) makes the useful point that the distribution of the largest preceramic sites not only correlates with areas of especially active upwelling and broad coastal shelf (which argues for the maritime hypothesis), but also correlates with that section of the Peruvian coast with the narrowest coastal plain. It is along this stretch of coast that the best fishing is closest to land that can be cropped without irrigation; for example, in valley mouth areas.

We would characterize the adaptation of late preceramic coastal populations in the central and northern Peruvian coast as one based primarily on the rich resources of the maritime province, with considerable but varying supplements from exogenous domesticated plants that appear to have increased in number over time and that were important sources of carbohydrates, vitamins, and minerals.

Considering the productivity (and, likely, the acceptable efficiency) of a maritime-based subsistence on the coast of Peru, and the availability of carbohydrates from wild plants of the lomas and coastal wetlands, why would populations have imported crop plants? It may be significant that the first documented appearance of a few crop plants between 7600 and 5000 B.P. at the Paloma village site occured when El Niño appears to have been developing on the coast and, especially, that the diversity of crop plants increases dramatically during the subsequent cotton preceramic period. By this time, all lines of evidence indicate that El Niño was operating at full capacity. We suggest that the dramatically changing environments on the Peruvian coast brought on by the probable emergence of the modern El Niño between 7000 and 5000 years ago may, in some substantial part, have contributed to the appearance and establishment of farming communities.

Drops in the productivity of marine resources that would have occurred, by modern standards, approximately every 4 or 5 years, and that may have lasted over much more protracted periods in the past (Martin *et al.,* 1993), would have dramatically lowered availability of marine foods of the cold-water province. Although yields of warm-water taxa would have increased and lomas vegetation would have bloomed (Moseley, 1992), El Niño introduced a high amount of unpredictability to the maritime resource base of the central and northern coast of Peru. Between 4600 and 4200 B.P., a pattern of marine-based subsistence along with use of a varied assemblage of cultivated plants, which were more predictable and controllable resources, was established. Thus, the late preceramic may represent a successful adaptation to life next to some of the most productive coastal waters in the world, and to fluctuations produced by El Niño.

What about the effect of El Niño on agriculture? As Moseley (1992) has argued, the increased erosion and devastation of coastal irrigation systems by heavy El Niño rains likely required a longer recovery time than did disruption of fishing for affected populations. A case in point is the suggestion by Wells (1987, 1990) that an El Niño-generated flood in the Casma Valley dated to AD 1330 correlates with "Nyamlap's Flood," which contributed to the breakdown of the dynasty of Nyamlap and the invasion of the region by Chimu. Historic floods in this valley buried much of the agricultural land as well as many habitation sites. We do not deny that El Niño can have negative effects on irrigation agriculture. Rather, we suggest that the advent of floodwater farming in the preceramic increased the stability of coastal adaptations in the face of El Niño. Washed out fields can be replanted, and newly flooded terrain provides the opportunity for increased farming.

Factoring El Niño into causation is not new (e.g, Osborn, 1977; Parsons, 1970; Wilson, 1981). Osborn (1977), for example, proposed that the ecological instability brought by El Niño may have led to the construction of facilities for food storage and redistribution. We propose a different response—the addition of predictable and controllable domesticated plants. As Raymond (1981) points out, El Niño does not produce frequent disasters—how would storage help? Furthermore, the phenomenon is episodic but not predictably so.

Eventually, however, the growing populations of coastal Peru went down the same path as their neighbors to the north in coastal Ecuador and, through the advent of irrigation agriculture, increased their reliance on cultivated foods, including maize and, especially, root crops and tree resources. That this happened later in time can probably be seen as a result of the greater richness of the Peruvian maritime province.

THE INTERIOR OF THE AMAZON BASIN

The Eastern Amazon

Recall that lake cores taken from near Santarem and Monte Alegre just north of the Amazon River had revealed no early Holocene human modification of the forests despite archeological evidence for early human occupation of the area (see Chapter 4). In one of these cores (Geral), located 15 km from the current main Amazon River channel, where deposits extend back 7500 radiocarbon years, human forest disturbance appears to commence in a core level directly dated to 5760 ± 90 B.P. (Figs. 5.2 and 5.3) (Bush et al., 1997). Prior to the first disturbance, there is virtually no charcoal, and the pollen and phytolith records are dominated by arboreal species of moist forest with little or no representation by herbaceous or other disturbance taxa.

The 5760 B.P. pollen spectra record a substantial rise of herbaceous taxa, includ-ing grasses, and woody secondary taxa (Bush *et al.*, 1997). The phytolith record is in agreement because levels of grasses, *Heliconia,* and charcoal increase greatly at this time (Piperno, unpublished data). Nineteen percent of the grass, 70% of the *Heliconia,* and 6% of the tree phytoliths are charred, which is also strongly suggestive of human activity. No maize or other cultigen fossils are present at this time. If the disturbance is a product of slash-and-burn cultivation, the particular crops involved are yet to be identified.

An intensification of vegetational disturbance occurs in the pollen and phytolith records at ca. 3350 B.P., an interpolated age based on radiocarbon determinations from above and below this core section (Bush *et al.*, 1997). At this time, levels of charcoal rise substantially, and 40% of the grass and 20% of the tree phytoliths are burnt (Piperno, unpublished data). This horizon represents slash-and-burn agriculture. A single pollen grain of maize is present together with maize phytoliths, although these are also rare. Scarce representation of maize would suggest either that maize was not important in the terra firme agricultural systems or that it was not being grown close enough to the lake edge to make an impact on the records. Given the high visibility of maize pollen and phytoliths in other paleoecological records discussed here, we suspect that the former alternative is the most likely.

It is noteworthy that during this period of maximal agricultural intensification represented by the pollen and phytolith records at Lake Geral, which occurred between ca. 3350 B.P. and the Conquest period, arboreal pollen and phytolith frequencies never decline and frequencies of pollen from grass and other herbaceous plants never increase to a point indicating large-scale destruction of the forest, as in many other sequences discussed here located outside of the Amazon Basin or on the fringes of it. Also, unlike in sequences from elsewhere, the diversity of arboreal taxa from the primary forest remains consistently high throughout the period of slash-and-burn agriculture.

As at Lake Ayauch[i], these patterns may relate to the presence of smaller popula-tions of shifting cultivators in the Geral watershed than in other places examined, a smaller scale and longer fallow system than practiced elsewhere, the relative unimportance of maize and possibly other kinds of soil-demanding seed crops in terra firme agricultural plots, or a combination of all three of these factors. It is also significant that the forest around Lake Comprida, 15 km from Geral, apparently was never subjected to human alteration through burning or slash-and-burn cultiva-tion (Bush *et al.*, 1997).

Toward the top of the Geral sequence, disturbance indicators, including char-coal, decline and several arboreal taxa return in significant quantities (Bush *et al.*, 1997). This probably marks the reduction of indigenous people in the area due to European influence during the past few centuries and consequent relaxation of agriculture and pressure on the forests.

At the Caverna de Pedra Pintada, a rock shelter less than 30 km from Lake Geral that had been occupied by human populations by at least 10,800 B.P., signs of agriculture become apparent by 3600 B.P. At this time, thick ceramic griddles of the type used today to cook bitter manioc, called budares, appear in the refuse. The pottery associated with the griddles relates to the Formative period elsewhere in lowland South America and is possibly associated with a rapid spread of bitter manioc-based agriculture in northern and eastern South America during the fourth millennium B.P. This horizon is probably also associated with the intensification of the landscape disturbance seen at Geral at 3350 B.P. and it suggests that bitter manioc was one of the principal crops grown in the terra firme slash-and-burn systems of that time and after. Above this horizon, the first cobs of maize occur along with fragments of bottle gourd. Eventually, people along the Amazon River floodplain near Pedra Pintada came to live in numerous, well-populated sedentary villages headed by chiefs and were probably supported by a maize-based economy (Roosevelt, 1987). These people are well described by the earliest Europeans that traveled the Amazon River (Carneiro, 1970a,b; Roosevelt, 1987).

The crop plants that may have caused the earliest forest disturbance at Geral at 5760 B.P. are not yet identified in either archeological or paleoecological records. They are most likely to be found somewhere in the complex of tuber and tree crops native to northern South America and the fringes of the Amazon Basin. Because most of the tuber crops cannot be recognized with phytoliths, and all of them produce very little pollen, they are very difficult to identify in paleoecological sequences. An exception, demonstrated in the cases of the Abeja site, discussed above, and paleoecological sequences from Belize, below, is when the coring location is in or on the edge of a field.

The Middle Amazon

Direct evidence for prehistoric changes in the central Amazonian terra firme forest is also available from a site located 70 km north of Manaus, Brazil (Piperno and Becker, 1996). Here, the "natural soils" occurring under a well-studied evergreen forest in a 1000-ha reserve (Lovejoy and Bierregaard, 1990) were sampled by digging small cores to a depth of 60 cm at systematic intervals across a 200-ha area and then analyzed for the presence of phytoliths and charcoal. Direct radiocarbon dating of phytolith assemblages from varying soil depths in the cores indicated that the soil sequences contained phytoliths of an age between 7800 B.P. and the modern era.

Although charcoal was common, the phytolith record contained no evidence for past human modification of the forest by swidden cultivation. Herbaceous plants were nearly absent in the records, and no signs of cultigens occurred. Also, there were no burnt phytoliths from either trees or herbaceous taxa. Fifteen charcoal samples covering the area and depth of deposits sampled at the site yielded radiocarbon dates of between 1800 and 550 B.P. The dates clustered between 1300 and 1100 B.P. Piperno and Becker (1996) concluded that in the absence of any evidence for human interference with this forest, the charcoal probably represented formerly drier times in the Amazon Basin possibly associated with strong El Niño events, during which natural fires occurred. It is significant that the period between 1300 and 1100 B.P. was also the driest time during the past 8000 years in the Yucatan Peninsula, where it coincided with the collapse of the classic Maya civilization (Hodell *et al.*, 1995). Thus, the past 1500 radiocarbon years may have seen particularly strong, relatively brief climatic events felt in both Central and South America.

The results of this study indicated that this one tract of terra firme forest was never significantly altered by humans. This finding is in accord with our belief that, for the most part (see Balée, 1989, for possible exceptions), the interfluve Amazonian forest in the interior of Brazil experienced later and less intensive forms of exploitation than other forests discussed here.[4] A similar viewpoint has been expressed by Denevan (1992).

Venezuela and the Orinoco Basin

Roosevelt's (1980) studies at Parmana in the middle Orinoco Basin defined a cultural sequence that began approximately 4000 B.P. and extended into the contact period. During the initial La Gruta phase (4000–3000 B.P.), site refuses are marked by numerous budares, indicating an intensive manioc-based agriculture. Also present are very small, longitudinal stone chips similar to ethnographically known "grater chips," which are inserted into wooden platforms, glued with resin, and used to shred the bitter manioc pulp. During the subsequent Corozal phase, beginning at approximately 2700 B.P., carbonized maize and *Phaseolus* beans appear for the first time, along with manos and metates. Isotope studies of human bone from the site indicated that maize became a staple plant during late Corozal times.

The location and study of plant remains from earlier sites in the region is needed, especially because one of the possible hearths of manioc domestication is nearby (see Fig. 3.18). We also note that the lithic remains at this and other sites with the same kinds of changes do not necessarily indicate that maize arrived late into the region but possibly only that intensive maize agriculture was practiced later than an intensive manioc agriculture based on bitter varieties of this crop.

Near the Caribbean littoral of western Venezuela at a large village site called Rancho Peludo, a similar lithic sequence suggests that intensive maize cultivation followed the development of vegeculture with bitter manioc as the staple crop

[4] Other Holocene pollen sequences from the humid lowlands of the Amazon Basin have been published (e.g., Liu and Colinvaux, 1988; Frost, 1988). However, these studies were undertaken before palynological techniques were standardized, when substantially fewer taxa could be identified, and before the analysis of phytoliths and charcoal began to be carried out. For these reasons, we do not include them in our discussion, although their interpretations are in accord with our beliefs about the relative scale of forest modification in the Amazon Basin.

(Rouse and Cruxent, 1963). Manos and metates occur stratigraphically above budares, suggesting an intrusion of maize agriculture into an area where vegeculture was first established, as at La Gruta. The earliest levels at Rancho Peludo with the budares are radicarbon dated to 3800 B.P. Their replacement by manos and metates occurred by approximately 3000–2000 B.P. (Rouse and Cruxent, 1963). Although this kind of lithic sequence occurred in the western and central parts of Venezuela where Rancho Peludo and La Gruta are located, in eastern Venezuela budares continued to be common into the contact period era, suggesting that bitter manioc remained the staple food (Rouse and Cruxent, 1963).

OTHER AREAS IN COLOMBIA

The Caribbean littoral of the Colombian lowlands has yielded several sites with early fiber- and mineral-tempered ceramics dating from the late sixth to the fifth millennium B.P. (Fig. 5.2). They include the well-known occupations of Puerto Chacho, Puerto Hormiga, and Monsú (Reichel-Dolmatoff, 1985; see Rodríguez, 1995, for a recent review). The subsistence orientations of the groups who inhabited these sites are still poorly understood. Puerto Hormiga and Puerto Chaco are shell middens located on former estuaries where mangrove swamps were conspicuous, so a significant portion of marine protein in the diet is likely. Grinding stones and hammerstones for the preparation of plant foods were also abundant in the middens. Whether the populations were practicing food production and the degree to which wild and tended plant sources may have been exploited are unknown.

Monsú is not a shell midden. The faunal remains from the earliest occupation of the site, possibly dating to 5275 B.P., point to an emphasis on freshwater resources and include deer, peccary, armadillos, monkeys, turtles, cayman, and fish (Reichel-Dolmatoff, 1985). Gradually, resources taken from the sea and estuaries increase through time. The tool kit contained large numbers of grinding stones dating from the earliest occupation onwards and stone axes dating from at least 4200 B.P. Also, large hoes made out of *Strombus* shells were present beginning approximately 3200 B.P. These suggest root crop cultivation.

The most recent reviews of possible subsistence orientations at these early ceramic sites favor the notion that the earliest occupants were full-time hunters and gatherers who relied on marine and terrestrial animal and plant resources and who probably shifted their settlements seasonally (Rodríguez, 1995; Oyuela-Caycedo, 1996). Unfortunately, the sites were in large part excavated before modern techniques of plant recovery were developed. The question of plant subsistence should probably be deferred until such analyses can be undertaken, but we consider it unlikely that the occupants of the sites were not practicing horticulture and/or were not in contact with people practicing horticulture further inland.

Regarding the latter point, the Caribbean littoral Colombian sites may suggest analogies with settlement patterns during the fifth and fourth millennium B.P. in central Pacific Panama (discussed later), where people occupied coastal locations on a seasonal basis and maintained horticultural settlements located farther in the interior of the country during other parts of the year.

An intriguing site, called San Jacinto 1., in northern Colombia that has recently been excavated by Oyuela-Caycedo and Bonzani (Oyuela-Caycedo, 1993, 1995, 1996; Bonzani, 1997) may shine some light on this question. Located further inland in the upper Magdalena Valley approximately 125 km from Monsú in a current savanna environment, it contains the earliest known fiber-tempered pottery, with an associated ^{14}C date of 5940 B.P. Occupation at the site lasted until approximately 5300 B.P.

Numerous carbonized plant remains were recovered from San Jacinto 1. (SJ-1) (Bonzani, 1997). They are still undergoing analysis, but grass seeds constitute the majority of the plant remains. Various fruits are also represented in the plant assemblage. No macrobotanical tuber remains have been reported to date. Some remains of terrestrial fauna were recovered, although apparently in smaller quantities than the plant remains. Grinding and nutting stones were present in high number, and hearths and earthen ovens were conspicuous on the numerous living floors identified by the careful excavation of the site.

No exotic exchange items were identified at SJ-1. The excavators believe that the total territory exploited by the occupants of the site may have been no larger than approximately 10 km in radius. The excavators' interpretation of SJ-1 is that it was a special-purpose site repeatedly occupied by hunters and gatherers for several months every year beginning at the end of the rainy season, and that the economy had as its focus the collection, processing, and cooking of wild grass seeds.

The final assessment of subsistence must await the complete analysis of the plant remains. However, we consider Oyuela-Caycedo's (1996, p. 83) belief that "emphasis on harvesting of wild grass seeds for food was under way at the site before horticulture began" unlikely for a number of reasons. Food production was under way in Colombia long before SJ-1 was occupied. Furthermore, the intensive harvesting of wild grasses is a very expensive strategy (see Chapter 2), one unlikely to have been a subsistence focus when less costly methods of food procurement involving food production were almost certainly available to these groups.

Following the growth of ceramic-using populations exploiting estuarine and mangrove environments in northern Colombia during the sixth and fifth millennium B.P., a number of sites were either founded in the region or, as discussed previously for the site of Monsú, demonstrate continuing occupation at or near the earliest sites containing ceramics. Only 5 km south of San Jacinto 1., Oyuela-Caycedo excavated a somewhat later occupation called San Jacinto 2. Radiocarbon dates place the occupation between 4565 and 3505 B.P., although the excavator considers the latter date to be anomalously young. The cultural remains at San Jacinto 2. are said to be more elaborate than those at SJ-1. They include numerous metates, ground stone axes, and ground "nutcracker" stones. Some of the pottery is now sand tempered, as at later occupations at Puerto Chacho and Puerto Hormiga, and it bears technological and stylistic affiliations to the pottery from the latter two sites.

Finally, we mention the records from a site called Momil, located in the lower Magdalena Valley. Momil was a small village that was occupied later in time and in which the now-familiar pattern of manos and metates stratigraphically above budares occurs (Reichel-Dolmatoff, 1957). Occupation of Momil probably occurred first at 3000 B.P. and ended around the time of Christ. Simple polished stone axes occurred throughout the sequence.

In summary, it appears that an intensive form of agriculture based on bitter manioc spread rapidly from its hearth, probably located somewhere in the southern Venezuelan and Guianan regions, north to the tip of South America and south to eastern Brazil by approximately 4000–3600 B.P. We are aware that budares may have been used to process kinds of plant material other than manioc (e.g., DeBoer, 1975), and that small pieces of stone identified as grater chips may similarly have been used for other purposes. Nevertheless, the weight of the evidence leads us to conclude that the budares did indeed function as budares at least some of the time and indicate the presence of a bitter form of manioc. Starch grain analysis may eventually provide significant information regarding this question. In some areas, maize subsequently came to be the dominant crop plant, as previously discussed. In eastern Venezuela, the Guianas, and many areas of the eastern Brazilian Amazon, societies continued to rely on bitter manioc (Rouse and Cruxent, 1963).

Current evidence indicates that ceramics associated with bitter manioc processing occur no earlier than approximately 4000 B.P. near the hearth of manioc in South America, and that a sweet form of manioc was likely grown in Panama by 7000–5000 B.P. (see Chapter 4 and the following section). It appears that the cultivation and spread of sweet varieties of manioc using simpler processing methods may have occurred much earlier than those of the bitter varieties, as first suggested by Sauer (1950).

In the following section, we discuss southern Central America and Mesoamerica where, as in western South America, systematic archeological survey/excavation and botanical recovery combined with paleoecological studies have provided substantial information on the growth of early food production and the development of agricultural systems.

CENTRAL PACIFIC PANAMA

In Chapter 4, we concluded discussion of this region with two highly important and interrelated events in its history. One was the introduction of an apparently primitive species of maize and a sweet form of manioc into the Aguadulce rock shelter, located on the coastal plain in deciduous forest, that probably occurred shortly before 7000 B.P. and by 6000 B.P., respectively. The other was the initiation of slash-and-burn agriculture at approximately 7000 B.P. at La Yeguada, a lake near Aguadulce at 650 m above sea level (a.s.l.) in wetter but still highly seasonal forest (Figs. 5.6 and 5.7). Early maize was documented with phytoliths, starch grains, and pollen from archeological and lake sediments. Manioc was identified by its characteristic starch grains, which were isolated from the surface of a plant grinding stone base and an edge ground cobble. At the Aguadulce shelter, maize appeared to have undergone substantial morphological change shortly after it arrived in Panama, possibly increasing its productivity.

The first manifestations of early slash-and-burn agriculture in the La Yeguada records were unmistakeable and clear. Arboreal taxa of primary forest experienced marked declines, woody secondary taxa rose substantially in frequency, and particulate charcoal continued to be very common (Figs. 5.8 and 5.9). Also remember that the ca. 7000 B.P. introduction of maize into Panama was preceded by the appearance of leren (*Calathea allouia*), bottle gourd, and probably a domesticated squash. The root crop leren is more likely to be an introduction than the squash, which may be a local product of the wild species of squash recently discovered near the sites. Archeological and paleoecological data indicate that this earliest horticultural expression was largely confined to house gardens and/or bits of alluvium of minor streams. Occupants of the region then moved out of this simple "kitchen garden" horticulture and into a regular pattern of burning and clearing forests for larger scale field cultivation at approximately 7000 B.P., shortly after maize arrived.

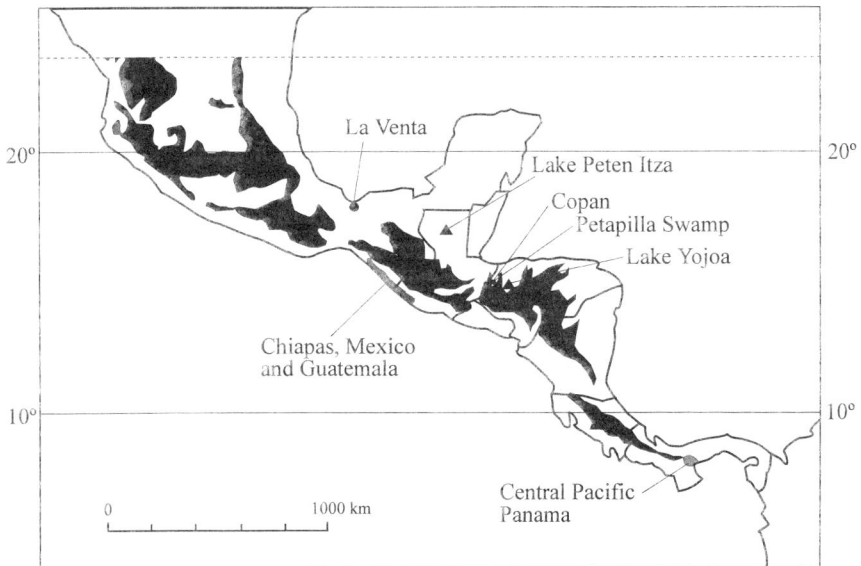

FIGURE 5.6   Map with the location of the sites in Middle and Central America discussed in the text.

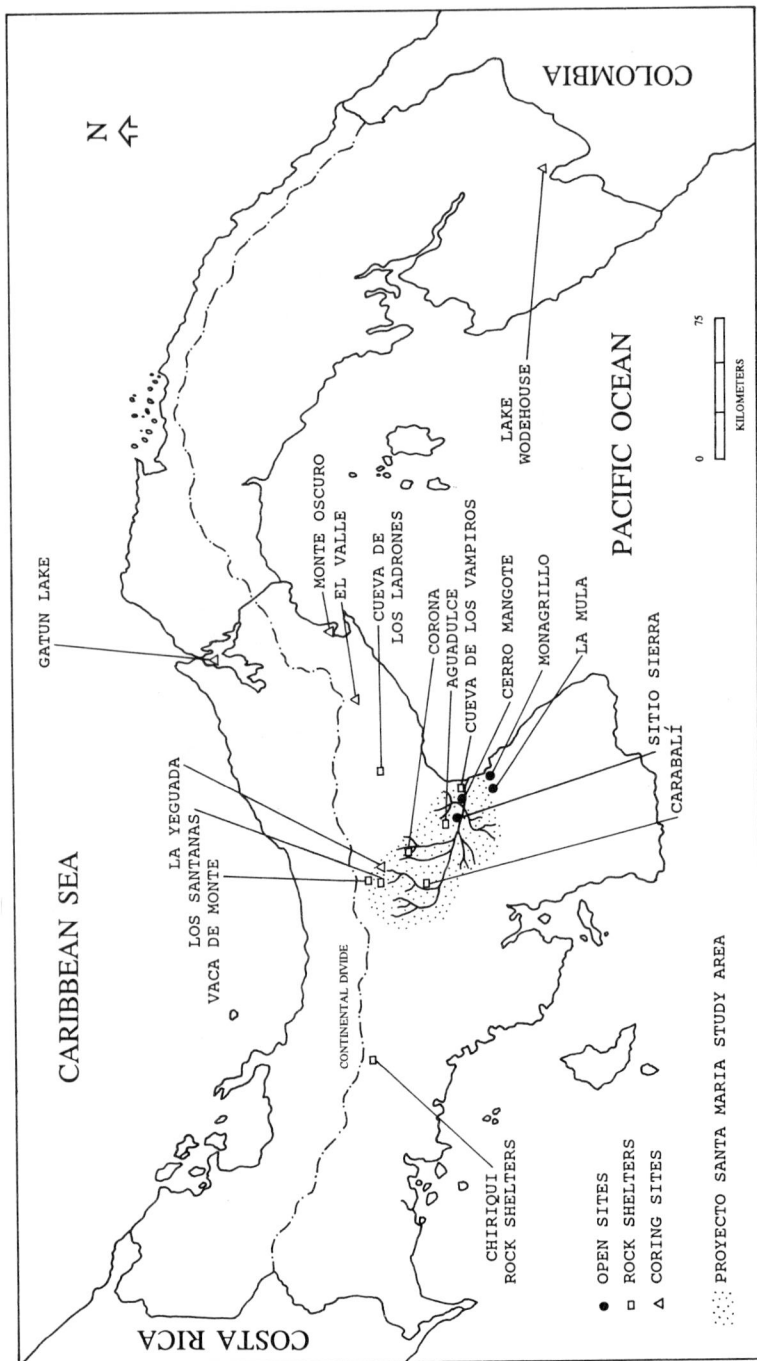

FIGURE 5.7 Map with the location of sites from Panama discussed in the text.

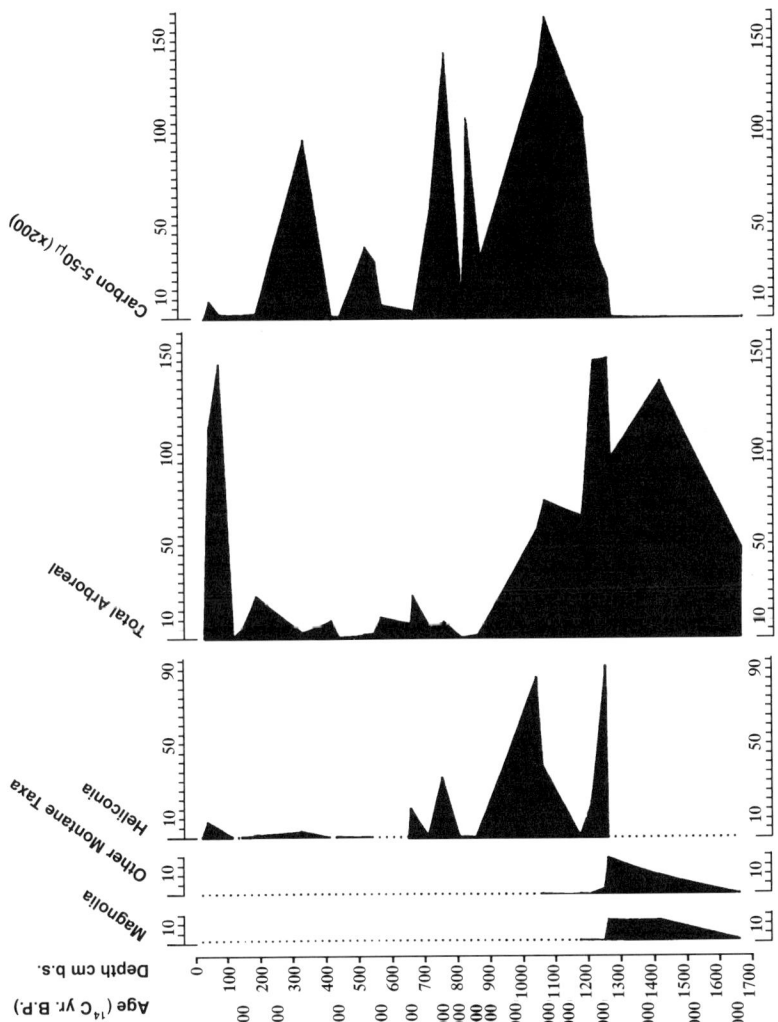

FIGURE 5.8 A summary phytolith and charcoal profile from Lake La Yeguada showing the decline of forest due to the onset and intensification of slash-and-burn agriculture through time. Phytolith and charcoal values are expressed in absolute rates of deposition ($20 \text{ cm}^{2-1} \text{ yr}^{-1}$). Total arboreal = the sum total of all arboreal and arboreal-associated phytoliths, which are largely from the primary forest. Carbon = charcoal

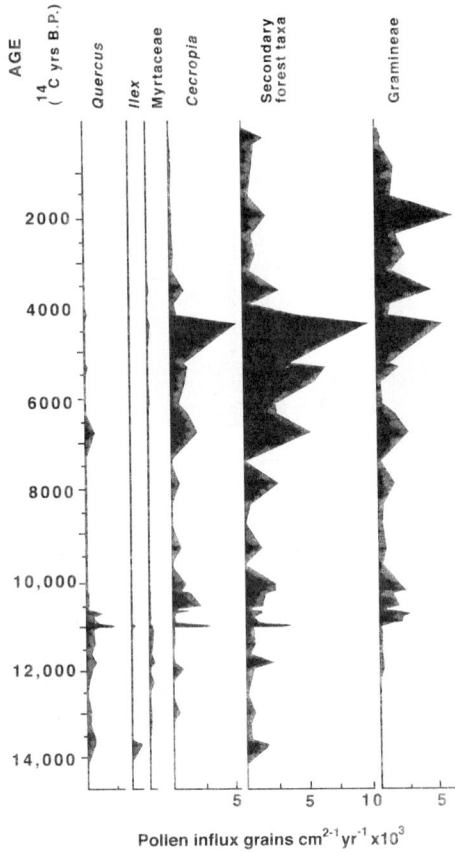

FIGURE 5.9   A summary pollen profile from Lake La Yeguada. Pollen analysis by Mark B. Bush, (See Piperno *et al.*, 1991b).

The Late Preceramic Period

In the cultural chronology of Panama, the time between 7000 and 5000 B.P. is the latest preceramic period (Period IIb; Table 5.3). During this time, slash-and-burn agriculture intensified in the Lake La Yeguada watershed (Piperno, 1995a, Piperno *et al.*, 1991b). Pollen from secondary forest trees (*Cecropia* and secondary forest taxa) becomes abundant between 6000 and 5000 B.P. Grass pollen also increases markedly, and phytoliths from primary forest trees remain in low frequencies (Figs. 5.8 and 5.9). Levels of burnt *Heliconia* and grass phytoliths reach the highest frequencies of the sequence during the 7000–5000 B.P. period, whereas burnt arboreal phytoliths continue to be common (Fig. 5.10). As discussed in Chapter 4, this intensification of slash-and-burn agriculture appears to have taken place during a climatic interval that was drier than today's. Such a drying would have increased the ease of burning and tree felling in this already highly seasonal environment.

During this time settlements continued to be small, generally less than 1 ha in size, and were situated on promontories overlooking streams or on interfluvial spurs. They suggest analogies with the organization of modern shifting cultivators—dispersed hamlet clusters of several families. The Proyecto Santa Maria settlement survey (Weiland, 1984; Cooke and Ranere, 1992a) indicated that there was a 15-fold increase in the number of sites occupied between 7000 and 5000 B.P. With little doubt, this was a result of an effective agricultural system, which used both major, domesticated seed and root crops. Growing numbers of people on the landscape, in turn, led to the intensification of the agricultural system.

Despite the loss of bifacial stone tool reduction techniques, other characteristics of the lithic inventories suggest considerable cultural continuity across the early preceramic–late preceramic boundary (Ranere and Cooke, 1996). Edge ground cobbles become the dominant plant-grinding implement represented in lithic assemblages from the sites, and they become more common after 7000 B.P. Other changes in plant subsistence at this time are evident in the botanical records from the archeological sites. Arrowroot begins to decline in importance at Aguadulce, Cueva de los Ladrones, and Vaca de Monte (Piperno, 1988a, pp. 188–193; 1995a). This probably occurred because it had serious competition from other suppliers of carbohydrates, such as more productive tuber crops such as manioc and maize. Sweet manioc would also require less processing than arrowroot.

Although few carbonized plant remains were recovered, macrobotanical data provide additional subsistence information for the late preceramic period. At the Aguadulce shelter, Corona, Vaca de Monte, and other sites in the region, a variety of tree fruits continued to be used. These included palms (*Elaeis,* the oil palm, and *Acrocomia mexicana*), arrowroot, nance (*Byrsonima*), *Hymenaea,* and hogplum (*Spondias*) (Cooke and Ranere, 1992a) (Table 5.3).

Bone isotope data from the burials at the preceramic coastal shellmound Cerro Mangote are available to assess the relative importance of maize and C_3 plants between 7000 and 5000 B.P. (Norr, 1989, 1995).[5] A mean bone collagen $\delta\ ^{13}C$ value of -13.7 from 18 individuals at the site is intermediate between C_3 and C_4 food chains. The $\delta\ ^{15}N$ values are low, indicating little consumption of protein from marine sources. The bone collagen results, then, indicate a predominantly terrestrial diet with consumption of significant amounts of C_3 plants but with input from C_4 plants, whose source is likely to be maize.

[5] Anomalously young AMS radiocarbon dates of $2630\ \pm\ 60$ and $2320\ \pm\ 50$ B.P. recently have been obtained on two skeletons from Cerro Mangote. They are rejected by the excavator, Anthony J. Ranere (personal communication, 1996), who notes that the midden is clearly late preceramic in age and that no ceramics have been found associated with the burials. Also, the burials are very similar to those at the preceramic Vegas site in Ecuador that date to the late preceramic period. We can only defer on this question until more data are obtained.

TABLE 5.3   Cultural Chronology and Plant Remains in Central American Regions

Cultural chronology	Dioscorea sp.	Maranta arundinacea (arrowroot)	Manihot esculenta (yuca)	Ipomoea batatas (sweet potato)	Cucurbita spp. (squash)
		Roots and tubers			Vegetables
Central Pacific Panama, Period IIb (late preceramic), 7000–5000 B.P.	SG	Phy[a], SG	SG		Phy
Period IIIa (early ceramic), 5000–3000/2500 B.P.		Phy[a]	SG	P	Phy
Period IIIb (middle ceramic), 2500 B.P.–A.D 230	SG	Phy[b]	SG		Phy, P
Chiapas, Mexico, Barra phase (early Formative), 3500–3300 B.P.					
Locona phase (early Formative), 3300–3100 B.P.					
Northern Belize, Swasey Phase (middle Formative), 3100–2000 B.P.			M		M

Note. M, macrobotanical; Phy, phytoliths, P, pollen; SG, starch grains.
[a] Arrowroot declines in importance.
[b] Arrowroot now virtually absent from records.

More precise isotopic evidence as to the source of calories and proteins in the diet is available from the bone apatite carbonate values from 12 individuals from Cerro Mangote (Norr, 1995). They indicate a diet with a protein source that was mainly terrestrial fauna and an energy source that was mainly C_4 and, thus, probably maize.

However, in light of the young AMS dates obtained on the human bone, we cannot conclude that people were drawing most of their calories from maize during the late preceramic period. A mixture of caloric sources from C_3 and C_4 plants at the time may be more likely. The fact that the isotope data point to a mainly terrestrial diet may also appear anomalous for this "coastal" site because considerable amounts of marine fish, shorebirds, crabs, mollusks, and racoons were recovered from the Cerro Mangote midden. However, isotope values for the protein sources are consistent with the other faunal remains from the site, which included deer, iguana, small reptiles, and birds, and that were present in quantities indicating they were also important in the diet (Cooke and Ranere, 1989). We note again that if the human bone dates are accurate, the isotope values may be reflecting much later subsistence strategies in the region.

Cultural chronology	Grains Zea mays (maize)	Legumes Phaseolus vulgaris (common bean)	Persea americana (avocado)	Tree fruits Palmae (palms)	Byrsonima (nance)	Spondias (hoplum)	Sapotaceae (sapotes)
Central Pacific Panama, Period IIb (late preceramic), 7000–5000 B.P.	Phy, P, SG			M, Phy	M		M
Period IIIa (early ceramic), 5000–3000/2500 B.P.	Phy, P, SG			M, Phy	M	M	
Period IIIb (middle ceramic), 2500 B.P.–A.D 230	Phy, M, P, SG	M		M, Phy			
Chiapas, Mexico, Barra phase (early Formative), 3500–3300 B.P.	M						
Locona phase (early Formative), 3300–3100 B.P.	M	M	M				
Northern Belize, Swasey Phase (middle Formative), 3100–2000 B.P.	M	M	M	M	M		

Regardless of how old the skeletons turn out to be, it is likely that Cerro Mangote represents a site that was occupied seasonally, perhaps during the dry season, by people who spent other parts of the year, most likely the wet season, concentrating on farming in more interior locations. Cerro Mangote is the only securely identified preceramic shell midden in central Panama. A heightened emphasis on coastal living with an increase in the number, size, and occupational density of sites, does not seem to have occurred until the onset of the ceramic period (Ranere, 1992).

The Early and Middle Ceramic Periods

The well-known Mongarillo ceramics (Willey and McGimsey, 1954), the earliest to appear in Central America, were first produced in Central Pacific Panama by 5000–4500 B.P. (Cooke, 1995). Cooke (1995) notes that although the idea of making pottery quite possibly spread to Panama from Colombia, Monagrillo ceramics are stylistically unique. Crop plants not recorded earlier are first in evidence during the early ceramic period (5000–3000 B.P.). At the Aguadulce shelter, pollen grains comparing favorably to sweet potato, plus maize pollen, are present in layer

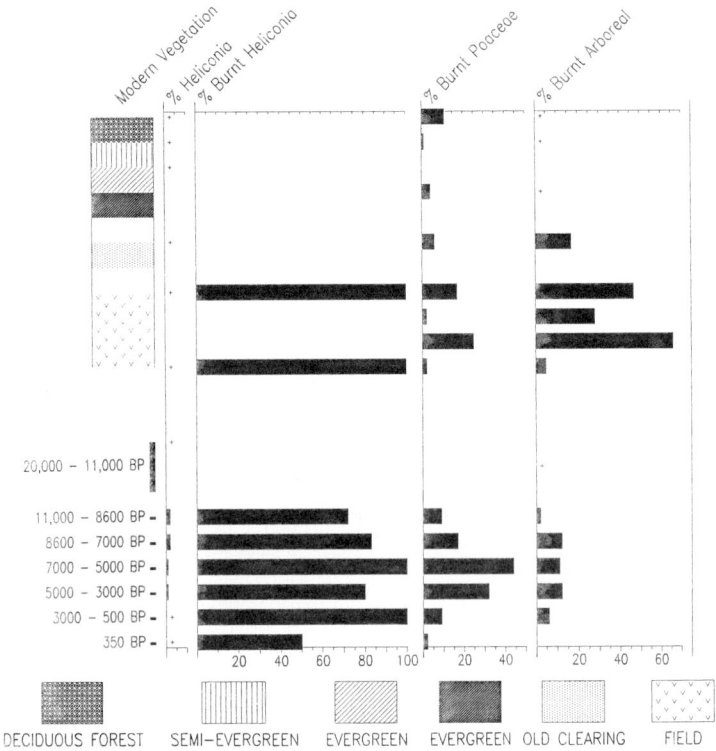

FIGURE 5.10   The frequencies of successional and burnt successional and arboreal phytoliths in modern forests and through time at Lake La Yeguada. For explanation, see Figure 4.4.

B1 containing early Monagrillo ceramics (J. Jones, personal communication, 1996). Associated phytoliths from this context produced a direct date of 4250 ± 60 B.P. Earlier sediments from Aguadulce contain little pollen.

An edge ground cobble recently recovered from just below the surface of the Monagrillo type-site that dates to between 5000 and 3300 B.P. contained the starch grains of manioc and maize on its grinding facet (Piperno and Holst, 1997).

A shift in settlement pattern appears to have commenced in the region during the early ceramic periods. Sites located on the coast become larger and more numerous. However, the sites do not appear to be permanently occupied, suggesting that the pattern of alternating settlement between the coast and interior zones continued (Cooke and Ranere, 1992b; Ranere, 1992). Nevertheless, this shift toward the coastline may represent the beginning of the major settlement transformation in the region, discussed below, whereby the earlier emphasis on foothill and interfluve coastal plain farming, represented in the La Yeguada and rockshelter records, was replaced by a focus on the alluvial bottomlands of the coastal plain by about 2000 B.P.

The likelihood of an early ceramic settlement shift is supported by its coincidence with another major boundary in the Lake La Yeguada record. At about 4200 B.P., frequencies both of particulate charcoal and pollen of woody secondary taxa decline sharply, and pollen and phytoliths from grasses increase substantially (Figs. 5.8 and 5.9). Meanwhile, phytoliths from primary trees remain in very low frequency. Saddle-shaped phytoliths from grasses, which are conspicuous components of present-day agricultural fields in Panama, attain their highest frequencies of the entire sequence. Even pollen of oak (*Quercus*), which had been lingering on the local upland peaks since the Pleistocene, is lost from the record.

These changes indicate a marked intensification of agriculture to a point where even secondary growth woody taxa became scarce on the landscape, accounting for the decline of charcoal. Fallow periods must have become shorter and shorter, and growing human populations were probably finding it more difficult to find productive land to farm, perhaps forcing some of them to seek more fertile localities nearer the coast.

Some of the vegetational records from archeological sites indicate that environmental modification was also accelerating on the coastal plain a few centuries before 4000 B.P. At the Aguadulce shelter, *Curatella americana*, a major indicator taxon for serious land degradation leading to the formation of anthropogenic savanna, first appears in sediments containing Monagrillo ceramics (Piperno, 1988a) (Plate 5.3). These sediments are 5 cm below a 5 cm level where the phytolith assemblage produced a direct date of 4250 ± 60 B.P. (Table 4.6).

Severe agricultural pressure on the landscape continued during the third millennium B.P. at La Yeguada. After 2000 B.P. at La Yeguada, agricultural pressure on the landscape appears to have ceased. Frequencies of grass pollen start to decline and never rise again. Maize is nearly lost from the record, being present in a single sample dating to approximately 1500 B.P.

The period between 3000 and 2000 B.P. was another crucial one for cultural development in the region. It is when an agricultural system based on a few productive crop plants, and with a more selective emphasis on maize, began to be practiced on a nonshifting basis on the alluvial soils of major coastal plain rivers and streams. As people abandoned hamlet living, the primary regional settlements became villages occupied by hundreds and up to a thousand individuals (Cooke and Ranere, 1992a,b). Tool technologies changed and involved greater effort in production. A ground and polished stone tool technology used for clearing the riverine forests appeared, suggesting that fire and simpler-made implements were now insufficient for the task in alluvial zones that have wetter soils and generally larger trees (Ranere, 1992). Manos and metates appeared and replaced edge ground cobbles as the dominant plant-grinding implements. Ceramics were better made and employed a wider range of shapes.

The first sedentary and nucleated village was founded by 2500 B.P. at the site of La Mula (Hansell, 1987). Starch grains found on a breadboard (legless) metate from the site and associated with a date of 2300 B.P. suggest that a variety of crop plants were still being prepared and consumed on a fairly regular basis. These plants include maize, manioc, some type of yam, and probably a *Calathea* species (Piperno and Holst, 1997). As with the tools from earlier time periods, *D. trifida* can be ruled out as the contributing yam species, suggesting that native species of yams were still being used in the subsistence system. By shortly after the time of Christ, villages were well-established along the alluvium of watercourses that cross the coastal plains. At one of these sites (Sitio Sierra), the presence for the first time in the region of numerous manos, table metates, and abundant macroremains of maize, plus isotope values for human bone, all indicate that maize was now the economic staple (Cooke and Ranere, 1992b; Norr, 1995).

Still, Briggs (1986) and Cooke and Ranere (1992b) believe that it was not until several hundred years later in the region, approximately AD 500–700—when "elite burials" are found—that status can first be described as being inherited and not achieved, and that some degree of "institutionalized social inequality" (Price, 1995, p. 142) was present. Starch grain analysis of a metate dating to approximately AD 700 from the site of Cerro Juan Diaz, located near La Mula (R. Cooke, personal communication, 1996), supports a dramatic narrowing of the subsistence base shortly after the time of Christ. The metate contains substantial (more than 20) numbers of maize, and only maize, starch grains, in contrast to lithics from earlier periods that typically yielded the remains of a variety of crop plants (Piperno and Holst, 1997).

In summary, the evidence from central Pacific Panama clearly indicates that swidden cultivation systems are sustainable for thousands of years in seasonal tropical forest growing on fertile soils and can support very substantial increases in population growth through time. Eventually, growing populations appear to have outstripped the productive potential of interfluve soils using slash-and-burn techniques and moved to alluvial bottomlands. This process took a long time (approximately 5000 years). With higher populations filling up circumscribed pieces of good agricultural land, denser and more permanent settlements, and increased probability of more irregular yields from highly productive but fewer crop plants, the processes leading to competition, social conflicts, and status-grabbing by a relatively few individuals, the hallmarks of "cultural complexity," were set in motion.

Other Sites in Panama

Data for the development of agriculture elsewhere in lowland Panama are available from two lake sequences to the east. In the Gatun (Chagres River) Basin near the present-day Panama Canal, where the potential vegetation is semi-evergreen tropical forest, slash-and-burn agriculture using maize was initiated by 5000 B.P.

(Piperno, 1985c; Fig. 5.7). This region receives an annual rainfall of 2600 mm
and has a long and marked dry season. Hence, it contains favorable areas for
niche expansions of crop plants through cutting and burning of vegetation. Such
expansions of maize, and probably other domesticated plants, appear to have taken
place under conditions drier than those of the present day. Pollen studies by Bartlett
and Barghoorn (1973) indicated that the development of slash-and-burn agriculture
in this region was coincident with a dry period that lasted from approximately
7300 to 4200 B.P. and that effectively expanded the available area of forests suitable
for such farming techniques.

There is little evidence from the Gatun Basin sequence of significant interference
with the tropical forest during Paleoindian times or for an early development of
food production and increasing intensification of forest modification over time, as
are present in the record from Lake La Yeguada. These findings are consistent
with our belief that the central Pacific watershed of Panama, encompassing the
Rio Santa Maria drainage and points nearby, was the nuclear area of food production
and cultural development in Panama. Maize may have started to expand into the
wetter forests of Panama before 5000 B.P.

By 3200 B.P. tree pollen nearly disappears from the Gatun Basin record (Bartlett
and Barghoorn, 1973). It must be noted that this event predates the appearance
of pottery in the region by more than 1000 years (Cooke *et al.,* 1996). Pollen
evidence signals the presence of manioc and sweet potato shortly after the time
of Christ, although, given the evidence from central Panama, it is likely that both
crops were incorporated earlier into the agricultural systems of the region.

Another paleoecological site that has been studied in Panama, Lake Wodehouse,
is situated in Darien province in wet but still seasonal forest near the Colombian
border. Here, maize pollen and phytoliths, together with evidence for deforestation
of the watershed, are present from the bottom of the core sediments, which are
dated to 4000 B.P. (Piperno 1994). Apparently, well-developed slash-and-burn
agricultural systems were present throughout the seasonally dry and low-elevation
zones of the Isthmus by this time.

MESOAMERICA

Honduras

In Honduras, data on the development of slash-and-burn cultivation and spread
of maize are available from pollen sequences from lake cores. Rue (1987, 1988)
examined the pollen records from a lake, Yojoa, and a peat bog, Petapilla, located
in western Honduras (Fig. 5.6). Petapilla is only 5 km from the archeological site of
Copan, a major Mayan urban center. Rue recovered earlier evidence for prehistoric
occupation of the region than had been revealed with archeological survey and
excavation, a finding that would become increasingly common as more paleoeco-

logical sequences were investigated in the lowland tropics. The presence of maize pollen along with other indicators of slash-and-burn agriculture, which were evidenced from the very beginning of the Yojoa sequence, ^{14}C dated to 4770 B.P., revealed the considerable antiquity of seed cropping and associated forest modification in the Maya area. These results also highlighted the important questions of Maya origins (extrusive or *in situ* development).

Rue's (1987, 1988) studies further demonstrated the significance of "paleoecological" research to archeological enquiry in the Maya area by shedding light on the cultural and demographic factors associated with the Maya collapse. The vegetational records from Petapilla, which start about 1000 B.P., showed that forest clearing, undertaken at least in part for maize agriculture (Abrams and Rue, 1988), was widespread around Copan at the time of the collapse. However, the records provided no evidence for forest regeneration that would be expected had the region been abandoned during the ninth century AD. It seemed that the rural lands outside of Copan continued to be occupied for several centuries after the collapse, a finding in accordance with the most recent archeological evidence (Paine and Freter, 1996). Rue's findings also support the idea that environmental degradation was an important factor in the collapse of the Classic Maya, at least around some Maya centers (e.g., Abrams and Rue, 1988; Paine and Freter, 1996).

Guatemala

An interesting new Holocene pollen record is available from Lake Peten Itza located in the Department of Peten, northern Guatemala (Islebe *et al.,* 1996; Fig. 5.6). The beginning of the sequence dates to the early Holocene, approximately 8800 B.P. During the first 3000 years of the lake's history no human disturbance is evident. A significant reduction of forest taxa and increase of grasses occurs just above a level of the core ^{14}C dated to 5600 ± 35 B.P. Also common at this time are taxa such as Melastomataceae, *Terminalia,* and members of the Burseraceae that might indicate considerable turnover of secondary forest, although these taxa could also find drier climates more to their liking. There is no maize pollen during the early forest reduction period. The investigators of the sequence note that maize pollen is heavy and very poorly dispersed, and possibly would not enter a lake record unless people were cultivating nearshore habitats (although cultivar and other pollen also regularly enter tropical lake records via stream flow, Piperno, 1993). Islebe and coinvestigators reasonably do not come down squarely on the side of either climatic change or human interference to account for the vegetational changes at 5600 B.P.

Modern pollen spectra from seasonally dry tropical forest in Costa Rica and Panama little disturbed by humans demonstrate significantly lower percentages of grasses than present in Islebe's diagrams at 5600 B.P. (Bush, 1991; Rodgers and Horn, 1996). This is one line of evidence in support of the notion that the vegetational changes at Lake Peten Itza were culturally induced. On the other hand, herbaceous taxa decrease and forest taxa increase again soon after their initial perturbations. This is a finding unlike every other sequence discussed here in which, after slash-and-burn cultivation is initiated, the system either intensifies, resulting in further loss of arboreal taxa, or proceeds unabated for a long time. Fluctuations such as those in Islebe's core may therefore be more in accord with short cycles of climatic drying. Charcoal and phytolith data would be a great help in resolving this issue.

After these fluctuations, the pollen record of an undated part of the core, but that clearly antedates 1880 B.P., shows a persistent decline of arboreal taxa and the presence of high numbers of sucessional, disturbance kinds of plant taxa. Shortly after, pollen spectra signal almost a complete removal of the forest. Arboreal taxa descend to extremely low levels, and indicators of treeless landscapes in the Maya area (*Ambrosia* and the Compositae) reach very high frequencies. *Zea* pollen is not only present but also common (2–4% of the pollen sum), probably indicating that riparian habitats were being used for planting maize. On the basis of age-depth regression lines and associations with similar changes recorded in other lakes of the region, Islebe and coworkers date this episode to 1880 B.P., the beginning of the Classic Maya period.

Belize

Exciting new information relating to the development of agriculture in the Maya lowlands is being uncovered in northern Belize. Here, allied archeological and paleoecological efforts are revealing what appear to be severe agricultural impacts on the landscape associated with a much earlier cultural occupation than had heretofore been recognized.

John Jones (1991, 1994) carried out the first studies to demonstrate such changes at a site called Cobweb Swamp. This large swamp is located next to the archeological site Colha, which is famous as a specialized lithic production site during Mayan times (Fig. 5.11) (Hester *et al.*, 1982). Jones' palynological investigations of the swamp sediments revealed that manioc pollen was present in an excavation unit dug at the site associated with a ^{14}C date of 4591 B.P. on worked wood. The size (greater than 200 μm) and morphological characteristics of the pollen indicate a domesticated and not a wild manioc. The stratum containing the manioc was the lowermost "field" unit of the site and is an easily identified horizon located stratigraphically below a moderately thick unit of sterile fill. It was determined elsewhere at Cobweb Swamp that this horizon predates 4400 B.P.

In a sediment core taken through Cobweb Swamp, a dramatic period of forest clearing associated with both maize and manioc pollen begins at approximately 4000 B.P. (Fig. 5.12). Dryland arboreal taxa (the Moraceae) decrease dramatically at this time and are replaced by high numbers of invasive, herbaceous taxa character-

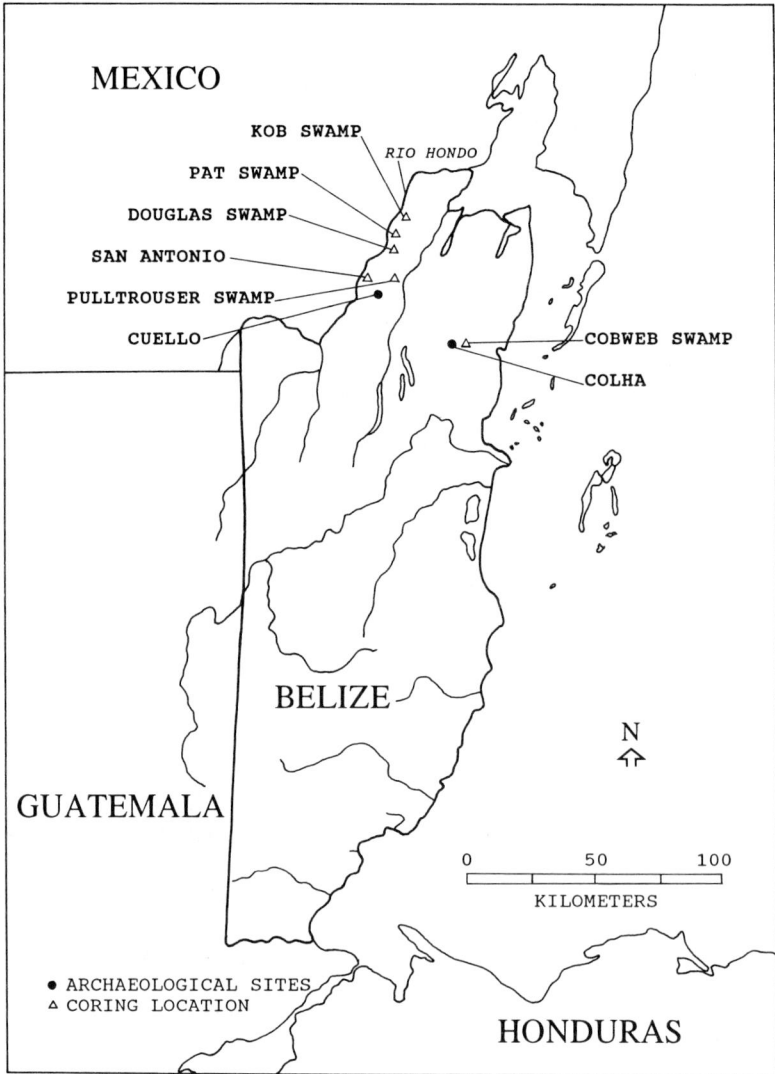

FIGURE 5.11   Map of Belize showing the locations of the sites discussed in the text. Modified from Jones (1994).

istic of agricultural fields in this area (Cheno-Ams). Levels of charcoal also attain high frequencies. These manifestations indicate that people were practicing intensive slash-and-burn agriculture in the semi-evergreen forests of the swamp's catchment area.

Jones also identified cotton and chile pepper at Cobweb Swamp associated with contexts dated to between 4000 and 3000 B.P. Interestingly, *Brosimum* (ramón), a

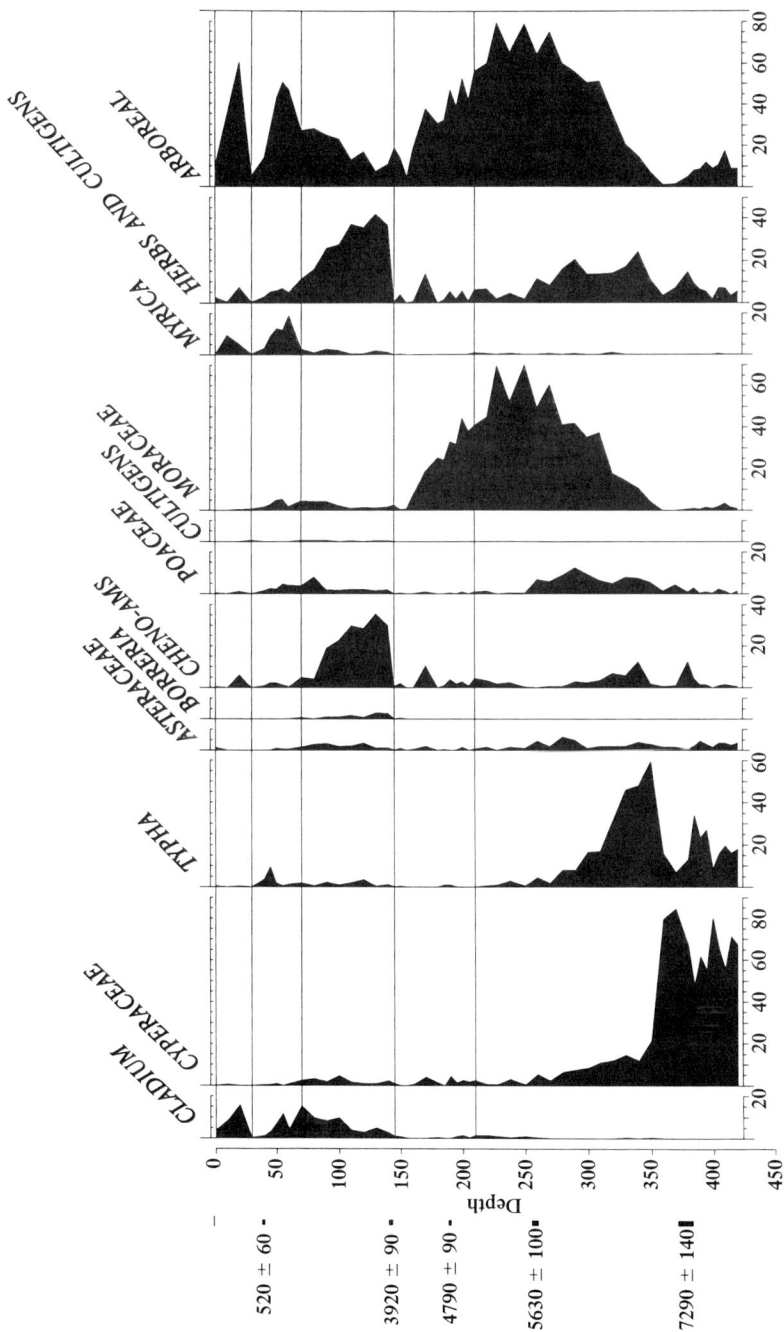

FIGURE 5.12 A summary pollen profile from Cobweb Swamp. Cultigens are maize, cotton, and chile pepper. Reproduced with permission from Jones, (1991).

tree whose fruits were postulated to have been a major source of food during Mayan times (Puleston and Puleston, 1971), experienced the same catastrophic decline as other noneconomic arboreal taxa at the beginning of the deforestation episode and did not subsequently increase artifically due to the selective removal of noneconomic trees as did other arboreal taxa (Sapotaceae and Palmae) at Cobweb Swamp (Fig. 5.12). These factors strongly suggest that it was not a primary food plant.

Although the bottom of the deposits at Cobweb predated 7290 B.P., the changes beginning during the middle fifth millennium B.P. appeared to be the first significant food production and forest disturbance occurring in the area. The pollen spectra from the bottom of the Cobweb core appear to represent an open, park-like savanna under a drier climate. In this region, the establishment of a high tropical forest may have occurred after 7300 years ago.

Recent interdisciplinary efforts directed by Mary Pohl, Kevin Pope, and John Jones have also focused on a program of excavation and coring of freshwater wetlands northwest of Cobweb Swamp, some of which had been used as fields and modified by ditching by prehistoric occupants (Pohl et al., 1990, 1996). Sites investigated included Pulltrouser, Kob (also called Cob), Pat, and Douglas Swamps (Fig. 5.11). Pulltrouser Swamp had previously been investigated by Turner and Harrison (1983), who believed that major mounding and modification at the site had been carried out by Classic Maya agriculturalists.

Pohl's team coring program penetrated deposits as old as 7200 B.P. The first indications of cultivars at the sites occur at 4750 ± 60 B.P., when manioc pollen is first recorded in a sediment core from Kob Swamp (Fig. 5.13). This is remarkably close to the earliest date registered for this domesticate at Cobweb Swamp. The manioc in the Kob swamp core is followed by the appearance of maize by approximately 4200 B.P. Maize pollen appears at 4610 B.P. in the Kob Swamp excavation unit near the coring locality. The pollen studies, carried out by John Jones, indicate that when manioc is first documented, the regional vegetation was still a high tropical forest with only minimal levels of disturbance evident in the pollen spectra (Fig. 5.13).

Then, coincident with the appearance of maize in the sediment core, the pollen data record an abrupt crash of forest species and dramatic increase of disturbance taxa, such as Cheno-Ams and the Compositae (Asteraceae). Secondary growth arboreal taxa (Trema) also increase. A sudden, major increase of charcoal fragments occurs at this time, indicating forest clearing using slash-and-burn techniques. Pohl et al. (1996) note that it is difficult to say whether deforestation was a regional or local event. On this matter, it is significant that the cores further south at Cobweb Swamp previously discussed showed maize and major forest clearing beginning at about the same time. The abrupt and marked increase of maize, disturbance plants, and charcoal at different sites suggest a swift spread of maize slash-and-burn agriculture through northern Belize.

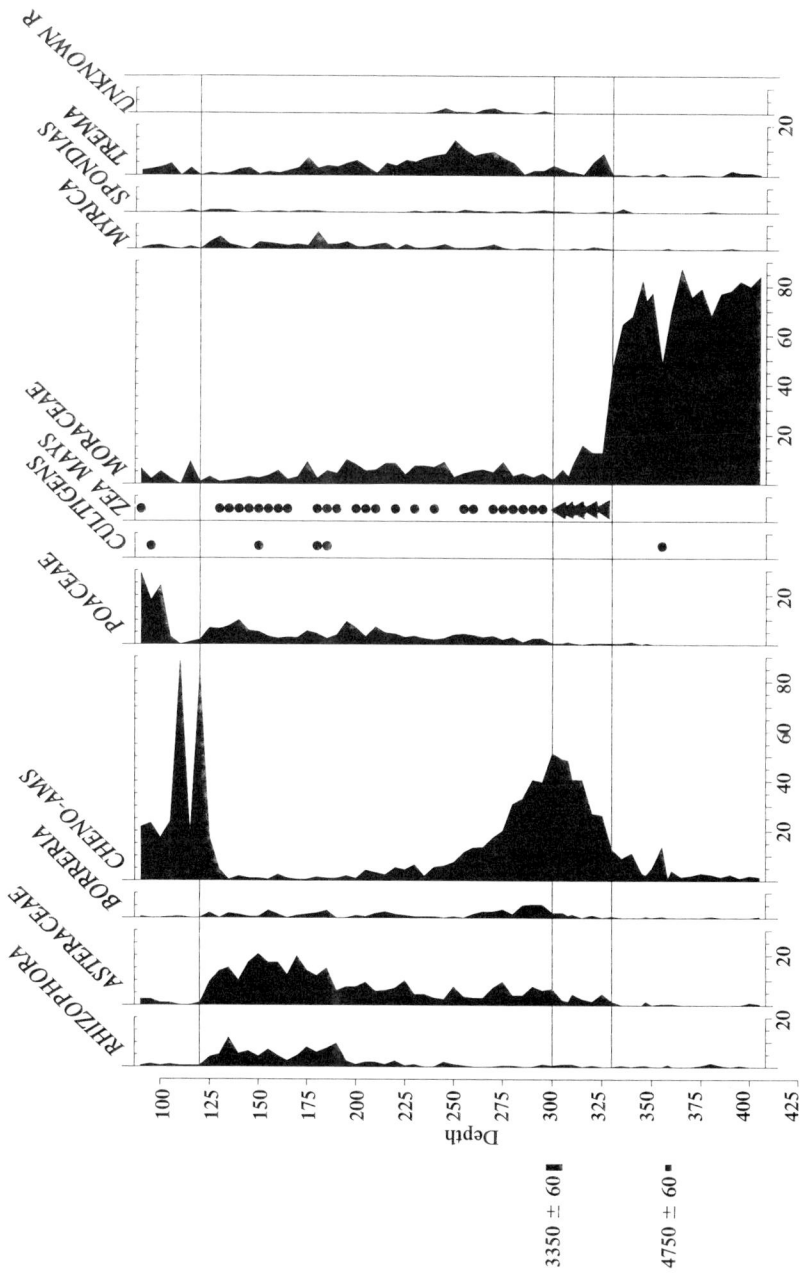

FIGURE 5.13 A summary pollen diagram from Kob Swamp. ▲, Old or Archaic (pre-Maya) maize pollen; ●, modern-type or Mayan maize. Cultigens are manioc and cotton pollen.

Whether this spread involved colonization of the region by new peoples or a rapid adoption of a new agricultural system by poeple already in place is difficult to determine. The Belize paleoecological profiles certainly display marked differences from those at Lake La Yeguada in Panama. There, significant forest disturbance using fire and involving marked increases of successional herbaceous taxa was occurring before 8600 B.P., and slash-and-burn agriculture with maize was evident from the seventh millennium B.P. onwards (for a comparison of these sequences, see Fig. 5.14). That an *in situ* intensification of the food production system over time does not appear to be evident in the Belize sequences suggests to us that new populations carrying advanced agricultural techniques may have moved into northern Belize. As in central Pacific Panama, the fertile upland soils of northern Belize endured thousands of years of intensive cultivation once swidden agriculture was initiated.

Jones' pollen data also suggest that maize may have been subjected to further experimentation and change by Belizean cultivators. The earliest maize pollen from northern Belize is relatively small and has much thicker exines compared with pollen from later, middle Formative times (Jones, 1996). Similar differences between early and middle Formative maize grains are noted by Rust and Leyden (1994), who studied the pollen record from La Venta in Tabasco, Mexico (discussed later). Whether these differences reflect an earlier presence of a more primitive variety of maize replaced by something more advanced and productive remains to be determined. This proposition would fit with emerging evidence that maize underwent substantial genetic and morphological change in various regions over a protracted period of time after it initially dispersed out of southwestern Mexico.

Pohl's team's studies are also generating important information about the land-use patterns associated with Preclassic and Classic Maya agriculture. The initial propositions about wetland use in the Maya lowlands stated that the transformation of them for agricultural purposes was widespread. Also, this transformation included the large area of karstic depressions or bajos in the Peten, Guatemala, which was the center of the Classic Maya culture (e.g., Adams, 1980; Adams *et al.,* 1981). Furthermore, it was thought that modifications of wetlands for agriculture did not begin until the late Preclassic period and involved transport of soil from the uplands to enrich the soils and build planting platforms (Turner and Harrison, 1983).

Two important implications followed from these views. First, as wetland cultivation assumed more prominence in agricultural production, slash-and-burn agriculture declined in stature as a likely contributor to the Maya economic base, especially for the burgeoning populations of the Classic period (Harrison and Turner, 1978). Second, the intensive labor postulated to have been associated with the movement of tons of upland earth into the swamps was seen as encouraging the development of corporate social and/or religious structures.

Bloom *et al.* (1983) first challenged the assumption that wetland agriculture was the preeminent factor in Maya agricultural production. They found that in a swamp in northern Belize along the Rio Hondo, called San Antonio, wetland cultivation

YEARS B.P.	PANAMA Lake La Yeguada	Gatun Lake	Lake Wodehouse	HONDURAS Lake Yojoa	GUATEMALA Lake Peten-Itza	BELIZE Cobweb Swamp	Kob Swamp
	Forest Regrowth		Forest Regrowth	Loss of Zea	Forest Regrowth after 950 B.P.	No forest recovery	No forest recovery
	--- 350 ---		--- 500 ---		---- 950 ----		
1000	Agricultural decline	Intensive S+B			Intensive agriculture		
2000	Intensive S+B		SLASH AND BURN	Agriculture intensifies	1850 ZP		
	Decline and disappearance						
3000	of secondary woody taxa	--3300 ZP--		SLASH AND BURN		Intensive S+B	Intensive S+B
					S+B (?)		
4000	--- 4300 ---	SLASH AND BURN	ZP, Phy BASE			ZP	ZP
		4850 ZPhy		-- 4770 ZP-- BASE		4591 MP	4750 MP
5000	S+B initiated in the watershed					Semi-evergreen forest. No disturbance recorded	
6000					---- 5760 ----		Undisturbed Semi-evergreen forest
7000	--- 6700 ZP---	Undisturbed Semi-evergreen forest.				--- 7290 ---	
	Small horticultural clearings in forest are				Undisturbed Semi-evergreen forest		BASE UNDATED
8000	indicated					Open Parkland	
9000	--- 8600 ---				---- 8840 ---- BASE	BASE UNDATED	
	Initiation and progressive intensification						
10000	of human forest disturbance						
11000	--- 11000 ---	--- 11300 ---					
12000	Forest with high percentage of montane elements						
13000							
14000	BASE						

FIGURE 5.14   Summary of pollen and phytolith data from various paleoecological sites from Middle and Central America discussed in the text. P, pollen: Phy, phytolith: Z and M, first appearance of Zea and manioc in pollen (P) and/or phytolith (Phy) records; S+B, slash-and-burn agriculture.

could be documented during the Preclassic but not during the Classic Maya period and required less labor than previously thought because it did not involve the construction of artificial platforms. They concluded that Preclassic Maya agricultural systems probably employed a combination of wet season upland (dryland) swiddening and dry season wetland farming.

Pope and Dahlin (1989) extended these findings to a much broader area. Through use of sophisticated satellite radar imagery and field research, they determined that wetland agriculture probably did not take place in the central Maya lowlands, where the seasonal water table variation was too extreme to support field complexes. Instead, it was primarily confined to the river margins of three regions, including northern Belize, southern Quintana Roo, Mexico, and along the upper Candelaria River in Campeche, Mexico. They also suggested that the wetlands they confirmed as fields may have been cultivated at an earlier date than was commonly believed.

Pohl's team (1990, 1996) subsequently established that in the northern Belize sites discussed previously, use of wetlands by farmers began approximately, 3400 B.P., much earlier than thought, when water table levels stabilized upon a temporary cessation of sea level rise. Excavations revealed that canals were not dug during these earliest periods of cultivation. Thus, swamps were probably first utilized during the dry season when the water table was at its lowest, and they may have formed an important complement to wet-season farming that was occurring in the upland forest.

Ditching or canal construction began at some swamps by 2900 B.P., at the beginning of the middle Preclassic (Formative) period, as a response to the return of rising water tables in the region. The continued rise of groundwater levels caused the submergence and abandonment of wetland fields during the late Preclassic period (approximately 2300 B.P.). The soil deposit at Pulltrouser Swamp, originally thought by Turner and Harrison (1983) to represent upland dirt transported by the Classic Maya for the construction of artificial platforms, postdates agricultural use of the fields. This deposit actually appears to be a combination of soils deposited naturally through a combination of the aggradation of river sediment injected by rising water levels, slope wash from upland soil erosion, and gypsum accumulation within the archeological sediments. The slope wash is probably a result of the intensification of upland farming during the Classic Maya period (Pohl *et al.*, 1996).

Thus, the agricultural techniques employed at the wetlands are more properly described as "flood recessional" than "raised field" agriculture. Some swamps investigated, including Pulltrouser, showed no evidence for ditching, indicating that much wetland agriculture involved simple modifications of areas on the margins of swamps. Also, it appears that the most important crop in the wetland fields was maize. Maize pollen is ubiquitous at all of the fields investigated, and at one site along the Rio Hondo in northern Belize carbonized maize stems comprised more than 60% of the macrobotanical remains (Pohl *et al.*, 1990).

Pohl *et al.* (1990) also considered that the year-round agricultural production and likely resulting surplus made possible by dry-season cultivation during Preclassic times stimulated political competition and allowed the establishment of an elite class with power.

Initially there was considerable reluctance about accepting a revised chronology and function for the wetland fields for two reasons. First, archeological sites dating to the late Archaic and early Preclassic periods (ca. 4400–3000 B.P.) had not been found near wetland areas. Archeologists not familiar or comfortable with the potential of paleoecology to reveal important information on human behavior wanted archeological data to confirm Jones' and Pohl's results. Second, and para-digmatically more important, is that slash-and-burn agriculture has traditionally been viewed as an inefficient method of food production. Scholars have considered it unlikely that it could have fueled the very high population densities recorded around Classic Maya sites (e.g., Harrison and Turner, 1978).

Consequent to Jones' (1991) work at Cobweb Swamp indicating that people living near the swamp practiced agriculture associated with dramatic forest clearing from 4200 B.P., Hester's team reexcavated Colha. This time, they established the presence of a preceramic occupation (Hester *et al.*, 1993; Iceland *et al.*, 1995). Associated with this early settlement is a distinctive tool called a "constricted uniface," probably used to fell trees and/or ditch soils.

Pohl's teams' excavations in the other swamps investigated also revealed late Archaic-age cultural materials, including manos and metates, biface axes, and Lowe projectile points. As it turns out, agricultural settlements at the edges of wetlands during the late fifth millennium B.P. may have been common. These findings made it even more clear that early and important preceramic occupations in the lowland tropics may often not be easily found using traditional archeological methods and emphasized that sites with specialized types of production should not be excluded as having been important loci of food production.

The results of Pohl's team bring slash-and-burn cultivation back into the fore-front as a primary subsistence mode that needs to be understood in considering cultural evolution and the fueling of population increase and very high population densities during the late Archaic and Maya periods. This seems particularly true in the Classic Maya heartland of northern Guatemala, where wetland agriculture may never have provided the agricultural product that originally seemed possible. Early slash-and-burn cultivation subsequently underwent intensification that culmi-nated in widespread soil erosion and, possibly, a landscape increasingly less capable of supporting large populations by Classic Mayan times (Deevey *et al.*, 1979; Islebe *et al.*, 1996; Vaughn *et al.*, 1985).

The Later Formative Period in Belize

If, in discussing the Maya, it seems we have gotten ahead of ourselves or exceeded the stated scope of this work, we felt that a consideration of the evolution of the agricultural techniques that supported growing populations and increasing social complexity through time in the Maya areas was important. As mentioned, the cultural affinity of the peoples who first cleared the forests near the northern Belize swamps is difficult to determine. They may have been Maya precursors, perhaps arriving from the Guatemalan highlands as suggested by evidence from glottochronology (discussed by Pohl *et al.,* 1996). The initial practitioners of intensive slash-and-burn agriculture may have arisen *in situ,* to be displaced by later arriving Maya foreigners. We feel this possibility is less likely, given the low to no amounts of forest modification in the region before slash-and-burn agriculture was initiated. Some archeologists believe that the Gulf coast of Mexico, whose Olmec culture is beginning to be better understood by archeologists, may have contributed the donor populations to northern Belize (J. Jones, personal communication, 1996).

During the middle Formative period, more substantial archeological evidence for occupations that are clearly ancestral to the splendid Classic Maya culture are present. The well-known site of Cuello in northern Belize is perhaps the best, and best-studied, example (Hammond, 1991; Hammond *et al.,* 1995). People first settled here approximately 3100 B.P., during the Swasey phase. They lived in settled villages, made technically proficient pottery, and grew corn, beans, manioc, and, no doubt, a large number of other domesticated crop plants. These and other villages of Maya forbearers flourished in Belize between 3000 and 1800 B.P.

OTHER AREAS OF MESOAMERICA

Coastal Chiapas, Mexico

As in Belize and Guatemala, archeological expressions of settlement and subsistence prior to the late Archaic period are extremely limited. Studies carried out recently along the Pacific coast of Chiapas, Mexico, have greatly increased knowledge of the late Archaic and early Formative periods in these regions (Blake *et al.,* 1992a,b, 1995; Kennett and Voorhies, 1996). The beginning of the cultural sequence, called the "Chantuto A" phase, is represented by one large shellmound site, Cerro de las Conchas, which was occupied between ca. 5700 and 4400 B.P. Subsistence at this time is very poorly known, although marine resources were clearly heavily used. Occupants of these and later Archaic-period sites are thought to have been shifting their locations seasonally and living in large residential camps in the more interior coastal plain localities of the region for part of the year. During the subsequent Chantuto B phase (4600–ca. 3700 B.P.), known occupations are still largely shell middens near estuaries.

Few artifacts belonging to the Chantuto phases were recovered, and cultigens are unknown. Stable carbon isotopes of remains from two Chantuto B individuals indicate high consumption of C_4 plants, supporting the possibility that camps dedicated to agricultural production were occupied inland on a seasonal basis. Kennett and Voorhies (1996) documented changes in the season of shellfish harvesting during the late Archaic period and argued that such changes were the result of scheduling conflicts brought on by the initiation of maize agriculture. Blake, Voorhies, and colleagues note that the degree of dependence on wild vs domesticated resources during Chantuto times is unknown. Given the time depth to known productive agriculture in Belize, we believe it is likely that Chantuto societies were growing a range of domesticated plants and, perhaps, placing considerable reliance on them.

The Formative period is considered to have begun in the region during the Barra phase (ca. 3500–3300 B.P.), when village life was first established. Ceramics are present for the first time. Some of them were imitations of bottle gourds, suggesting these were grown. A few charred maize kernels were recovered from Barra sites but preservation of macroremains appears to be poor and little else has been identified to date. By the Locona phase (3300–3100 B.P.), several large villages were present. Some of these had fairly substantial architecture tied to considerable labor investments, along with elaborate ceramics and figurines. Site number and size indicate that Locona-phase population may have been two or three times greater than that of the Barra phase. Macrobotanical records contained small amounts of maize, beans, and avocado. Blake et al., (1992a,b) believes that this period marks the transition from fairly egalitarian societies to chiefdoms in the region.

Blake et al. (1992a,b) considered whether nonagricultural, especially marine resources, supported the growth of population and social complexity. Stable carbon isotopes of skeletal remains from the oldest to the youngest of the cultural phases indicated little maize consumption during the early Formative period; less, in fact, than during the late Archaic. These results have been questioned by Ambrose and Norr (1992) on the grounds that insufficient collagen was present in the skeletal populations to support isotopic analysis.

We believe that little reliance on agriculture after 5000–4000 B.P., and especially when village life was established, is unlikely given the advanced development of agriculture in other areas of Mesoamerica such as Belize by this time. Also, the coastal Chiapas Formative peoples were obviously interacting through trade and other mechanisms with societies elsewhere in Mesoamerica that may have been practicing food production.

After the Locona phase, the trend of increasing sociopolitical complexity in the region accelerates. By approximately 3000 B.P., large regional centers with monumental architecture are present. In summary, there are many unanswered questions about the origins and growth of food-producing economies in lowland Mesoamerica. These may, in part, be revealed by ongoing survey and excavation by Voorhies, Blake, and other investigators.

The Mexican Gulf Coast

We end our discussion of Mesoamerica with the Olmec culture of the south Gulf coast of Mexico. Olmec society was one of the few primary civilizations known in the New World. That it occurred in a wet, lowland tropical forest makes it all the more interesting to archeologists. What was known about the Olmec was, until recently, restricted to the age and distribution of ceramics, art styles, and monumental architecture (Sharer and Grove, 1989). However, significant information has been revealed recently concerning the early settlement patterns and subsistence base of these people and their relationships to the emergence of cultural complexity (e.g., Rust and Sharer, 1988; Rust and Leyden, 1994).

The site of La Venta was a major Olmec center and one of two (San Lorenzo being the other) that has seen considerable archeological study (Fig. 5.6). Analysis of botanical remains from deep test pits at La Venta revealed the presence of *Zea* pollen and charred maize cobs in strata dating to ca. 4200–3800 B.P. (Rust and Leyden, 1994). At this time, called the early Bari period, people were making agricultural clearings in mangrove vegetation near their sedentary settlements located on the river levees, and they were exploiting the rich aquatic resources of the mangrove swamps and coastal rivers. The maize pollen grains of this period are small, suggesting a small-eared variety of maize.

During the subsequent middle and late Bari periods (3700–3000 B.P.), macromaize remains become much more common. Their morphologies indicate a 10- to 14-rowed race of maize with tiny ears and kernels. The cobs bear overall similarities to the modern race Argentine popcorn. Rust and Leyden (1994) suggest that this small-eared popcorn may have evolved as an adaptation to a perennially wet environment. They note that smaller plants with tiny ears would be less susceptible to damage from mold-producing dampness.

After 3000 B.P., there appears to be a dramatic increase in maize use, as indicated by the recovery of substantially more carbonized maize remains from archeological sites, and settlement expands in the area. Thereafter, social complexity increases to culminate in the Classic-period Olmec civilization. Much like Pohl *et al.* (1990, 1996), Rust and Sharer (1988) and Rust and Leyden (1994) suggest that population growth resulting from productive agriculture, plus competition for limited amounts of good agricultural land, may have fueled the development of social stratification and complexity in the area.

SUMMARY AND DISCUSSION

Cultural Continuities and Intensification of Food Production

A number of substantial themes and trends have emerged from the records presented here. First, it is worth repeating that considerable cultural continuity is evident or very possible between the early and middle Holocene periods in at least five regions where early food production occurred; southwestern Ecuador, the middle Cauca Valley in Colombia, Amazonian Colombia, northwest Peru, and central Pacific

Panama. In each of these areas, the practice of food production was initiated by 10,000–8000 B.P., and subsequently new crop plants were added and the food production system was intensified in different ways. In central Pacific Panama, the middle Cauca Valley, and Amazonian Colombia, swidden cultivation developed.

Although this was the dominant pattern, there were regional differences. In coastal Ecuador, following the small-scale horticulture of the Vegas period, the Valdivia peoples chose to organize themselves in sedentary villages and farm the bottomlands of river valleys. It seems that only after the river valleys were filled with settlements did slash-and-burn agriculture of interfluve forest take place. Whether the post-Vegas developments in this region represent an incursion of Valdivia-culture people from the Guayas Basin to the east, who had already moved from a shifting to a permanent mode of planting, or some particularistic feature of the ecology of southwest Ecuador is unclear. We believe that these independent trends for regional continuity, and the expansion and intensification of food production during the late-early to middle Holocene, are difficult to explain in the absence of an early Holocene development or acceptance of food production in the regions in question.

Elaboration of the social sphere as food production intensified and developed into more productive systems is also apparent in many of these regions. This would be expected if significant social pressures and constraints on subsistence largely followed, rather than preceded, the practice and intensification of food production. What can be called "big men" societies, with some degree of social stratification and ascribed stature from birth, are not evident in many records until shortly after the time of Christ.

The Paleoecological Record and Early Archeological Sites

Another important feature of the records we discussed is that paleoecological evidence for human occupation of some regions, such as Honduras and Amazonian Ecuador, predates available archeological evidence. Apparently, the environmental correlates of early foraging and farming in tropical forest are more visible than the stones, wood, and other implements used by people in their daily lives. Many regions have not been studied with a systematic archeological survey, so the antiquity of settlement typically is based on the excavation of one or two large sites, which reached their zenith later in time. It is the case, however, that an efficient and inexpensive method of archeological "survey" in the lowland tropics may be to core available lakes and swamps in areas of investigation and look for evidence of a human presence and settlement trends over time using plant micro-fossils.

Of course, the low visibility of archeological sites before the Formative period was reached and/or ceramics were used is probably due to the fact that people practicing swidden agriculture generally shifted locations frequently and used many implements made out of wood that do not survive over time. Sites of preceramic age in Panama and elsewhere occupied by early horticulturists typically exhibit surface manifestations of small "lithic scatters" located on small spurs overlooking minor streams. At first glance, such sites provide little hint that people living in houses, cutting the forests, and growing a variety of domesticated plants once lived there.

It is noteworthy in this regard that sites such as the Maya center Copan contained some cultural debris well below ceramic levels, consisting of small amounts of lithics, charcoal, and charred nuts (Rue, 1988). Because the material was excavated before the advent of radiocarbon dating, its age was never assessed. One can also imagine that, paling in comparison to the splendor of the Mayan occupation of the site, this debris was not considered to be of much importance.

The Development and Spread of Effective Food Production

It is clear from the evidence that people living in simple hamlets and moving their living sites often utilized slash-and-burn techniques of agriculture with fully domesticated plants in the lowland and mid-elevational forests throughout tropical America between 7000 and 4500 radiocarbon years ago. Some small-scale clearing of forest for the preparation of plant-growing plots probably commenced during the middle of the ninth millennium B.P. in Panama. This was detectable because a combination of pollen, phytolith, and charcoal analyses were employed, techniques not used in every paleoecological sequence we reviewed.

By the seventh to sixth millennium B.P., significant modification of the primary forest for fields is evident in Panama, the Cauca Valley, Colombia, and western Amazonia (the Ecuadorian Ayauch[i] sequence). Maize is present in all these sequences. Swidden cultivation also possibly starts by 5600 B.P. in the eastern Amazon Basin (Lake Geral) in a form less demanding on, and destructive of, the forest. In Panama, populations grew substantially after shifting cultivation was initiated (it is not possible to make statements about demographic trends in other areas because systematic archeological surveys have not been carried out). The demographic trends in Panama can reasonably be taken to indicate that truly effective agriculture capable of supporting substantial increases in population number and density was being practiced.

We believe that beginning during the seventh millennium B.P., there was a shift in agricultural techniques as maize and, most likely, other crop plants, such as manioc and sweet potato, became increasingly available to populations in southern Central American and northern South America. Occupants of many regions began to move out of simple kitchen garden horticulture and into a regular pattern of burning and clearing forests for larger scale field cultivation. Thus, the roots of slash-and-burn agriculture appear to go back at least 7000 radiocarbon years in the lowland tropical forest.

By the fifth millennium B.P., severe forest clearance was manifested over a very wide area, including, in addition to the previously mentioned areas, eastern Panama, northern Belize, and the Colombian Amazon, and it is clearly associated with maize agriculture. The development and spread of slash-and-burn agriculture may have been enhanced by drier mid-Holocene climates that have been evidenced in several regions we considered. Longer and/or drier dry seasons would have expanded the range of environments suitable for slash-and-burn techniques using crops such as maize.

In the eastern Amazon Basin (Lake Geral), maize is not present until much later in time, and the removal of the forest by swidden cultivators never achieved the scale and intensity indicated in the other sequences. In one small tract of terra firme forest located 70 km north of Manaus in the central Amazon, neither human alteration of the forest nor food production could be demonstrated during the past 7200 years. Whether these patterns were true over broader areas of the Amazonian terra firme forest will be revealed by future research. If so, they may have been substantially a result of regional variation in the acceptance and use of maize and other soil-demanding crop plants in agricultural systems in and outside of the interior of the Amazon Basin, and this variation was probably largely structured by the fertility of the soils.

We discussed how the Valdivia culture of southwestern Ecuador appeared to represent the first expression of the Formative way of life in the New World, and that it predated similar developments in highland areas of Central and South America. At this point, it is instructive to compare existing highland and lowland paleoecological sequences because they offer a major source of data on the question of where truly effective agricultural techniques may have first emerged. We find that productive agriculture, indicated by major impacts on vegetation associated with the presence of cultivars, occurs earlier in Panama, Amazonian Ecuador, Amazonian and lower mid-elevational Colombia, Honduras, and Belize than it does in the arid and subarid Central Mexican highlands.

For example, in pollen records from five lake basins at elevations between 1700 and 2575 m a.s.l. in central Mexico (Metcalfe *et al.*, 1989; see also Street-Perrott *et al.*, 1989; O'Hara *et al.*, 1993), three of which extend back to the Pleistocene, a marked decline of arboreal species and the presence of maize do not occur until 3500–3000 B.P. Only one site has disturbance this old. This is a full 1000 years later than the manifestations in northern Belize, which occur toward the later end of the time spectrum among low-elevation sites. It is important to note that the highland Mexican lakes form what constitutes a long transect from the Pacific to the Atlantic side, and they cover marked gradients in climate and vegetation. Therefore, they adequately represent the past conditions of vegetation, and human impacts on them, on the important central Mexican plateau.

From these data and others presented in this and Chapter 4, we conclude that the humid tropical lowlands were the major settings for the origin and development of agricultural systems and associated increases of population settlement and density. The time has come for the Tehuacán Valley plant remains to relinquish their role as the core database for evaluating the course of agricultural development in the New World. Tehuacán (along with the Oaxaca Valley) was probably not a pristine area of food development. The early and middle Holocene sequences from the Tehuacán Valley should be considered on the basis of what they probably represent—encampments of small and mobile groups who were essentially hunters and gatherers and who received plants from their lowland neighbors that were already domesticated and grew them in small quantities.

The earliest clear signals of slash-and-burn agriculture occur in central Pacific Panama, where they are recorded early in the seventh millennium B.P. This region seems to have been precocious in a number of developments relating to tropical forest settlement and agricultural development. Systematic archeological surveys have shown that it was occupied from the beginning of the Paleoindian period (11,000 years ago). People interfered with the forest during terminal Pleistocene times and grew squash, bottle gourd, and leren in small house gardens at an early date. This portion of the isthmus is narrow, and it once supported considerable expanses of tropical deciduous forest on fertile soils, on which grew a close wild relative of at least one important domesticated plant, *Cucurbita moschata*. It experienced major vegetational changes at the close of the Pleistocene that favored the development of food-producing strategies. It contains a large and productive estuarine zone that was intensively exploited starting 7000 years ago. Thus, all the factors seemingly necessary for an early *in situ* development and intensification of food production converged in the region and made it, we are increasingly led to believe, a nuclear area for the origins and development of food production.

It may also be highly significant that in this region maize phytoliths were shedding some of their primitive characteristics and beginning to look "modern" in basic cob morphology (probably indicating, in part, the softening of a hard glume), if probably not in size, between the seventh and sixth millennium B.P. (see Chapter 4). At this point, slash-and-burn agriculture took off and never looked back.

Significant improvements in the productivity of maize allowing productive agriculture have been tied largely to significant increases in the size of the cob, which, given the small-sized kernels recovered from early Preclassic sites, probably occurred after the earliest manifestations of slash-and-burn agriculture in the humid tropics. For example, Kirkby (1973) suggests that in the Oaxaca Valley maize did not reach a threshold of productivity warranting considerable emphasis on its cultivation at the expense of wild resource collection until 3400 B.P. At this time, cobs grew to a size of approximately 6 cm, and the mesquite forests may have been cut to plant maize for the first time.

A scenario such as this is inappropriate for explaining change in the humid and lowland areas for two reasons. First, individual plants with many small ears, typical of "primitive" maize races, would yield as much as plants with a few larger ears if kernels were naked and could be fairly efficiently ground or otherwise prepared and consumed. Second, Kirkby's (1973) analysis relates to choices between alternatives of resource exploitation in a highly productive wild resource zone, the thorny scrub environments of Oaxaca, where maize had many wild competitors contributing substantial inputs into the food base.

Wild plant productivity is dramatically different in the tropical forest, where from its earliest introduction maize was probably a far more efficient source of calories than many plants in the natural flora and was probably a focus of experimentation, leading to improvement. The storeability and transportability of maize also ensured its early use and spread in the humid tropics. As discussed in Chapter 4, the late development of effective maize agriculture in the Mexican highlands may, in part, relate to the relative ease of harvesting productive wild plants and the low population densities, which resulted in a limited amount of selection pressure placed on maize during earlier periods.

We believe that an evolution of cob and kernel morphology leading to productive varieties of maize may have occurred in the humid tropical lowlands outside of the Balsas River valley. Such developments helped to fuel the initiation and spread of slash-and-burn agriculture throughout the tropical forest between 7000 and 5000 B.P. Maize was not the only crop plant spreading at that time. Manioc probably arrived in Panama by 7000–6000 B.P. It was present in the Colombian Amazon by approximately 5000 B.P. By this time, it had moved into Belize, possibly as part of an expanding slash-and-burn agricultural system.

Given the time-transgressive nature from south to north of the appearance of sites evidencing food production and then slash-and-burn agriculture during the early and middle Holocene, it is tempting to place the origins of effective agricultural systems in southern Central America and northern South America. Our information regarding this issue is severely limited because paleoecological studies have not been widely carried out. Also, the middle Archaic period (ca. 8000–5000 B.P.) is essentially missing in archeological records from Mesoamerica. Therefore, no continuum of cultural change associated with early food production and agricultural intensification can be evidenced from archeological data, as is possible in central Pacific Panama, northwest Peru, the Colombian Amazon, and southwest Ecuador if, indeed, it occurred.

Is the low visibility of the middle (and, in some regions, early) Archaic period in lowland Mesoamerica compared to southern Central America and northern South American an artifact of insufficient archeological survey and attention to paleoecological data? Is it a result of effective food production systems having been developed first farther south, to then make their appearance in Mesoamerica during the late Archaic Period? These are among several very meaningful questions that will have light shed on them by future studies.

The Spread of Food Production and Individual Crop Plants

Why did food production spread and intensify? We basically take a Darwinian approach to this problem. The substantially increased yield of food from cultivation compared to foraging raised the carrying capacity of local environments. Food production spread because in any area people practicing it raised more and/or better fit children than people not practicing it. This is an argument similar to one used by Rindos (1984). With an increasing number of people on the landscape filling up the low-cost and circumscribed locales of early food production, more labor-intensive forms of food production that could be practiced in the forest were developed. These practices then increased aggregate food yields and continued to support population growth, which, in turn, demanded further intensification (Boserup, 1965). This, perhaps, was the first time that population pressure played a significant role in agricultural dynamics.

It should be noted that although the creation of larger scale field systems in the forest almost certainly involved a per capita increase of labor relative to the return of food when compared with garden horticulture, a simple and inexpensive technology (fire) and simple tools apparently were sufficient in many regions to clear the forest. The time-consuming task of making and refurbishing stone axes was often a much later development than slash-and-burn cultivation itself.

We have seen how many plants were cultivated and domesticated in both hemispheres of the Neotropics and how some of them were spread far outside of their areas of origin by the time of Christ. They include maize, manioc, sweet potato, achira, peanuts, squash, beans, cotton, certain palms, and coca. How particular plants spread is a different question than how food production as a subsistence strategy was disseminated. Large-scale population movements or colonization are possible. We suggested that these processes might account for the seemingly rapid onset and spread of slash-and-burn agriculture into certain regions of Mesoamerica. Similarly, bitter manioc-based agriculture probably spread through population movements in northeastern South America and the eastern Amazon Basin.

However, we generally find this explanation much less appealing for the early and middle Holocene periods of southern Central America and northwestern South America, when population densities were low but long-distance transmissions of crops, such as maize, leren, manioc, squash, and sweet potato, took place against signals of significant cultural continuity. Pickersgill and Heiser (1977) and Harlan (1986) commented that known patterns of transfer seemed to indicate plant-by-plant transfer rather than movement of whole plant complexes. For some regions, the evidence presented here does little to change this view, which also implies plant movement without substantial and permanent population expansion.

Exchange of plants within regional exchange networks is a fundamental characteristic of tropical forest societies today, and with little doubt it was practiced in the past (Lathrap, 1973b). Simple forms of down-the-line exchange may have been among the earliest types of exchange networks (Pearsall, 1977). It is equally likely that early plant transmission was not rooted in any stated or structured system but perhaps had much to do with the nature of early and middle Holocene settlement characteristics and reproductive networks. The cultivators of this time were spread over the landscape at low densities, and they were likely to have interacted with other groups and to have chosen mates who lived in rather far-flung areas (Wobst, 1978). Such behavior would have had positive political and economic implications (Turnbull, 1986), and it also served to maintain viable reproductive networks (Wobst, 1974). The Pygmy, for example, who have a prescription to "marry far" (Turnbull, 1986), choose mates who, on average, live 53 km away (Hewlett *et al.,* 1986). The mean mating distance for the Kung is 66 km (Harpending, 1976). Under these social conditions, plants surely would spread as far and as wide as they would in structured exchange systems.

Early exchange may have been "neighborly" (Ford, 1984) and casual (e.g., an early central Panamanian cultivator may have said to a visitor, "Have you seen the new plant we are growing") or merely entailed semibounded populations and changes in residence of a husband or a wife at marriage. All in all, the evidence indicates that at least before 4000–3000 B.P. the mechanism of transmission was largely cultural, not demic, in the areas better known to us at the current time.

Why particular plants moved and then were permanently accepted by populations at various times is another issue that we believe involves a return to evolutionary ecology for an explanation. The most deterministic form of the diet breadth model predicts that new crop plants will be added when and only when they increase the efficiency of the existing resource system (Gremillion, 1996). We note that the new crop plant may initially be low ranked but will be accepted by horticulturists anyway if higher ranked resources are in short supply (Gremillion, 1996). We believe it is likely that the early spread and widespread adoption of plants such as maize, squash, and certain tubers were in large part related to their efficiency when compared with naturally available resources. This is likely why they were among the earliest plants to have been taken under cultivation.

When sensitivity to risk is built into foraging models some interesting insights also emerge with regard to crop spread (Gremillion, 1996). Winterhalder's (1986, 1990) simulations, discussed at length in Chapter 4, indicate that people might benefit by accepting new resources to mitigate risk at the expense of some loss of energetic efficiency when expected average food returns are much higher or much lower than the minimum needed for survival. This finding does not support the intuitive notion that populations will not experiment with new resources when under resource stress. They might if the new resources are high ranked, and under these circumstances the diet would narrow, not broaden. On the other hand, "affluent farmers" could be expected to diversify their crop inventories with the prospect of further minimizing risk, or to just try a new plant because it looks interesting, because they can tolerate some loss of efficiency in crop production. Perhaps, the tendency to add new crop plants irrespective of their return to labor (various tree fruits and tuber crops demanding more costly processing) came somewhat later in time, when horticultural systems with a fairly high-yielding product were established.

Remember also that when expected food returns are anywhere near what is needed for survival, the energetically optimal diet also minimizes risk (see Chapter 4). Given the reliable food supplies typical of horticultural production by small and shifting groups of people in the humid tropics today (e.g., Carneiro, 1983; Hames, 1990; Johnson, 1983), the proclivity of tropical horticultural people to share resources with high variance of return, and the low-cost "storage" options provided by tubers that would allow for continuous production (Lathrap, 1973b), we might expect that factors other than risk aversion were generally structuring the subsistence behaviors of many cultivators before the emergence of social complexity.

Current Evidence for Tropical Food Production Evolution in Light of Traditional Archeological Evidence and Theory

Paleoecological data formed an essential part of our study of the evolution of agricultural systems. Such data are also meaningful when they are viewed in the context of prehistoric change in the way that it has been evaluated by investigators using traditional archeological data and concepts. For example, central Panamanian populations began to make ceramics at a relatively early (5000 B.P.) time. However, the introduction of a ceramic technology seems to have little changed other aspects of the material culture, settlement patterns and densities, and demographic trajectories (Cooke, 1995; Linares, 1979; Ranere and Hansell, 1978). Instead, these attributes appear to have been affected more by the much earlier incorporation of exogenous domesticates, such as maize and manioc, into subsistence economies and the subsequent demands of these crops on labor and on the environment.

In eastern Panama, Honduras, Belize, and the Colombian Cauca Valley and Amazon, the emergence of ceramic traditions occurred several thousand years later in time, but between 6000 and 4000 B.P. each of these regions witnessed an intensification of agriculture and land-use changes parallel to those recorded in central Panama. Intensive forest modification and clearance in all these regions appears to have occurred without a reliance on ground and polished stone implements, which do not appear as components of the lithic inventories until after 3000 B.P. Furthermore, the clearance of forests was carried out not by the occupants of sedentary farming villages but by people who were probably still organized as shifting cultivators.

These factors emphasize the difficulties of using only traditional archeological correlates (e.g., the presence/absence of ceramics, polished stone implements, and settled villages) in studying the evolution of settlement and agriculture in the lowland tropics. They make it clear that recent attempts to correlate the onset of ceramic manufacture with the emergence of horticulture (e.g., Hoopes, 1995) may be fruitless exercises in many regions of the lowland tropics.

If the correlation between the origin and spread of ceramics and the appearance and development of food production has been found to be a weak one, a possibly stronger link can be established between pots and the improvement and spread of specific crop plants. For example, many tubers, fleshy fruits, and plant vegetative structures were easily prepared by simple roasting on coals and baking in pits, and maize and other hard seeds were prepared by a somewhat more difficult process involving grinding before cooking. However, new cooking techniques made possible by pottery may have led to the use of the lima and common bean as important carbohydrate and protein sources and not just as green vegetables (see Chapter 3). Also, surely the availability of griddles to bake bitter manioc cakes hastened the development and spread of this important crop. Thus, we view ceramics as having importance for the emergence of certain crop plants as important resources and not for the beginnings and subsequent intensification of food production.

Although slash-and-burn agriculture was, and is, the dominant technique of food production in the humid tropics, it generally has not earned much respect among prehistorians, who sometimes consider it to be a primitive system. Swidden cultivation does not leave the terraces, mounds, and ditches that attend other systems of food production and is thus less materially impressive. Also part of the problem is that today slash-and-burn cultivation is seen as a primary cause of the disappearance of the remaining forests by well-intentioned conservationists and is viewed as a "wasteful" and "destructive" technique. We need to be reminded from time to time that swidden cultivation was, and still is, an enormously successful adaptation to the rigors and constraints of the tropical forest. It allowed effective food production to be practiced in highly diverse, and sometimes exceptionally poor, environments. It could be adjusted as needed to varying conditions of climate. In areas where dry seasons were long and marked, it could be practiced without the need for sophisticated and labor-intensive tool technologies. If practiced on fertile soils at high levels of intensity (with short fallow periods), it could, for a time, support high levels of population, and it did just that in at least two regions we know about—central Pacific Panama and the Maya heartland and periphery.

Prehistorians interested in the development of agriculture and social complexity in the New World should also not lose sight of the fact that slash-and-burn cultivation is associated with a social organization characterized by household autonomy in decision making, relatively small and shifting settlements often composed of a few related families, and low population densities. These features, plus the very success of the slash-and-burn mode of planting, help explain why people in the Neotropics adopted a lifestyle oriented around sedentary and nucleated villages with increased social complexity long after food production was initiated.

This lifestyle is documented in many areas of the humid lowlands by 2000 B.P. or shortly thereafter. The agricultural subsistence base had clearly narrowed by this time. Available bone isotope records indicate high maize consumption, and carbonized maize becomes ubiquitous in the plant records of many regions of Central America and northern South America. In other areas, such as central and eastern Amazonia, bitter manioc played the role of staple crop, as testified to by the presence of pieces of stone and ceramics that probably were used to grate the poisonous pulp of the tuber and bake the bread.

Christenson (1980) notes that agricultural intensification leading to a substantial narrowing of the food base are features of prehistoric life not easily explained by current optimal foraging models, a point to which we agree. Similarly, the development of social complexity is not likely to be satisfactorily sought in simple models from evolutionary ecology. Other factors no doubt fueled increasing economic specialization and social complexity. They include population pressure arising from the long-term growth of human populations subsisting on the substantial yields from cultivated foodstuffs during the preceding 7000–8000 years, the development of very high-yielding varieties of maize and other crops (actually, this factor follows the diet breadth model because resources of high rank should narrow the diet if they are abundant), competition over good agricultural lands, the opportunity for power-hungry individuals to accumulate wealth and status by manipulating a large agricultural product, and other elements embedded in social relationships (e.g., Cooke and Ranere, 1992b; Pohl *et al.,* 1996; Rust and Leyden, 1994).

The allure of the Formative and later periods of New World prehistory, when individuals eschewed a simpler and egalitarian existence and began to accumulate material trappings of brightly painted ceramics, carefully polished stone celts, exotic exchange items, and monumental architecture, has been a potent attraction for many archeologists. In light of the evidence presented here, the role of the early shifting cultivators in influencing the later courses of New World prehistory deserves appreciation.

These simpler people were enormously successful in their own right. Not long after arriving in the Neotropics, they skillfully negotiated a period of immense environmental upheaval in a new habitat that cannot not have been easy to live in at first, and they went on to develop the productive systems of agriculture that made everything that followed possible. We agree with Dillehay *et al.* (1989, p. 756), who noted that many of the later, more materially elaborate cultural developments owe their existence to these small-scale societies who are "less conspicuous and more ephemeral because they never achieved social and economic unity over wider areas." Nevertheless, they achieved a stature such as to have left abundant evidence of the first effective agricultural systems in the New World that we study and admire today.

CHAPTER 6

The Relationship of Neotropical Food Production to Food Production from Other Areas of the World

COMMON CHARACTERISTICS OF INCIPIENT FOOD-PRODUCING SOCIETIES AROUND THE WORLD: HOW DOES THE NEOTROPICAL EXAMPLE COMPARE?

As a result of research undertaken in other areas of the world during the past 10–15 years, a number of ecological, demographic, and cultural factors thought to be closely associated with the development of food production have been proposed by various investigators. In this section, we place Neotropical food production in a broader context by evaluating how many of these factors appear to have been important in the Neotropics. The factors are (i) sedentary, or at least semisedentary, living; (ii) social complexity, often accompanied by some level of social circumscription; (iii) abundant resources; (iv) diets characterized by high resource diversity; (v) sufficient population numbers to make resource intensification possible; (vi) the presence of good potential domesticates; (vii) a technology to use plants effectively; and (viii) a long period of availability of cultivated plants before the full-fledged emergence of agriculture (e.g., Gebauer and Price, 1992a; Hayden, 1995; Price and Gebauer, 1995).

The first expectation, sedentary or semisedentary living, is met by the Neotropical data. Many sites with early food production were probably occupied on a semisedentary basis. Their visibility and location, alongside lakes or small parcels of alluvium of watercourses, indicate they were occupied for considerable periods during a year and/or visited regularly. Some of the Vegas settlements of southwest Ecuador may have been occupied year-round. We do distinguish the short-term sedentism that seems to be characteristic of tropical sites from the longer term sedentism possibly associated with Natufian sites in southwest Asia (Bar Yosef and Belfer-Cohen, 1989, 1992).

The second factor, social complexity and social circumscription, is not evidenced by the Neotropical data. As we have discussed in other chapters, the earliest groups that practiced food production appear to have been organized at the level of the single family or hamlet and were probably egalitarian. This should come as no surprise because similar levels of social organization still characterize many horticultural groups in the American tropics today. Also, in apparent contrast to some other areas of the world, such as southwest Asia, early Holocene people in the Neotropics do not appear to have been socially circumscribed. They would have been able to "vote with their feet" (Gebauer and Price, 1992a, p. 9) and migrate to new areas for new sources of food. However, given the widespread distribution of forest on the landscape, this generally would not have improved the efficiency of their subsistence system and their food supply.

The human tendency to experiment with food resources until better return rates are achieved would, at any rate, have generally lessened the value of migration to human foragers (and it ultimately would have been disadvantageous to the forager to delay food production in this manner). We conclude that social complexity and other social factors commonly cited (e.g., Hayden, 1990, 1992; Price, 1995) were not important in the lowland Neotropics.

In relation to this issue, we also note that social factors are often cited as having substantial importance in areas of secondary developments of food production. In such regions, crop plants occurred later in time, populations may have grown substantially after the termination of the Pleistocene, and a demic diffusion of a well-developed agriculture may have taken place (e.g., Price, 1995, Price and Gebauer, 1995). We have very little information on the relationship of social factors to food production from several important, and probably pristine, areas of food production, such as New Guinea (Haberle, 1994; Golson, 1991a,b) and mainland Southeast Asia (Sauer, 1952). We suspect that in these areas, Neotropical kinds of social relations on the eve of food production may have been more typical than others commonly discussed. In evaluating the significance of social factors, we should be careful not to draw too many conclusions from nonpristine areas or a single pristine area (the southern Levant).

The third factor thought to be important in the emergence of food production, food abundance, is tricky. Tropical forests are not areas with abundant and stable wild resources, so we might reject this factor for the Neotropics. However, those areas where plant tuber and seed production were first developed were in the most optimal zones of the most optimal types of forests and, in this sense, they represent resource abundance. Nonetheless, as long as people are not starving or otherwise experiencing frequent and severe shortfalls of food, food abundance per se may have little relevance. Rather, we believe that the crucial factor relating to food procurement is its energetics at any given time in relation to those of the past and what potentially can be gained by trying alternative strategies, such as food production.

We make a similar comment with regard to the factor of having a sufficient human population to promote resource intensification. Under the optimal diet model, resource intensification can be driven largely by changes in foraging efficiencies and diet breadth. In the Neotropics, these changes were primarily responses to natural shifts in the abundance and distribution of resources and had little to do with human population pressure (we discuss the possibility of a somewhat different scenario in another region of pristine food production later in this chapter). The lowland Neotropics may have had the lowest population densities of any region shown to have supported the emergence of food production during the early Holocene.

As in examples from other parts of the world (e.g., Bar-Yosef and Meadow, 1995; Cohen, 1977a, 1987; Cowan and Watson, 1992; Flannery, 1969; Gebauer and Price, 1992b; Hillman *et al.,* 1989), Neotropical foragers appeared to have had diverse diets on the eve of food production and, as elsewhere, they probably broadened their food base and incorporated more costly resources into the diet shortly before cultivation began. Estimations of diet breadth based on archeological data can be tenuous (Broughton and Grayson, 1993; Madsen, 1993), particularly when very few pre-10,000 B.P. subsistence data are available. However, as discussed in Chapter 4, the characteristics of the pre-10,000 B.P. resource base in the Neotropics, combined with the available archeological evidence and expectations from optimal foraging theory, make it likely that the Neotropical diet diversified substantially approximately 10,500–10,000 B.P.

In relation to the sixth factor, availability of good potential domesticates, the presence of suitable plant domesticates on the humid and forested Neotropical landscape is undisputed. Many important crop plant ancestors are now understood to have been naturally distributed in tropical deciduous and semi-evergreen forests, where the dry season is long and marked. As discussed in Chapter 4, grinding implements to effectively use these plants are present in archeological assemblages by 10,000 B.P.

As for the final factor, a long period of availability of cultigens before the full emergence of agriculture, the Neotropical case follows the trends reported for other regions. According to the current evidence, the earliest Neotropical cultigens were supplements to the existing diet. It took at least 3000 years for small-scale horticulture practiced in gardens and other small plots near houses to develop into larger scale field systems using slash-and-burn techniques. It took another several thousand years before the subsistence base narrowed sharply and a few domesticated plants dominated subsistence systems. Both of these processes took a long time in the Neotropics.

In addition to these factors, we add another factor that is highly relevant to the Neotropical example of food production and quite possibly to many others— vegetational disturbance by humans, with or without fire. Because fire was the chief method used to modify vegetation by early humans, we would expect that early vegetational disturbance very often included burning in those environments where fires could easily be started. Keeley (1995) notes that setting fires is highly correlated with ethnographically known hunters and gatherers who practice proto-cultivation. McCorriston and Hole (1991) argue that vegetational disturbance by humans in the southern Levant before the beginning of food production was important in sustaining faunal densities high enough for exploitation.

Anthropogenic disruption of the tropical forest using fire is apparent in several paleoecological records from the Neotropics dating to the earliest Holocene. In the Neotropics, it not only increased the yield of wild plant and animal products but also enhanced the opportunity for decreased residential mobility before the practice of cultivation began.

Lastly, at the beginning of this book we suggested that an important element that has been missing in discussions of food production origins was the ability of past human ancestors to exploit resources in an energetically optimal manner. We believe there is increasing empirical evidence indicating that populations capable of modern human cognition and behavior, including forms of resource exploitation and management characteristic of modern foragers and incipient farmers, evolved a little more than 100,000 years ago and perhaps by only 35,000 years ago (e.g., Brooks *et al.*, 1995; Foley, 1988; Klein, 1995; Mithen, 1996). The important issues of why previous glacial/interglacial environmental perturbations did not lead to the propagation of plants and why it took so long for humankind to practice plant propagation become much less problematical if one accepts this proposition (see Chapter 1). We believe that the factors necessary for the emergence of food production probably did not converge until the end of the Pleistocene.

THE RELEVANCE OF EVOLUTIONARY ECOLOGY TO THE ORIGINS OF FOOD PRODUCTION WORLDWIDE

As is probably obvious by now, we believe that models of optimal foraging and others derived from evolutionary ecology have substantial promise for the study of agricultural origins in other areas of the world. Russell (1988) has already provided such a study for the Near East that is consistent with the available archeological evidence. He concluded that return rates from cultivating wild cereals in optimal habitats were likely to have been higher than those from the same cereals collected in the wild. Wright (1994) feels that Russell underestimated the costs of processing wild cereals. However, this objection may have little relevance to the heart of the matter, which is the estimation of overall foraging return rates before and after the close of the Pleistocene and the efficiency of cultivation in relation to the post-Pleistocene resources. In Chapter 4, we discussed how these factors probably favored the long-term persistence of foraging rather than the early emergence of farming in the arid mountain valleys of Mexico.

Russell (1988) also notes that only a very few wild members of the large grass family, notably including the ones that were domesticated in the southern Levant, appear to be capable of providing the return rates probably needed for the selection of food-producing strategies to occur. Hawkes and O'Connell (1992) and Hawkes *et al.* (1997) add that rapid improvements in handling efficiency (e.g., development of stiffer rachises and larger seeds) may have been very important in the domestication process and that few plants probably possessed the genetic and reproductive characteristics that predisposed them to such improvements. These factors may help explain why only a small number of grasses and other kinds of plants were domesticated and why broad spectrum economies did not become pristine food producing economies everywhere in the world following the close of the Pleistocene.

We end this book by returning to two issues that have attracted considerable attention in discussions of food production origins: the changes in dietary breadth and human demography that seemed to have preceded the shift from foraging to food production by a relatively short time. These issues provide a useful example of the potentially broad applicability of evolutionary ecology to the issue of agricultural origins.

The trends toward dietary generalization before the emergence of food production, first termed the "broad spectrum revolution" (Flannery, 1969), have been commonly explained (i) by population pressure (e.g., Cohen, 1977a; Christenson, 1980), (ii) by a combination of population growth in favorable habitats leading to population/resource imbalances in more marginal areas (Binford, 1968), (iii) as a mechanism to reduce resource unpredictability in dynamic post-Pleistocene environments (Flannery, 1986a), or (iv) as a response to the new availability of resources following the post-Pleistocene environmental changes (McCorriston and Hole, 1991). In the Near East, population growth appears to have occurred in concert with resource diversification, and it led to increasing territoriality during the Natufian period, just before the first evidence for domesticated plants appears (e.g., Bar-Yosef and Belfer-Cohen, 1992; Bar-Yosef and Meadow, 1995). It appears that this series of events did not occur in the Neotropics.

We argued that resource diversification can be more parsimoniously and robustly explained by the diet breadth model, whereby an expansion of diet breadth occurs in response to an increasing scarcity of highly ranked resources and a decrease in the foraging return rate. Models using a combination of foraging theory and population ecology can provide even sharper insights into the process. They can also clarify why significant growth of human populations, and the various implications for cultural development that follow, may or may not take place in pristine areas before the advent of food production.

For example, Winterholder and Goland (1993) embedded the diet breadth model into a computer simulation that considered population models for foragers and their resources. In other words, in addition to the standard components of a diet breadth analysis, they included the resources' response to continued exploitation as a variable in evaluating the relationships between diet breadth, foraging efficiency, and human population response.

They found that an increase in dietary breadth arising from decreasing return rates could increase, lower, or effect no change on forager population density. The outcome is largely determined by the sustainable yield (density and intrinsic rate of increase (r), or recovery rate) of resources being added to the diet, although resource density seems to be more important. Incorporation of new and low-ranked resources that are densely distributed on the landscape and have a high r (such as wild cereals) likely will result in an increase of human population density. On the other hand, incorporation of resources that are low ranked but that occur in low densities on the landscape (such as tubers, squash, and many other Neotropical plants) will likely cause a decline in human population density. Also, increasing human population density can deplete prey and cause subsistence strategies to further broaden.

Therefore, changes in the density of human foragers do not depend on changes in mobility or territoriality. They can result solely from the spatial and reproductive characteristics of newly exploited resources, which largely determine how these resources respond to predation and how many human foragers they can support (Winterhalder and Goland, 1993). Processes such as these may explain the significant increase in human populations in the Near East before the Neolithic period, and they open the possibility that such increases started to occur before sedentary life. They suggest that increasing diet breadth in the Neotropical forest at the close of the Pleistocene may have led to declining human population density before the emergence of food production.

Furthermore, because increases in human population subsequent to the expansion of diet breadth could have lowered the foraging return rate of certain resources, and generally depleted others, increasing territoriality, resource circumscription, and demographic pressure may have been influential in food production origins in certain nuclear areas, such as the Near East (e.g., Bar-Yosef and Belfer-Cohen, 1992; Bar Yosef and Meadow, 1995; McCorriston and Hole, 1991; Wright, 1994), but of negligible importance elsewhere, such as the Neotropics.

It has been suggested that the records of the latest foraging and earliest food production from different regions of the world contain so much "local" variability that searching for a single explanation for agricultural origins might be fruitless (Flannery, 1986a; Gebauer and Price, 1992a; McCorriston and Hole, 1991). In contrast, we believe that these local differences may substantially depend on the variation among the natural, exploitable resources of the regions in question and can be parsimoniously accounted for without sacrificing much universality of explanation on the basis of a theory that relies on evolutionary ecology.

This example illustrates how evolutionary theory can help to explain the diversity of human cultural behaviors while providing unified explanations for some signal events that took place during human prehistory. Darwinian reasoning has much to offer to the growing studies of the origins and development of food production.

References

Abrams, E. M., and D. J. Rue (1988). The causes and consequences of deforestation among the prehistoric Maya. *Human Ecol.* **16,** 377–395.

Absy, M. L., A. Cleef, M. Fournier, L. Martin, M. Servant, A. Sifeddine, M. F. Da Silva, F. Soubies, K. Suguio, B. Turcq, and T. Van der Hammen (1991). Mise en évidence de quatres phases d'ouverture de la foret dans le sud-est de l'Amazonia au cours des 60.000 derniéres années: Premiere comparaison avec d'autres régions tropicales. *Comptes Rendus de l'Académie des Sciences, Paris, Ser.* 2 **312,** 673–678.

Aceituno, P. (1988). On the functioning of the Southern Oscillation in the South American sector. Part I: Surface climate. *Monthly Weather Rev.* **116,** 505–524.

Adams, R. E. W. (1980). Swamps, canals, and the location of ancient Maya cities. *Antiquity* **54,** 206–214.

Adams, R. E. W. (1991). *Prehistoric Mesoamerica,* 2nd. Ed. Univ. of Oklahoma Press, Norman.

Adams, R. E. W., W. E. Brown, Jr., and T. P. Culbert (1981). Radar mapping, archaeology, and ancient Maya land use. *Science* **213,** 1457–1463.

Alexander, J. (1971). The domestication of yams: A multi-disciplinary problem. In *Science in Archaeology* (E. S. Higgs and D. Brothwell, Eds.), 2nd. ed., pp. 229–234. Thames and Hudson, London.

Alexander, J., and D. G. Coursey (1969). The origins of yam cultivation. In *The Domestication and Exploitation of Plants and Animals* (P. J. Ucko and G. W. Dimbleby, Eds.), pp. 405–425. Aldine, Chicago.

Allem, A. C. (1987). *Manihot esculenta* is a native of the Neotropics. *FAO/IBPGR Plant Genetic Resour. Newslett.* **71,** 22–24.

Alvard, M. S. (1993). Testing the "ecologically noble savage" hypothesis: Interspecific prey choice by Piro hunters of Amazonian Peru. *Human Ecol.* **21,** 355–387.

Alvard, M. S. (1995). Intraspecific prey choice by Amazonian hunters. *Curr. Anthropol.* **36,** 789–818.

Ambrose, S. H., and L. Norr (1992). On stable isotopic data and prehistoric subsistence in the Soconusco region. *Curr. Anthropol.* **33,** 401–407.

Anderson, A. B., and D. A. Posey (1989). Management of a tropical scrub savanna by the Gorotire Kayapó of Brazil. In *Resource Management in Amazonia: Indigenous and Folk Strategies* (D. A. Posey and W. Balée, Eds.), Advances in Economic Botany, Vol. 7, pp. 159–173. The New York Botanical Garden, Bronx, NY.

Anderson, E. (1952). *Plants, Man and Life.* Univ. of California Press, Berkeley.

Andres, T. C. (1990). Biosystematics, theories on the origin, and breeding potential of *Cucurbita ficifolia.* In *Biology and Utilization of the Cucurbitaceae* (D. M. Bates, R. W. Robinson, and C. Jeffrey, Eds.), pp. 102–119. Cornell Univ. Press, Ithaca, NY.

Andres, T. C. (1995). Complexities in the infraspecific nomenclature of the *Cucurbita pepo* complex. In *Acta Horticulturae: Second International Symposium on the Taxonomy of Cultivated Plants* (H. B. Tukey, Ed.), Vol. 413, pp. 65–91. Cornell Univ. Press, Ithaca, New York.

Andres, T. C., and D. R. Piperno (1995). New evidence on the past and present distribution of *Cucurbita*. Paper presented at the annual meeting of the Society for Economic Botany, Ithaca, NY.

Athens, J. S. (1990). Prehistoric agricultural expansion and population growth in northern highland Ecuador: Interim report for 1989 fieldwork. International Archaeological Research Institute, Inc., Honolulu.

Austin, D. F. (1978). The *Ipomoea batatas* complex—I. Taxonomy. *Bull. Torrey Botanical Club* **105**, 114–129.

Austin, D. F., R. L. Jarret, C. Tapia, and F. de la Puente. (1992). Collecting tetraploid *I. batatas* (Linnaeus) Lamarck in Ecuador. *FAO/IBPGR Plant Genetic Resour. Newslett.* **91/92**, 33–35.

Ayensu, E. S., and D. G. Coursey (1972). Guinea yams. *Econ. Bot.* **26**, 301–318.

Bahuchet, S., D. McKey, and I. de Garine (1991). Wild yams revisited: Is independence from agriculture possible for rain forest hunter–gatherers? *Human Ecol.* **19**, 213–243.

Bailey, L. H. (1947). *The Standard Cyclopedia of Horticulture*. Macmillan, New York.

Bailey, R. C., and T. N. Headland (1991). The tropical rain forest: Is it a productive environment for human foragers? *Human Ecol.* **19**, 261–285.

Bailey, R. C., G. Head, M. Jenike, B. Owen, R. Rechtman, and E. Zechenter (1989). Hunting and gathering in tropical rain forest: Is it possible? *Am. Anthropol.* **91**, 59–82.

Baksh, M., and A. Johnson (1990). Insurance policies among the Machiguenga: An ethnographic analysis of risk management in a non-western society. In *Risk and Uncertainty in Tribal and Peasant Economies* (E. Cashdan, Ed.), pp. 193–227. Westview, Boulder, CO.

Balée, W. (1988). Indigenous adaptation to Amazonian palm forests. *Principes* **32**, 47–54.

Balée, W. (1989). The culture of Amazonian forests. In *Resource Management in Amazonia: Indigenous and Folk Strategies* (D. A. Posey and W. Balée, Eds.), Advances in Economic Botany, Vol. 7, pp. 1–21. The New York Botanical Garden, Bronx, NY.

Balée, W., and A. Gély (1989). Managed forest succession in Amazonia: The Ka'apor case. In *Resource Management in Amazonia: Indigenous and Folk Strategies* (D. A. Posey and W. Balée, Eds.), Advances in Economic Botany, Vol. 7, pp. 129–158. The New York Botanical Garden, Bronx, NY.

Balick, M. J. (1984). Ethnobotany of palms in the Neotropics. In *Ethnobotany in the Neotropics* (G. T. Prance and J. A. Kallunki, Eds.), Advances in Economic Botany, Vol. 1, pp. 9–23. The New York Botanical Garden, Bronx, NY.

Balick, M. J. (1986). *Systematics and Economic Botany of the Oenocarpus-Jessenia (Pamae) Complex*. Advances in Economic Botany, Vol. 3. The New York Botanical Garden, Bronx, NY.

Balick, M. J. (Ed.) (1988). *The Palm-Tree of Life: Biology, Utilization and Conservation*. Advances in Economic Botany, Vol. 6, The New York Botanical Garden, Bronx, NY.

Balick, M. J. (1989). Neotropical oil palms. In *Oil Crops of the World. Their Breeding and Utilization* (G. Robbelen, R. Keith Downey, and A. Ashri, Eds.), pp. 505–517. McGraw-Hill, New York.

Banks, D. J. (1990). Peanut (*Arachis hypogaea* L.) collecting in coastal Peru. *FAO/IBPGR Plant Genetic Resour. Newslett.* **81/82**, 45.

Barnola, J. M., D. Raynaud, Y. S. Korotkevich, and C. Lorius (1987). Vostoc ice core provides 160,000-year record of atmospheric CO_2. *Nature* **329**, 408–414.

Barse, W. P. (1990). Preceramic occupations in the Orinoco River valley. *Science* **250**, 1388–1390.

Barse, W. P. (1995). El período arcaico en el Orinoco y su contexto en el norte de Sud América. In *Ambito y Ocupaciones Tempranas de la America Tropical* (I. Cavelier and S. Mora, Eds.), pp. 99–113. Instituto Colombiano de Antropología, Fundación Erigaie, Santa Fe de Bogotá.

Bartlein, P. J., M. E. Edwards, S. L. Shafer, and E. D. Barker, Jr. (1995). Calibration of radiocarbon ages and the interpretation of paleoenvironmental records. *Quat. Res.* **44**, 417–424.

Bartlett, A. S., and E. S. Barghoorn (1973). Phytogeographic history of the Isthmus of Panama during the past 12,000 years. In *Vegetation and Vegetational History of Northern Latin America* (A. Graham, Ed.), pp. 203–299. Elsevier, Amsterdam.

Bar-Yosef, O., and A. Belfer-Cofen (1989). The origins of sedentism and farming communities in the Levant. *J. World Prehistory* **3**, 447–498.

Bar-Yosef, O., and A. Belfer-Cofen (1992). From foraging to farming in the Mediterranean Levant. In *Transitions to Agriculture in Prehistory* (A. B. Gebauer and T. D. Price, Eds.), pp. 21–48. Prehistory Press, Madison, WI.

Bar-Yosef, O., and R. H. Meadow (1995). The origins of agriculture in the Near East. In *Last Hunters–First Farmers: New Perspectives on the Prehistoric Transition to Agriculture* (T. D. Price and A. B. Gebauer, Eds.), pp. 39–94. School of American Research Press, New Mexico.

Baudoin, J. P. (1988). Genetic resources, domestication, and evolution of lima bean, *Phaseolus lunatus*. In *Genetic Resources of Phaseolus Beans: Their Maintenance, Domestication, Evolution, and Utilization* (P. Gepts, Ed.), pp. 393–407. Kluwer, Dordrecht.

Baudoin, J. P., J. P. Barthelemy, and V. Ndungo (1991). Variability of cyanide contents in the primary and secondary genepools of the lima bean, *Phaseolus lunatus* L. *FAO/IBPGR Plant Genetic Resour. Newslett* **85**, 5–9.

Beadle, G. W. (1972). The mystery of maize. *Field Mus. Nat. History Bull.* **43**(10), 2–11.

Beadle, G. W. (1977). The origins of *Zea mays*. In *Origins of Agriculture* (C. A. Reed, Ed.), pp. 615–636. Mouton, The Hague, The Netherlands.

Beadle, G. W. (1980). The ancestry of corn. *Sci. Am.* **242**, 112–119.

Beard, J. S. (1944). Climax vegetation in tropical America. *Ecology* **25**, 127–158.

Becerra Valásquez, V. L., and P. Gepts (1994). RFLP diversity of common bean (*Phaseolus vulgaris*) in its centres of origin. *Genome* **37**, 256–263.

Beckerman, S. (1979). The abundance of protein in Amazonia: A reply to Gross. *Am. Anthropol.* **81**, 533–560.

Beckerman, S. (1993). Major patterns in indigenous Amazonian subsistence. In *Tropical Forests, People and Food* (C. M. Hladik, A. Hladik, O. F. Linares, H. Pagezy, A. Semple, and M. Hadley, Eds.), pp. 411–424. UNESCO/Parthenon Publishing Group, Paris.

Behling, H. (1996). First report on new evidence for the occurrence of *Podocarpus* and possible human presence at the mouth of the Amazon during the late-glacial. *Vegetation History Archaeobot.* **5**, 241–246.

Bender, B. (1978). Gatherer–hunter to farmer: A social perspective. *World Archaeol.* **10**, 204–222.

Bender, B. (1985). Prehistoric developments in the American midcontinent and in Brittany, northwest France. In *Prehistoric Hunter–Gatherers: The Emergence of Cultural Complexity* (T. Price and J. Brown, Eds.), pp. 21–57. Academic Press, Orlando.

Benfer, R. A. (1982). Proyecto Paloma de la Universidad de Missouri y el Centro de Investigaciones de Zonas Aridas. *Zonas Aridas* **2**, 34–73.

Benfer, R. A. (1984). The challenges and rewards of sedentism: The preceramic village of Paloma, Peru. In *Paleopathology at the Origins of Agriculture* (M. N. Cohen and G. J. Armelagos, Eds.), pp. 531–558. Academic Press, New York.

Benfer, R. A. (1990). The Preceramic period site of Paloma, Peru: Bioindications of improving adaptation to sedentism. *Latin Am. Antiquity* **1**, 284–318.

Benfer, R. A., B. Ojeda, and G. H. Weir (1987). Early water management strategies on the coast of Peru. In *Risk Management and Arid Land Use Strategies in the Andes* (D. Browman, Ed.), pp. 195–206. Westview, Boulder, CO.

Benz, B. F., and H. Iltis (1990). Studies in archaeological maize. I. The "wild" maize from San Marcos cave reexamined. *Am. Antiquity* **55**, 500–511.

Best, R. C. (1984). The aquatic mammals and reptiles of the Amazon. In *The Amazon: Limnology and Landscape Ecology of a Mighty Tropical River and Its Basin* (H. Sioli, Ed.), pp. 371–412. Junk, Dordrecht.

Bettinger, R. L. (1991). *Hunter–Gatherers: Archaeological and Evolutionary Theory.* Plenum, New York.

Binford, L. R. (1968). Post-pleistocene adaptations. In *New Perspectives in Archaeology* (S. R. Binford and L. R. Binford, Eds.), pp. 313–341. Aldine, Chicago.

Binford, M. W., J. P. Bradbury, D. G. Frey, B. Leyden, R. M. Lewis, Jr., M. Salgado-Laboriau, C. Schubert, F. H. Weibezahn, and D. R. Whitehead (1981). Late Quaternary environmental history of Lake Valencia, Venezuela. *Science* **214**, 1299–1305.

Bird, J. B., and R. G. Cooke (1978). The occurrence in Panama of two types of Paleoindian projectile points. In *Early Man in America from a Circum-Pacific Perspective* (A. L. Bryan, Ed.), Occasional Papers No. 1, pp. 263–272. Department of Anthropology, University of Alberta, Edmonton, Canada.

Bird, J. B., J. Hyslop, and M. D. Skinner (1985). *The Preceramic Excavations at the Huaca Prieta, Chicama Valley, Peru,* Anthropological Papers No. 62, Part 1. American Museum of Natural History, New York.

Blake, M., B. S. Chisholm, J. E. Clark, and E. Mudar (1992a). Non-agricultural staples and agricultural supplements: Early formative subsistence in the Soconusco region, Mexico. In *Transitions to Agriculture in Prehistory,* (A. B. Gebauer and T. D. Price, Eds.), pp. 133–151. Prehistory Press, Madison, WI.

Blake, M., B. S. Chisholm, J. E. Clark, B. Voorhies, and M. W. Love (1992b). Prehistoric subsistence in the Soconusco region. *Curr. Anthropol.* **33,** 83–94.

Blake, M., J. E. Clark, B. Voorhies, G. Michaels, M. W. Love, M. E. Pye, A. A. Demarest, and B. Arroyo (1995). Radiocarbon chronology for the late Archaic and Formative periods on the Pacific coast of southeastern Mesoamerica. *Ancient Mesoamerica* **6,** 161–183.

Bloom, P. R., M. Pohl, C. Buttleman, F. Wiseman, A. Covich, C. Miksicek, J. Ball, and J. Stein (1983). Prehistoric Maya wetland agriculture and the alluvial soils near San Antonio Rio Hondo, Belize. *Nature* **301,** 417–419.

Bohack, J. R., P. D. Dukes, and D. F. Austin (1995). Sweet potato. In *Evolution of Crop Plants* (J. Smartt and N. W. Simmonds, Eds.), pp. 57–62. Longman, Essex, UK.

Bonavia, D. (1982) *Precerámico peruano. Los Gavilanes. Mar, desierto, y oasis en la história del hombre.* Corporación Financiera de Desarrollo S. A. (COFIDE). Oficina de Asuntos Culturales; Instituto Arqueológico Alemán, Comisión de Arqueología General y Comparada. Editorial Ausonia-Talleres Gráficos S. A., Lima.

Bonnichsen, R., and K. L. Turnmire (1991). *Clovis: Origins and Human Adaptation.* (R. Bonnichsen and K. Fladmark, Eds.) Center for the Study of the First Americans, Oregon State Univ., Corvallis.

Bonzani, R. M. (1997). Learning from the present: The constraints of plant seasonality on foragers and collectors. In *Recent Advances in the Archaeology of the Northern Andes* (A. Oyuelo-Caycedo and J. S. Raymond, Eds.), in press.

Boom, B. (1987). *Ethnobotany of the Chácobo Indians, Beni, Bolivia.* Advances in Economic Botany, Vol. 4, The New York Botanical Garden, Bronx, New York.

Boserup, E. (1965). *The Conditions of Agricultural Growth.* Aldine, Chicago.

Bozarth, S. R. (1987). Diagnostic opal phytoliths from rinds of selected *Cucurbita* species. *Am. Antiquity* **52,** 607–615.

Bradbury, J. P. (1989). Late Quaternary lacustrine palaeoenvironments in the Cuenca de Mexico. *Quat. Sci. Rev.* **8,** 75–100.

Braidwood, R. J. (1951). *Prehistoric Men* (second edition). Chicago Natural History Museum, Chicago.

Braidwood, R. J. (1960). The agricultural revolution. *Sci. Am.* **203,** 130–148.

Braun, A. (1968). *Cultivated Palms of Venezuela.* The International Palm Society, Lawrence, KS.

Bray, W. (1995). Searching for environmental stress: Climatic and anthropogenic influences on the landscape of Colombia. In *Archaeology in the Lowland American Tropics* (P. W. Stahl, Ed.), pp. 96–112. Cambridge Univ. Press, Cambridge, UK.

Bray, W. B., L. Herrera, and M. Cardale Schrimpff (1985). Report on the 1982 field season in Calima. *Pro Calima* **4,** 2–26.

Bray, W. B., L. Herrera, M. Cardale Schrimpff, P. Botero, and J. G. Monsalve (1987). The ancient agricultural landscape of Calima, Colombia. In *Pre-Hispanic Agricultural Fields in the Andean Region* (W. M. Denevan, K. Mathewson, and G. Knapp, Eds.), Vol. 359, pp. 443–481. British Archaeological Reports International Series, Oxford, UK.

Bray, W. B., L. Herrera, and M. Cardale Schrimpff (1988). Report on the 1984 field season in Calima. *Pro Calima* **5,** 2–42.

Bretting, P. K. (1990). New perspectives on the origin and evolution of the New World domesticated plants: Introduction. *Econ. Bot.* **44**(3, Suppl.), 1–5.

Bretting, P. K., M. M. Goodman, and C. W. Stuber (1987). Karyological and isozyme variation in West Indian and allied American mainland races of maize. *Am. J. Bot.* **74,** 1601–1613.

Briggs, P. S. (1986). Pre-conquest mortuary arts and status in the central region of Panama. Ph.D. dissertation, University of New Mexico. University Microfilms, Ann Arbor, MI.

Broecker, W. S. (1994). Massive iceberg discharges as triggers for global climate change. *Nature* **372,** 421–424.

Broecker, W. S., and G. H. Denton (1990). What drives glacial cycles? *Sci. Am.* **270,** 49–55.

Brokaw, N. V. L. (1982). Treefalls: Frequency, timing, and consequences. In *The Ecology of a Tropical Forest: Seasonal Rhythms and Long-Term Changes* (E. G. Leigh, Jr., A. S. Rand, and D. M. Windsor, Eds.), pp. 101–108. Smithsonian Institution Press, Washington, DC.

Brooks, A. S., D. M. Helgren, J. S. Cramer, A. Franklin, W. Hornyak, J. M. Keating, R. G. Klein, W. J. Rink, H. Schwarcz, J. N. L. Smith, K. Stewart, N. E. Todd, J. Verniers, and J. E. Yellen (1995). Dating and context of three middle stone age sites with bone points in the upper Semliki valley, Zaire. *Science* **268,** 548–556.

Broughton, J. M., and D. K. Grayson (1993). Diet breadth, adaptive change, and the White Mountains faunas. *J. Archaeol. Sci.* **20,** 331–336.

Brubaker, C. L., and J. F. Wendel (1994). Reevaluating the origin of domesticated cotton (*Gossypium hirsutum;* Malvaceae) using nuclear restriction fragment length polymorphisms (RFLPs). *Am. J. Bot.* **81,** 1309–1326.

Brücher, H. (1988). The wild ancestor of *Phaseolus vulgaris* in South America. In *Genetic Resources of Phaseolus Beans: Their Maintenance, Domestication, Evolution, and Utilization* (P. Gepts, Ed.), pp. 185–214. Kluwer, Dordrecht.

Brücher, H. (1989). *Useful Plants of Neotropical Origin and Their Wild Relatives.* Springer-Verlag, Berlin.

Bryan, A. (1989). *The Central American Filter Funnel: Earliest American Adaptation to Tropical Forests. Proceedings of the Circum-Pacific Prehistory Conference.* Washington State Univ. Press, Olympia.

Buckler, E. S., IV, T. B. Holtsford, and D. Pearsall (1995). Maize domestication: The molecular evidence with respect to biogeography and early Holocene climate change. Paper presented at the 60th Annual Meeting of the Society for American Archaeology, Minneapolis, MN.

Buckler, E. S., IV, and T. P. Holtsford (1996). *Zea* systematics: Ribosomal ITS evidence. *Mol. Biol. Evol.* **13,** 612–622.

Buckler, E. S., IV, D. M. Pearall, and T. P. Holtsford (1997). The impact of early Holocene climate and plant ecology on central Mexican archaic subsistence. *Curr. Anthropol.,* **38.**

Bullen, R. P., and W. W. Plowden, Jr. (1963). Preceramic Archaic sites in the highlands of Honduras. *Am. Antiquity* **28,** 382–385.

Bullock, S. H., H. A. Mooney, and E. Medina (Eds.) (1995). *Seasonally Dry Tropical Forests.* Cambridge Univ. Press, Cambridge, UK.

Burger, R. L. (1988). Unity and heterogeneity within the Chavín horizon. In *Peruvian Prehistory* (R. W. Keatinge, Ed.), pp. 99–144. Cambridge Univ. Press, Cambridge, UK.

Burger, R. L., and L. Salazar-Burger (1991). Recent investigations at the Initial period center of Cardal, Lurín valley. *J. Field Archaeol.* **18,** 275–296.

Bush, M. B. (1991). Modern pollen-rain data from South and Central America: A test of the feasibility of fine-resolution lowland tropical palynology. *Holocene* **1,** 162–167.

Bush, M. B. (1992). A simple yet efficient pollen trap for use in vegetation studies. *J. Vegetation Sci.* **3,** 275–276.

Bush, M. B. (1995). Neotropical plant reproductive strategies and fossil pollen representation. *Am. Nat.* **145,** 594–609.

Bush, M. B., and P. A. Colinvaux (1988). A 7,000 year-old pollen record from the Amazon lowlands. *Vegetatio* **76,** 141–154.

Bush, M. B., and P. A. Colinvaux (1990). A pollen record of a complete glacial cycle from lowland Panama. *J. Vegetation Sci.* **1,** 105–118.

Bush, M. B., D. R. Piperno, and P. A. Colinvaux (1989). A 6,000 year history of Amazonian maize cultivation. *Nature* **340,** 303–305.

Bush, M. B., P. A. Colinvaux, M. Wiemann, K. B. Liu, and D. R. Piperno (1990). Late Pleistocene temperature depression and vegetation change in Ecuadorian Amazonia. *Quat. Res.* **34,** 330–345.

Bush, M. B., D. R. Piperno, P. A. Colinvaux, P. De Oliveira, L. Krissek, M. Miller, and W. Rowe (1992). A 14,300 year paleoecological profile of a lowland tropical lake in Panama. *Ecol. Monogr.* **62,** 251–275.

Bush, M. B., P. A. Colinvaux, and M. C. Miller (1997). Two Holocene lake and forest histories from the Amazon lowlands. *J. Paleolimnol.,* submitted.

Butzer, K. W. (1971). *Environment and Archeology: An Ecological Approach to Prehistory,* 2nd ed. Aldine, Chicago.

Bye, R. A., Jr. (1981). Quelites—Ethnoecology of edible greens—Past, present and future. *J. Ethnobiol.* **1,** 109–123.

Byers, D. (Ed.) (1967). *The Prehistory of the Tehuacán Valley. Vol. I. Environment and Subsistence.* Univ. of Texas Press, Austin.

Byrd, B. (1992). The dispersal of food production across the Levant. In *Transitions to Agriculture in Prehistory* (A. B. Gebauer and T. D. Price, Eds.), pp. 49–61. Prehistory Press, Madison, WI.

Byrd, K. M. (1976). Changing animal utilization patterns and their implications: Southwest Ecuador (6500 BC–A.D. 1400). Unpublished Ph.D. dissertation, Department of Anthropology, University of Florida, Gainesville.

Byrd, K. M. (1996). Subsistence strategies in coastal Ecuador. In *Case Studies in Environmental Archaeology* (E. J. Reitz, L. A. Newsom, and S. J. Scudder, Eds.), pp. 305–316. Plenum, New York.

Campbell, K. E., Jr. (1982). Late Pleistocene events along the coastal plain of northwestern South America. In *Biological Diversification in the Tropics* (G. T. Prance, Ed.), pp. 423–440. Columbia Univ. Press, New York.

Cane, S. (1989). Australian aboriginal seed grinding and its archaeological record: A case study from the western desert. In *Foraging and Farming: The Evolution of Plant Exploitation* (D. R. Harris and G. C. Hillman, Eds.), pp. 99–119. Unwin Hyman, London.

Cardale de Schrimpff, M., W. Bray, T. Gähwiler-Walder, and L. Herrera (1992). *Calima: Diez Mil Años de Historia en el Suroccidente de Colombia.* Fundación Pro-Calima, Santa Fé de Bogotá.

Carneiro, R. L. (1961). Slash and burn cultivation among the Kuikuru and its Implications for cultural development in the Amazon Basin. In *The Evolution of Horticultural Systems in Native South America: Causes and Consequences, A Symposium* (J. Wilbert, Ed.), *Antropológica.* Supplement Publication No. 2., pp. 46–67. Sociedad de Ciencias Naturales La Salle, Caracas.

Carneiro, R. L. (1970a). The transition from hunting to horticulture in the Amazon Basin. Proceedings of the VIII International Congress of Anthropological and Ethnological Sciences, Vol. 3, pp. 144–148. Kyoto, Tokyo.

Carneiro, R. L. (1970b). A Theory of the Origin of the State. *Science* **169,** 733–738.

Carneiro, R. L. (1983). The cultivation of manioc among the Kuikuru of the upper Xingú. In *Adaptive Responses of Native Amazonians* (R. B. Hames and W. T. Vickers, Eds.), pp. 65–108. Academic Press, New York.

Cartelle, C., and W. C. Hartwig (1996). A new extinct primate among the Pleistocene megafauna of Bahia, Brazil. *Proc. Natl. Acad. Sci. USA* **93,** 6405–6409.

Casas, A., M. del Carmen Vásquez, J. J. Viveros, and J. Caballero (1996). Plant management among the Nahua and the Mixtec in the Balsas River Basin, Mexico. An ethnobotanical approach to the study of plant domestication. *Human Ecol.* **24,** 455–478.

Cashdan, E. (1982). The ecology of human subsistence. Review of *Hunter-Gatherer Foraging Strategies. Science* **216,** 1308–1309.

Castiñeiras, L., N. P. Nasser, and D. Piñero (1994). The origin of *Phaseolus vulgaris* L. in Cuba: Phaseolin patterns and their relationship with morpho-agronomical traits. *Plant Genetic Resour. Newslett.* **99,** 25–28.

Cavelier, I., C. Rodríguez, L. F. Herrera, G. Morcote, and S. Mora (1995). No solo de caza vive el hombre: Ocupación del bosque Amazónico, holoceno temprano. In *Ambito y Ocupaciones Tempranas de la América Tropical* (I. Cavelier and S. Mora, Eds.), pp. 27–44. Instituto Colombiano de Antropología, Fundación Erigaie, Santa Fé de Bogotá.

Chauchat, C. (1988). Early hunter–gatherers on the Peruvian coast. In *Peruvian Prehistory* (R. W. Keatinge, Ed.), pp. 41–66. Cambridge Univ. Press, Cambridge, UK.

Chikwendu, V. E., and C. E. A. Okezie (1989). Factors responsible for the ennoblement of African yams: Inferences from experiments in yam domestication. In *Foraging and Farming. The Evolution of Plant Exploitation* (D. R. Harris and G. C. Hillman, Eds.), pp. 344–357. Unwin Hyman, London.

Childe, V. G. (1952). *New Light on the Most Ancient East.* Praeger, New York.

Christenson, A. L. (1980). Change in the human food niche in response to population growth. In *Modeling Change in Prehistoric Subsistence Economies* (T. K. Earle and A. L. Christenson, Eds.), pp. 31–72. Academic Press, New York.

Churcher, C. S. (1966). The insect fauna from the Talara tarseeps, Peru. *Can. J. Zool.* **44,** 985–993.

Clark, J. S. (1988). Particle motion and the theory of charcoal analysis: Source area, transport, deposition, and sampling. *Quat. Res.* 30, 67–80.

Clement, C. R. (1988). Domestication of the pejibaye (*Bactris gasipaes*): Past and present. In *The Palm-Tree of Life* (M. J. Balick, Ed.), Advances in Economic Botany, Vol. 6, pp. 163–180. The New York Botanical Garden, Bronx, NY.

Clement, C. R. (1989). A center of crop genetic diversity in western Amazonia. *BioScience* **39,** 624–631.

Clement, C. R. (1993). Native Amazonian fruits and nuts: Composition, production and potential use for sustainable development. In *Tropical Forests, People and Food* (C. M. Hladik, A. Hladik, O. F. Linares, H. Pagezy, A. Semple, and M. Hadley, Eds.), pp. 139–152. UNESCO/ Parthenon Publishing Group, Paris.

Clement, C. R. (1994). Crops of the Amazon and the Orinoco regions: Their origin, decline, and future. In *Neglected Crops. 1492 From a Different Perspective* (J. E. H. Bermejo and J. León, Eds.), pp. 195–203. Food and Agriculture Organization of the United Nations, Rome.

Clement, C. R. (1997). Fruit trees and the origins of agriculture in the Neotropics. Proceedings of the Second International Congress on Ethnobiology, pp. 1–8. Kunming Institute of Botany, The Chinese Academy of Sciences, Kunming, Yunnan, China.

Clement, C. R., and J. E. Mora Urpi (1987). Pejibaye palm (*Bactris gasipaes,* Arecaceae): Multi-use potential for the lowland humid tropics. *Econ. Bot.* **41,** 302–311.

CLIMAP Project Members (1976). The surface of the Ice-Age earth. *Science* **191,** 1131–1137.

CLIMAP Project Members (1981). Seasonal reconstruction of the earth's surface at the last glacial maximum. Geological Society of America Map and Chart Series MC-36.

Coates, A. G. (1997). The forging of Central America. In *Central America: A Natural and Cultural History* (A. G. Coates, Ed.), in press. Yale Univ. Press, New Haven, CT.

Cochrane, T. T., and P. G. Jones (1981). Savannas, forests and wet season potential evapotranspiration in tropical South America. *Tropical Agric.* **6,** 185–190.

Cohen, M. N. (1977a). *The Food Crisis in Prehistory.* Yale Univ. Press, New Haven, CT.

Cohen, M. N. (1977b). Population pressure and the origins of agriculture: An archaeological example from the coast of Peru. In *Origins of Agriculture* (C. A. Reed, Ed.), pp. 135–177. Mouton, The Hague, The Netherlands.

Cohen, M. N. (1987). The significance of long-term changes in human diet and food economy. In *Food and Evolution* (M. Harris and E. B. Ross, Eds.), pp. 261–282. Temple Univ. Press, Philadelphia.

Cole, M. (1986). *The Savannas: Biogeography and Geobotany.* Academic Press, London.

Coley, P. D. (1983). Herbivory and defensive characteristics of tree species in a lowland tropical forest. *Ecol. Monogr.* **53,** 209–233.

Coley, P. D., and J. Barone (1996). Herbivory and plant defenses in tropical forests. *Annu. Rev. Ecol. Systematics* **27,** 305–335.

Coley, P. D., J. P. Bryant, and F. S. Chapin, III (1985). Resource availability and plant antiherbivore defense. *Science* **230,** 895–899.

Colinvaux, P. (1987). Amazon diversity in light of the paleoecological record. *Quat. Sci. Rev.* **6,** 93–114.

Colinvaux, P. (1993). Pleistocene biogeography and diversity in tropical forests of South America. In *Biological Relationships between Africa and South America* (P. Goldblatt, Ed.), pp. 473–499. Yale Univ. Press, New Haven, CT.

Colinvaux, P. A., and M. B. Bush (1991). The rain-forest ecosystem as a resource for hunting and gathering. *Am. Anthropol.* **93,** 153–160.

Colinvaux, P. A., P. E. De Oliveira, P. E. Moreno, M. C. Miller, and M. B. Bush. (1996a). A long pollen record from lowland Amazonia: Forest and cooling in glacial times. *Science* **274,** 85–88.

Colinvaux, P. A., K-b. Liu, P. De Oliveira, M. B. Bush, M. C. Miller, and M. Steinitz-Kannan (1996b). Temperature depression in the lowland tropics in Glacial times. *Climatic Change* **32,** 19–33.

Colinvaux, P. A., M. B. Bush, M. Steinitz-Kannan, and M. C. Miller (1997). Glacial and Postglacial pollen records from the Ecuadorian Andes and Amazon. *Quart. Res.* **48,** 69–78.

Connell, J. H. (1978). Diversity in tropical rain forests and coral reefs. *Science* **199,** 1302–1310.

Contreras, J., D. F. Austin, F. de la Puente, and J. Díaz. (1995). Biodiversity of sweet potato (*Ipomoea batatas,* Convolvulacae) in Southern Mexico. *Econ. Bot.* **49,** 286–296.

Cooke, R. G. (1992). Prehistoric nearshore and littoral fishing in the eastern tropical Pacific: An ichthyological evaluation. *World Archaeol.* **6,** 1–49.

Cooke, R. G. (1995). Monagrillo, Panama's first pottery: Summary of research, with new interpretations. In *The Emergence of Pottery: Technology and Innovation in Ancient Societies,* (W. K. Barnett and J. W. Hoopes, Eds.), pp. 169–184. Smithsonian Institution Press, Washington, DC.

Cooke, R. G. (1997). Human settlement of Central America and northern South America. *Quart. Int.,* in press.

Cooke, R. G. and A. J. Ranere (1984). The "Proyecto Santa María": A multidisciplinary analysis of prehistoric adaptations to a tropical watershed in Panama. In *Recent Developments in Isthmian Archaeology* (F. W. Lange, Ed.), British Archaeological Reports International Series 212, pp. 3–30. Oxford, UK.

Cooke, R. G., and A. J. Ranere (1989). Hunting in prehistoric Panama: A diachronic perspective. In *The Walking Larder: Patterns of Domestication, Pastoralism and Predation* (J. Clutton-Brock, Ed.), pp. 295–315. Unwin Hyman, London.

Cooke, R. G., and A. J. Ranere (1992a). Prehistoric human adaptations to the seasonally dry forests of Panama. *World Archaeol.* **24,** 114–133.

Cooke, R. G., and A. J. Ranere (1992b). The origins of wealth and hierarchy in the central region of Panama, with observations on its relevance to the phylogeny of Chibchan-speaking polities in Panama and elsewhere. In *Wealth and Hierarchy in the Intermediate Area* (F. W. Lange, Ed.), pp. 243–316. Dumbarton Oaks, Washington, DC.

Cooke, R. G., and A. J. Ranere (1992c). Human influences on the zoogeography of Panama: An update based on archaeological and ethnohistorical evidence. In *Biogeography of Mesoamerica* (S. P. Darwin and A. L Welden, Eds.), Special publication of the Mesoamerican Ecology Institute, pp. 21–58. Tulane Univ. of Louisiana, New Orleans.

Cooke, R. G., and A. J. Ranere (1997). The relation of fish resources to the location, diet breadth, and procurement technology of preceramic and ceramic sites in an estuarine embayment on the Pacific coast of Panama. Washington State Univ. Press, Olympia.

Cooke, R. G., Norr, L., and D. R. Piperno (1996). Native Americans and the Panamanian landscape. In *Case Studies in Environmental Archaeology* (E. J. Reitz, L. A. Newsom, and S. J. Scudder, Eds.), pp. 103–122. Plenum Press, New York.

Correal Urrego, G. (1981). *Evidencias Culturales y Megafauna Pleistocénica en Colombia.* Publicación de la Fundación de Investigaciones Arqueológicas Nacionales 12, Banco de la República, Bogotá, Colombia.

Correal Urrego, G. (1986). Apuntes sobre el medio ambiente Pleistocénico y el hombre prehistórico en Colombia. In *New Evidence for the Pleistocene Peopling of the Americas* (A. L. Bryan, Ed.), pp. 115–131. Center for the Study of Early Man, Univ. of Maine, Orono.

Correal Urrego, G., and T. Van der Hammen (1977). *Investigaciones Arqueológicas en los Abrigos Rocosos del Tequendama: 12,000. Años de Historia del Hombre y de su Medio Ambiente en la Altiplanicie de Bogotá.* Biblioteca Banco Popular, Bogotá, Colombia.

Correal Urrego, G., T. Van der Hammen, and J. C Lerman (1966–1969). Artefactos líticos de abrigos rocosos en El Abra, Colombia: Informe preliminar. *Revista Colombiana de Arqueología* **14,** 12–46.

Coursey, D. G. (1967). *Yams*. Longman, London.

Coursey, D. G. (1976). Yams. *Dioscorea* spp. (Dioscoreaceae). In *Evolution of Crop Plants* (N. W. Simmonds, Ed.), pp. 70-74. Longman, London.

Cowan, C. W., and P. J. Watson (Eds.) (1992). *The Origins of Agriculture: An International Perspective.* Smithsonian Institution Press, Washington, DC.

Croat, T. (1978). *The Flora of Barro Colorado Island.* Stanford Univ. Press, Stanford, CA.

Cruxent, J. M. (1970). Projectile points with Pleistocene mammals in Venezuela. *Antiquity* **44,** 223–225.

Cuevas, A. Axayacatl (1994). Spanish plum, red mombin (*Spondias purpurea*). In *Neglected Crops. 1492 from a Different Perspective* (J. E. H. Bermejo and J. León, Eds.), pp. 111–115. Food and Agriculture Organization of the United Nations, Rome.

Damp, J. E. (1984a). Architecture of the early Valdivia village. *Am. Antiquity* **49,** 573–585.

Damp, J. E. (1984b). Environmental variation, agriculture, and settlement processes in Coastal Ecuador (3300–1500 B.C.). *Curr. Anthropol.* **25,** 106–111.

Damp, J. E. (1988). *La primera ocupación Valdivia de Real Alto. Patrones económicos, arquitectónicos, e ideológicos.* Biblioteca Ecuatoriana de Arqueología, 3. Escuela Politécnica del Litoral, Centro de Estudios Arqueológicos y Antropológicos, Guayaquil.

Damp, J. E., and D. M. Pearsall (1994). Early cotton from coastal Ecuador. *Econ. Bot.* **48,** 163–165.

Damp, J. E., and L. P. Vargas S. (1995). The many contexts of early Valdivia ceramics. In *The Emergence of Pottery: Technology and Innovation in Ancient Societies* (W. K. Barnett and J. W. Hoopes, Eds.), pp. 157–168. Smithsonian Institution Press, Washington, DC.

Damp, J. E., D. M. Pearsall, and L. T. Kaplan (1981). Beans for Valdivia. *Science* **212,** 811–812.

Darwin, C. (1845). *The Voyage of the Beagle* (L. Engel, Ed.), Doubleday Anchor (This edition published in 1962.) The American Museum of Natural History, Garden City, NY.

DeBoer, W. (1975). The archaeological evidence for manioc cultivation: A cautionary note. *Am. Antiquity* **40,** 419–433.

Debouck, D. G. (1989). Early beans (*Phaseolus vulgaris* L. and *P. lunatus* L.) domesticated for their aesthetic value? *Annu. Rep. Bean Improvement Cooperative* **32,** 62–63.

Debouck, D. G. (1994). Beans (*Phaseolus* spp.). In *Neglected Crops. 1492). from a Different Perspective* (J. E. Hernández Bermejo and J. León, Eds.), pp. 47–62. Food and Agriculture Organization of the United Nations, Rome.

Debouck, D. G., J. H. Liñan Jara, A. Campana Sierra, and J. H. de la Cruz Rojas (1987). Observations on the domestication of *Phaseolus lunatus* L. *FAO/IBPGR Plant Genetic Resour. Newslett.* **70,** 26–32.

Debouck, D. G., R. Castillo T., and J. M. Tohme (1989a). Observations on little-known *Phaseolus* germplasm of Ecuador. *FAO/IBPGR Plant Genetic Resour. Newslett.* **80,** 15–21.

Debouck, D. G., A. Maquet, and C. E. Posso (1989b). Biochemical evidence for two different gene pools in lima beans, *Phaseolus lunatus* L. *Annu. Rep. Bean Improvement Cooperative* **32,** 58–59.

Debouck, D. G., O. Toro, O. M. Paredes, W. C. Johnson, and P. Gepts (1993). Genetic diversity and ecological distribution of *Phaseolus vulgaris* (Fabaceae) in northwestern South America. *Econ. Bot.* **47,** 408–423.

Decker-Walters, D. S., T. W. Walters, U. Posluszny, and P. G. Kevan (1990). Genealogy and gene flow among annual domesticated species of *Cucurbita*. *Can. J. Bot.* **68,** 782–789.

Decker-Walters, D. S., T. W. Walters, C. W. Cowan, and B. D. Smith (1993). Isozymic characterization of wild populations of *Cucurbita pepo*. *J. Ethnobiol.* **13,** 55–72.

Deevey, E. S., D. S. Rice, P. M. Rice, H. H. Vaughan, M. Brenner, and S. Flannery (1979). Mayan urbanism: Impact on a tropical karst environment. *Science* **206,** 298–306.

De Oliveira, P. (1992). A palynological record of late Quaternary vegetation and climatic change in southeastern Brazil. Unpublished Ph.D. thesis. The Ohio State University, Columbus.

Delgado, S., A. A. Bonet, and P. Gepts (1988). The wild relative of *Phaseolus vulgaris* in Middle America. In *Genetic Resources of Phaseolus Beans: Their Maintenance, Domestication, Evolution and Utilization* (P. Gepts, Eds.), pp. 163–184. Kluwer, Dordrecht.

Denevan, W. M. (1992). Stone vs metal axes: The ambiguity of shifting cultivation in prehistoric Amazonia. *J. Steward Anthropol. Soc.* **20,** 153–165.

Denevan, W. M., J. M. Treacy, J. B. Alcorn, C. Padoch, J. Denslow, and S. F. Paitán (1984). Indigenous agroforestry in the Peruvian Amazon: Bora Indian management of swidden fallows. *Interciencia* **9,** 346–357.

Dering, P., and G. H. Weir (1979). Analysis of plant remains from the Preceramic site of Paloma, Chilca Valley, Peru. Manuscript on file, Univ. of Missouri, Columbia.

Dering, P., and G. H. Wier (1981). Preliminary plant macrofossil analysis of La Paloma deposits. Manuscript on file, Univ. of Missouri, Columbia.

DeVries, T. J. (1987). A review of geological evidence for ancient El Niño activity in Peru. *J. Geophys. Res.* **92**(C13), 14,471–14,479.

DeVries, T. J., and L. E. Wells (1990). Thermally-anomalous Holocene molluscan assemblages from coastal Peru: Evidence for paleogeographic, not climate change. *Palaeogeogr. Palaeoclimatol. Palaeoecol.* **81,** 11–32.

Díaz, J., F. de la Puente, and D. F. Austin (1992). Enlargement of fibrous roots in *Ipomoea* section *batatas* (Convolvulaceae). *Econ. Bot.* **26,** 322–329.

Dillehay, T. (1989). *Monteverde: A Late Pleistocene Settlement in Chile,* Vol. 1. Smithsonian Institution Press, Washington, DC.

Dillehay, T. (1991). Disease ecology and initial human migration. In *The First Americans: Search and Research* (T. D. Dillehay and D. J. Meltzer, Eds.), pp. 231–266. CRC Press, Boca Raton, FL.

Dillehay, T. D., P. J. Netherly, and J. Rossen (1989). Middle preceramic public and residential sites on the forested slope of the Western Andes, Northern Peru. *Am. Antiquity* **54,** 733–759.

Dillehay, T. D., G. A. Calderón, G. Politis, and M. da C. Coutinho Beltrâo (1992). Earliest hunters and gatherers of South America. *J. World Prehistory* **6,** 145–204

Dillehay, T. D., J. Rossen, and P. J. Netherly (1997). The Nanchoc tradition: The beginnings of Andean civilization. *Am. Sci.* **85,** 46–55.

Dippery, J. K., D. T. Tissue, R. B. Thomas, and B. R. Strain (1995). Effects of low and elevated CO_2 on C_3 and C_4 annuals: I. Growth and biomass allocation. *Oecologia* **101,** 13–20.

Doebley, J. (1990). Molecular evidence and the evolution of maize. *Econ. Bot.* **44**(3, Suppl.), 6–27.

Doebley, J. (1992). Mapping the genes that made maize. *TIG* **8,** 302–307.

Doebley, J. (1994a). Morphology, molecules, and maize. In *Corn and Culture in the Prehistoric New World* (S. Johannessen and C. A. Hastorf, Eds.), pp. 101–112. Westview Press, Boulder, CO.

Doebley, J. (1994b). Genetics and the morphological evolution of maize. In *The Maize Handbook,* (M. Freeling and V. Walbot, Eds.), pp. 66–76. Springer-Verlag, New York.

Doebley, J., and A. Stec (1991). Genetic analysis of the morphological differences between maize and teosinte. *Genetics* **129,** 285–295.

Dorweiler, J., A. Stec, J. Kermicle, and J. Doebley (1993). Teosinte glume architecture. 1. A genetic locus controlling a key step in maize evolution. *Science* **262,** 233–235.

Drennan, R. D. (1995). Chiefdoms in northern South America. *J. World Prehistory* **9,** 301–340

Dufour, D. (1988). Cyanide content of Cassava (*Manihot esculenta,* Euphorbiaceae) cultivars used by Tukanoan indians in northwest Amazonia. *Econ. Bot.* **42,** 255–266.

Dunnell, R. C. (1980). Evolutionary theory and archaeology. In *Advances in Archaeological Method and Theory,* (M. B. Schiffer, Ed.), Vol. 3, pp. 38–99. Academic Press, New York.

Dunnell, R. C. (1989). Aspects of the application of evolutionary theory in archaeology. In *Archaeological Thought in America* (C. Lamberg-Karlovsky, Ed.), pp. 35–49. Cambridge University Press, New York.

Durham, W. (1981). Overview: Optimal foraging analysis in human ecology. In *Hunter–Gatherer Foraging Strategies* (B. Winterhalder and E. A. Smith, Eds.), pp. 218–231. Univ. of Chicago Press, Chicago.

Earle, T. (1980). A model of subsistence change. In *Modelling Change in Prehistoric Subsistence Economies* (T. K. Earle and A. Christenson, Eds.), pp. 1–29. Academic Press, New York.

Eder, J. (1978). The caloric returns to food collecting: Disruption and change among the Batak of the Philippine tropical forest. *Human Ecol.* **6,** 55–69.

Egawa, Y., and M. Tanaka (1984). Cytogenetical relationships among three species of chili peppers, *Capsicum chinense, C. frutescens,* and *C. baccatum. Jpn. J. Breed.* **34**, 50–56.

Eisenberg, J. F. (1989). *Mammals of the Neotropics: The Northern Neotropics, Volume 1.* Univ. of Chicago Press, Chicago.

Eisenberg, J. F., and R. W. Thorington, Jr. (1973). A preliminary analysis of a Neotropical mammal fauna. *Biotropica* **5**, 150–161.

Eiten, G. (1972). The cerrado vegetation of Brazil. *Bot. Rev.* **38**, 201–341.

Ellen, R. (1988). Foraging, starch extraction and the sedentary lifestyle in the lowland rainforest of Central Seram. In *Hunters and Gatherers 1: History, Evolution and Social Change,* (T. Ingold, D. Riches, and J. Woodburn, Eds.), pp. 117–134. Berg, Oxford, UK.

Emiliani, C. (1992). Pleistocene paleotemperatures. *Science* **257**, 1462.

Emiliani, C., and D. B. Erickson (1991). The glacial/interglacial temperature range of the surface water of the oceans at low latitudes. In *Stable Isotope Geochemistry: A Tribute to Samuel Epstein* (H. P. Taylor, Jr., R. O'Neil, and I. R. Kaplan, Eds.), The Geochemical Society Spec. Publ. No. 3. pp. 223–228.

Endicott, K., and P. Bellwood (1991). The possibility of independent foraging in the rain forest of peninsular Malaysia. *Human Ecol.* **19**, 151–185.

Enfield, D. B. (1992). Historical and prehistorical overview of El Niño/Southern Oscillation. In *El Niño: Historical and Paleoclimatic Aspects of the Southern Oscillations,* (H. F. Diaz and V. Markgraf, Eds.), pp. 95–117. Cambridge University Press, Cambridge, UK.

Engel, F. (1988). *Ecología Prehistórica Andina, I and II.* Centro de Investigaciones de Zonas Aridas (CIZA) de la Universidad Nacional Agraria, Lima, Peru.

Ericson, J. E., M. West, C. H. Sullivan, and H. W. Krueger (1989). The development of maize agriculture in the Viru valley, Peru. In *The Chemistry of Prehistoric Human Bone* (T. D. Price, Ed.), pp. 68–104. Cambridge University Press, Cambridge, UK.

Eshbaugh, W. H. (1976). Genetic and biochemical systematic studies of chili peppers (*Capsicum*-Solanaceae). *Bull. Torrey Bot. Club* **102**, 396–403.

Eshbaugh, W. H. (1979). Biosystematic and evolutionary study of the *Capsicum pubescens* complex. In *National Geographic Society Research Reports* (P. H. Oehser and J. S. Lea, Eds.), pp. 143–162. National Geographic Society, Washington, DC.

Eshbaugh, W. H., S. L. Guttman, and M. J. McLeod (1983). The origin and evolution of domesticated *Capsicum* species. *J. Ethnobiol.* **3**, 49–54.

Esquivel, M., L. Castiñeiras, L. Lioi, and K. Hammer (1993). The domestication of lima bean (*Phaseolus lunatus* L.) in Cuba: Morphological and biochemical studies. *FAO/IBPGR Plant Genetic Resour. Newslett.* **91/92**, 21–22.

Eubanks, M. (1995). A cross between two maize relatives: *Tripsacum dactyloides* and *Zea diploperennis* (Poaceae). *Econ. Bot.* **49**, 172–182.

Faegri, K. (1966). Some problems of representativity in pollen analysis. *Paleobotanist* **15**, 135–140.

Fairbanks, R. G. (1989). A 17,000-year glacio-eustatic sea level record: Influence of glacial melting rates on the Younger Dryas event and deep-ocean circulation. *Nature* **342**, 637–642.

Feldman, R. A. (1980). Aspero, Peru: Architecture, subsistence economy, and other artifacts of a preceramic chiefdom. Unpublished Ph.D. dissertation, Department of Anthropology, Harvard University, Cambridge, MA.

Fiedel, S. J. (1996). Paleoindians in the Brazilian Amazon. *Science* **274**, 1820–1825.

Fischer, A. (1960). Latitudinal variations in organic diversity. *Evolution* **14**, 64–81.

Flannery, K. V. (1967). Vertebrate fauna and hunting patterns. In *The Prehistory of the Tehuacán Valley Vol. 1: Environment and Subsistence* (D. S. Byers, Ed.), pp. 132–177. Univ. of Texas Press, Austin.

Flannery, K. V. (1969). Origins and ecological effects of early domestication in Iran and the Near East. In *The Domestication and Exploitation of Plants and Animals* (P. J. Ucko and G. W. Dimbleby, Eds.), pp. 73–100. Duckworth, London.

Flannery, K. V. (1973). The origins of agriculture. *Annu. Rev. Anthropol.* **2**, 271–310.

Flannery, K. V. (1986a). The research problem. In *Guilá Naquitz: Archaic Foraging and Early Agriculture in Oaxaca, Mexico* (K. V. Flannery, Ed.), pp. 3–18. Academic Press, Orlando.

Flannery, K. V. (1986b). Wild food resources of the Mitla caves: Productivity, seasonality, and annual variation. In *Guilá Naquitz: Archaic Foraging and Early Agriculture in Oaxaca, Mexico* (K. V. Flannery, Ed.), pp. 255–264. Academic Press, Orlando.

Flannery, K. V. (1986c). (Ed.) *Guilá Naquitz: Archaic Foraging and Early Agriculture in Oaxaca, Mexico.* Academic Press, Orlando.

Foley, R. (1988). Hominids, humans and hunter–gatherers: An evolutionary perspective. In *Hunters and Gatherers 1: History, Evolution and Social Change* (T. Ingold, D. Riches, and J. Woodburn, Eds.), pp. 207–221. Berg, Oxford, UK.

Food and Agriculture Organization FAO (1971). *Inventariacion y Demonstraciones Forestales, Panamá. Zonas de Vida.* (Based on the work of Joseph Tosi), FO:SF Pan 6, Technical Report 2. FAO, Rome.

Ford, R. I. (1984). Prehistoric phytogeography of economic plants in Latin America. In *Pre-Columbian Plant Migration* (D. Stone, Ed.), Papers of the Peabody Museum of Archaeology and Ethnology, pp. 177–183. Harvard Univ. Press, Cambridge, MA.

Ford, R. I. (1985). The processes of plant food production in prehistoric North America. In *Prehistoric Food Production in North America* (R. I. Ford, Ed.), Anthropological Paper No. 75, pp. 1–18. Museum of Anthropology, Univ. of Michigan, Ann Arbor.

Forsyth, A., and K. Miyata (1984). *Tropical Nature.* Scribner, New York.

Foster, R. B. (1982). Famine on Barro Colorado Island. In *The Ecology of a Tropical Rainforest: Seasonal Rhythms and Long-Term Changes* (E. G. Leigh, Jr., A. S. Rand, and D. M. Windsor, Eds.), pp. 201–212. Smithsonian Institution Press, Washington, DC.

Foster, R. B. (1990). The floristic composition of the Rio Manu floodplain forest. In *Four Neotropical Rainforests* (A. H. Gentry, Ed.), pp. 99–111. Yale Univ. Press, New Haven, CT.

Foster, R. B., and S. P. Hubbell (1990). The floristic composition of the Barro Colorado Island forest. In *Four Neotropical Rainforests* (A. H. Gentry, Ed.), pp. 85–98. Yale Univ. Press, New Haven, CT.

Frankie, G., H. Baker, and P. Opler (1974). Comparative phenological studies of trees in tropical wet and dry forests in the lowlands of Costa Rica. *J. Ecol.* **62**, 881–919.

Freeland, W. J., and D. H. Janzen (1974). Strategies in herbivory by mammals: The role of plant secondary compounds. *Am. Nat.* **108**, 269–289.

Fregene, M. A., J. Vargas, J. Ikea, F. Angel, J. Tohme, R. A. Asiedu, M. O. Akoroda, and W. M. Roca (1994). Variability of chloroplast DNA and nuclear ribosomal DNA in cassava (*Manihot esculenta* Crantz) and its wild relatives. *Theor. Appl. Genetics* **89**, 719–727.

Freyre, R., R. Ríos, L. Guzmán, D. G. Debouck, and P. Gepts (1996). Ecogeographic distribution of *Phaseolus* spp. (Fabaceae) in Bolivia. *Econ. Bot.* **50**, 195–215.

Fritz, G. J. (1994). Are the first American farmers getting younger? *Curr. Anthropol.* **35**, 305–309.

Fritz, G. J. (1995). New dates and data on early agriculture: The legacy of complex hunter–gatherers. *Ann. Missouri Bot. Garden* **82**, 3–15.

Frost, I. (1988). A Holocene sedimentary record from Anañgucocha in the Ecuadorian Amazon. *Ecology* **69**, 66–73.

Furnier, G. R., M. P. Cummings, and M. T. Clegg (1990). Evolution of the avocados as revealed by DNA restriction fragment variation. *J. Heredity* **81**, 183–188.

Gade, D. W. (1966). Achira, the edible Canna, its cultivation and use in the Peruvian Andes. *Econ. Bot.* **20**, 407–415.

Galinat, W. C. (1975). The evolutionary emergence of maize. *Bull. Torrey Bot. Club* **102**, 313–324.

Galinat, W. C. (1985). Domestication and diffusion of maize. In *Prehistoric Food Production in North America* (R. I. Ford, Ed.), Anthropological Papers No. 75, pp. 245–278. Museum of Anthropology, Univ. of Michigan, Ann Arbor.

Galinat, W. C. (1988). The origin of corn. In *Corn and Corn Improvement,* Agronomy Monograph No. 18, pp. 1–31. ASA-CSSA-SSSA, Madison, WI.

Galinat, W. C. (1992). Evolution of corn. *Adv. Agron.* **47**, 203–231.

Galinat, W. C. (1995). El origen del maíz. El grano de la humanidad; The origin of maize: Grain of humanity. *Econ. Bot.* **49,** 3–12.

García-Bárcena, J. (1982). *El Precerámico de Aguacatenango.* Colección Científica 110, INAH. Chiapas, Mexico.

Gebauer, A. B., and T. D. Price (1992a). Foragers and farmers: An introduction. In *Transitions to Agriculture in Prehistory* (A. B. Gebauer and T. D. Price, Eds.), pp. 1–10. Prehistory Press, Madison, WI.

Gebauer, A. B., and T. D. Price (Eds.) (1992b). *Transitions to Agriculture in Prehistory.* Prehistory Press, Madison, WI.

Gentry, A. H. (1988a). Changes in plant community diversity and floristic composition on environmental and geographical gradients. *Ann. Missouri Bot. Garden,* **75,** 1–34.

Gentry, A. H. (1988b). Tree species richness of upper Amazonian forests. *Proc. Nat. Acad. Sci. USA* **85,** 156–159.

Gentry, A. H. (1990). Floristic similarities and differences between Southern Central America and upper Central Amazonia. In *Four Neotropical Rainforests* (A. H. Gentry, Ed.), pp. 141–157. Yale Univ. Press, New Haven, CT.

Gentry, A. H. (1993). *A Field Guide to the Families and Genera of Woody Plants of Northwest South America (Colombia, Ecuador, Peru) with Supplementary Notes on Herbaceous Taxa.* Conservation International, Washington, DC.

Gentry, A. H. (1995). Diversity and floristic composition of Neotropical dry forests. In *Seasonally Dry Tropical Forests,* (S. H. Bullock, H. A. Mooney and E. Medina, Eds.), pp. 146–194. Cambridge Univ. Press, Cambridge, U.K.

Gentry, A. H., and J. Terborgh (1990). Composition and dynamics of the Cocha Cashu "Mature" floodplain forest. In *Four Neotropical Rainforests* (A. H. Gentry, Ed.), Yale Univ. Press, pp. 542–564. New Haven, CT.

Gepts, P. (1988). Phaseolin as an evolutionary marker. In *Genetic Resources of Phaseolus Beans: Their Maintenance, Domestication, Evolution, and Utilization* (P. Gepts, Eds.), pp. 215–241. Kluwer, Dordrecht.

Gepts, P. (1990). Biochemical evidence bearing on the domestication of *Phaseolus* (Fabaceae) beans. *Econ. Bot.* **44**(3, Suppl.), 28–38.

Gepts, P. (1991). Biotechnology sheds light on bean domestication in Latin America. *Diversity* **7,** 49–50.

Gepts, P., and F. A. Bliss (1986). Phaseolin variability among wild and cultivated common beans (*Phaseolus vulgaris*) from Colombia. *Econ. Bot.* **40,** 469–478.

Gepts, P., T. C. Osborn, K. Rashka, and F. A. Bliss (1986). Phaseolin–protein variability in wild forms and landraces of the common bean (*Phaseolus vulgaris*): Evidence for multiple centers of domestication. *Econ. Bot.* **40,** 451–468.

Giacometti, D. C., and J. León (1994). *Tannia, Yautia (Xanthosoma sagittifolium).* In *Neglected Crops. 1492 from a Different Perspective.* (J. E. Hernández Bermejo and J. León, Eds.), pp. 253–258. Food and Agriculture Organization of the United Nations, Rome.

Gibbons, A. (1996). First Americans: Not mammoth hunters, but forest dwellers? *Science* **272,** 346–347.

Glanz, W. E. (1982). The terrestrial mammal fauna of Barro Colorado Island: Censuses and long-term changes. In *The Ecology of a Tropical Forest: Seasonal Rhythms and Long-Term Changes* (E. G. Leigh, Jr., A. S. Rand, and D. M. Windsor, Eds.), pp. 455–468. Smithsonian Institute Press, Washington, DC.

Glanz, W. E. (1990). Neotropical mammal densities: How unusual is the community on Barro Colorado Island, Panama? In *Four Neotropical Rainforests* (A. H. Gentry, Ed.), pp. 287–313. Yale Univ. Press, New Haven, CT.

Glassow, M. A. (1978). The concept of carrying capacity in the study of culture process. In *Advances in Archeological Method and Theory,* (M. B. Schiffer Ed.), Vol. 1, pp. 31–48. Academic Press, New York.

Glynn, P. W. (1988). El Niño-Southern Oscillation 1982–1983: Nearshore population, community, and ecosystem responses. *Ann. Rev. of Ecol. Systematics* **19,** 309–345.

Gnecco, Valencia C. (1994). The Pleistocene–Holocene Boundary in the Northern Andes: An archaeological perspective. Unpublished Ph.D. dissertation, Washington University, St. Louis.

Gnecco, Valencia C. (1995). Movilidad y acceso a recursos de cazadores recolectores prehispánicos: el caso del Valle de Popayán. In *Ambito y Ocupaciones Tempranas de la América Tropical* (I. Cavelier and S. Mora, Eds.), pp. 59–72. Instituto Colombiano de Antropología, Fundación Erigaie, Santa Fe de Bogotá.

Gnecco, Valencia C., and A. Mohammed (1994). Tecnología de cazadores-recolectores subandinos: Análisis funcional y organización tecnológica. *Rev. Colombiana Antropol.* **31,** 7–31.

Gnecco, Valencia C., and H. Salgado López (1989). Adaptaciones precerámicas en el suroccidente de Colombia. *Museo del Oro Boletín* **24,** 35–53.

Goloubinoff, P., S. Paabo, and A.C. Wilson (1993). Evolution of maize inferred from sequence diversity of an *Adh2* gene segment from archaeological specimens. *Proc. Natl. Acad. Sci. USA* **90,** 1997–2001.

Golson, J. (1991a) Bulmer Phase II: Early Agriculture in the New Guinea Highlands. In *Man and a Half: Essays in Pacific Anthropology and Ethnobiology in Honor of Ralph Bulmer* (A. Pawley, Ed.), pp. 484–491. The Polynesian Society, Auckland, New Zealand.

Golson, J. (1991b). The New Guinea highlands on the eve of agriculture. In *Indo-Pacific Prehistory 1990* (P. Bellwood, Ed.), Vol. 2, Bull. No.11. pp. 82–91. Indo-Pacific Prehistory Association, Canberra, Australia.

Goodman, M. M., and C. W. Stuber (1983). Races of maize. VI. Isozyme variation among races of maize in Bolivia. *Maydica* **XXVIII,** 169–187.

Goslar, T., M. Arnold, E. Bard, T. Kuc, M. F. Pazdur, M. Ralska-Jasiewiczowa, K. Rózáński, N. Tisnerat, A. Walanus, B. Wicik, and K. Więckowski (1995). High concentration of atmospheric ^{14}C during the Younger Dryas cold episode. *Nature* **377,** 414–417.

Gragson, T. L. (1993). Human foraging in lowland South America: Pattern and process of resource procurement. *Res. Econ. Anthropol.* **14,** 107–138.

Gregory, W. C., and M. P. Gregory (1976). Groundnut. *Arachis hypogaea* (Leguminosae-Papilionatae). In *Evolution of Crop Plants* (N. W. Simmonds, Ed.), pp.151–154. Longman, London.

Gremillion, K. J. (1989). The development of a mutualistic relationship between humans and maypops: *Passiflora incarnata* L. in the southeastern United States. *J. Ethnobiol.* **9,** 135–155.

Gremillion, K. J. (1996). Diffusion and adoption of crops in evolutionary perspective. *J. Anthropol. Archaeol.* **15,** 183–204.

Grieder, T. (1988). Radiocarbon measurements. In *La Galgada, Peru. A Preceramic Culture in Transition* (T. Grieder, A. B. Mendoza, C. Earle Smith, and R. M. Malina) pp. 68–72. Univ. of Texas Press, Austin.

Grieder, T., A. B. Mendoza, C. Earle Smith, and R. M. Malina (1988). La Galgada in the world of its time. In *La Galgada. A Late Preceramic Culture in Transition* (T. Grieder, A. B. Mendoza, C. E. Smith, and R. M. Malina) pp. 192–203. Univ. of Texas Press, Austin.

Grobman, A., and D. Bonavia (1978). Pre-ceramic maize on the north central coast of Peru. *Nature* **276,** 386–387.

Gross, D. (1975). Protein capture and cultural development in the Amazon Basin. *Am. Anthropol.* **77,** 526–549.

Groube, L. (1989). The taming of the rainforest: A model for Late Pleistocene forest exploitation in New Guinea. In *Foraging and Farming. The Evolution of Plant Exploitation* (D. R. Harris and G. C. Hillman, Eds.), pp. 292–304. Unwin Hyman, London.

Gruhn, R., and A. L. Bryan (1984). The record of Pleistocene megafaunal extinction at Taima-Taima, Northern Venezuela. In *Quaternary Extinctions: A Prehistoric Revolution* (P. S. Martin and R. G. Klein, Eds.), pp. 128–137. Univ. of Arizona Press, Tucson.

Gruhn, R. D., A. L. Bryan, and J. D. Nance (1977). Los Tapiales: A Paleo-Indian campsite in the Guatemalan highlands. *Proc. Am. Philos. Soc.* **121,** 235–273.

Guidon, N. (1989). On stratigraphy and chronology at Pedra Furada. *Curr. Anthropol.* **30,** 641–642.

Guidon, N., and G. Delibrias (1986). Carbon-14 dates point to man in the Americas 32,000 years ago. *Nature* **321,** 769–771.

Guilderson, T. P., R. G. Fairbanks, and J. L. Rubenstone (1994). Tropical temperature variations since 20,000 years ago: Modulating interhemispheric climate change. *Science* **263,** 663–665.

Gutiérrez Salgado, A. P. Gepts, and D. G. Debouck (1995). Evidence for two gene pools of the Lima bean, *Phaseolus lunatus* L., in the Americas. *Genetic Resourc. Crop Evol.* **42,** 15–28.

Haberle, S. (1994). Anthropogenic indicators in pollen diagrams: Problems and prospects for late Quaternary palynology in New Guinea. In *Tropical Archaeobotany: Applications and New Developments* (J. G. Hather, Ed.), pp. 172–201. Routledge, London.

Haberle, S. (1997). Late Quaternary vegetation and climate history of the Amazon Basin: Correlating marine and terrestrial pollen records. In *Proceedings of the Ocean Drilling Program* (R. Flood, D. Piper, L. Peterson, and A. Klaus, Eds.), Vol. 155, in press. Texas A&M Univ., College Station.

Haffer, J. (1969). Speciation in Amazonian forest birds. *Science* **165,** 131–137.

Haffer, J. (1974). *Avian Speciation in Tropical South America,* Publ. No.14. Nuttal Ornithological Club, Cambridge, MA.

Hahn, S. K. (1995). Yams. In *Evolution of Crop Plants* (J. Smartt and N. W. Simmonds, Eds.), pp.112–120. Longman, Essex, UK.

Hall, D. A. (1995). Stone-tool tradition endures radical environmental change. *Mammoth Trumpet* **10,** 1–5.

Hallam, S. J. (1989). Plant usage and management in Southwest Australian aborigine societies. In *Foraging and Farming: The Evolution of Plant Exploitation* (D. R. Harris and G. C. Hillman Eds.), pp. 136–151. Unwin Hyman, London.

Halward, T. M., H. T. Stalker, E. A. LaRue, and G. Kochert (1991). Genetic variation detectable with molecular markers among unadapted germ-plasm resources of cultivated peanut and related wild species. *Genome* **34,** 1013–1020.

Halward, T. M., H. T. Stalker, E. A. LaRue, and G. Kochert (1992). Use of single-primer DNA amplifications in genetic studies of peanut (*Arachis hypogaea* L.). *Plant Mol. Biol.* **18,** 315–325.

Hamann, A., D. Zink, and W. Nalg (1995). Microsatellite fingerprinting in the genus *Phaseolus. Genome* **38,** 507–515.

Hames, R. (1980). Game depletion and hunting zone rotation among the Ye'kwana and Yanomamö of Amazonas, Venezuela. In *Working Papers on South American Indians, Vol. 2* (R. Hames, Ed.), pp. 24–62. Bennington College, Bennington, VT.

Hames, R. (1990). Sharing among the Yanomamö: Part I, The effects of risk. In *Risk and Uncertainty in Tribal and Peasant Economies* (E. Cashdan, Ed.), pp. 89–105. Westview, Boulder, CO.

Hames, R. B., and W. T. Vickers (1982). Optimal diet breadth as a model to explain variability in Amazonian hunting. *Am. Ethnol.* **9,** 358–378.

Hammel, B. (1990). The distribution of diversity among families, genera, and habit types in the La Selva Flora. In *Four Neotropical Rainforests* (A. H. Gentry, Ed.), pp. 75–84. Yale Univ. Press, New Haven, CT.

Hammond, N. (Ed.) (1991). *Cuello: An Early Maya Community in Belize.* Cambridge Univ. Press, Cambridge, UK.

Hammond, N., A. Clarke, and, S. Donaghey (1995). The long goodbye: Middle Preclassic Maya archaeology at Cuello, Belize. *Latin Am. Antiquity* **6,** 120–128.

Hancock, J. F. (1992). *Plant Evolution and the Origin of Crop Species.* Prentice Hall, Englewood Cliffs, NJ.

Hansell, P. (1987). The Formative in Pacific central Panama: La Mula-Sarigua. In *Chiefdoms of America* (R. D. Drennan and C. Uribe, Eds.), pp. 119–139. University Press of America, Lanham, MD.

Hardy, K. (1996). The preceramic sequence from the Tehuacán Valley: A reevaluation. *Curr. Anthropol.* **37,** 700–716.

Harlan, J. R. (1971). Agricultural origins: Centers and noncenters. *Science* **174,** 468–474.

Harlan, J. R. (1986). Plant domestication: Diffuse origins and diffusions. In *The Origin and Domestication of Cultivated Plants* (C. Barigozzi, Ed.), pp. 21–34. Elsevier, Amsterdam.

Harlan, J. R. (1992). *Crops and Man,* 2nd ed. American Society of Agronomy and Crop Science Society of America, Madison, WI.

Harpending, H. (1976). Regional variation in Kung populations. In *Kalahari Hunter–Gatherers* (R. B. Lee and I. DeVore, Eds.), pp. 152–165. Harvard Univ. Press, Cambridge, MA.

Harris, D. R. (1969). Agricultural systems, ecosystems and the origins of agriculture. In *The Domestication and Exploitation of Plants and Animals* (P. J. Ucko and G. W. Dimbleby, Eds.), pp. 3–14. Duckworth, London.

Harris, D. R. (1971). The ecology of swidden cultivation in the Upper Orinoco rain forest, Venezuela. *Geogr. Rev.* **61,** 475–495.

Harris, D. R. (1972). The origins of agriculture in the tropics. *Am. Sci.* **60,** 180–193.

Harris, D. R. (1973). The prehistory of tropical agriculture: An ethnoecological model. In *The Explanation of Cultural Change: Models in Prehistory* (C. Renfrew, Ed.), pp. 391–417. Duckworth, London.

Harris, D. R. (1977a). Alternate pathways toward agriculture. In *Origins of Agriculture* (C. A. Reed, Ed.), pp. 173–249. Mouton, The Hague, The Netherlands.

Harris, D. R. (1977b). Settling down: An evolutionary model for the transformation of mobile bands into sedentary communities. In *The Evolution of Social Systems* (J. Friedman and M. L. Rowlands, Eds.), pp. 401–417. Duckworth, London.

Harris, D. R. (1987). Aboriginal subsistence in a tropical rain forest environment: Food procurement, cannibalism, and population regulation in Northeastern Australia. In *Food and Evolution* (M. Harris and E. B. Ross, Eds.), pp. 257–285. Temple Univ. Press, Philadelphia.

Harris, D. R. (1989). An evolutionary continuum of people–plant interaction. In *Foraging and Farming: The Evolution of Plant Exploitation* (D. R. Harris and G. Hillman, Eds.), pp. 11–26. Unwin Hyman, London.

Harris, D. R. (1996). (Ed.) *The Origins and Spread of Agriculture and Pastoralism in Eurasia.* Smithsonian Institution Press, Washington, DC.

Harrison, P. D., and B. L. Turner, II (1978). *Pre-Hispanic Maya Agriculture.* Univ. of New Mexico Press, Albuquerque.

Hart, T., and J. Hart (1986). The ecological basis of hunter-gatherer subsistence in African rain forests: The Mbuti of Eastern Zaire. *Human Ecol.* **14,** 29–55.

Hartshorn, G. S. (1990). An overview of Neotropical forest dynamics. In *Four Neotropical Rainforests* (A. H. Gentry, Ed.), pp. 585–595. Yale Univ. Press. New Haven, CT.

Hartshorn, G. S. and L. J. Poveda (1983). Plants: Checklist of trees. In *Costa Rican Natural History* (D. H. Janzen, Ed.), pp. 158–183. Univ. of Chicago Press, Chicago.

Hassan, F. (1977). The dynamics of agricultural origins in Palestine: A theoretical model. In *Origin of Agriculture* (C. A. Reed, Ed.), pp. 589–609. Mouton, The Hague, The Netherlands.

Hassan, F. (1981). *Demographic Archaeology.* Academic Press, New York.

Hastenrath, S. (1985). *Climate and Circulation of the Tropics.* Reidel, Dordrecht.

Hastenrath, S. (1991). *Climate Dynamics of the Tropics.* Kluwer, Dordrecht.

Hawkes, J. G. (1989). The domestication of roots and tubers in the American tropics. In *Foraging and Farming: The Evolution of Plant Exploitation* (D. R. Harris and G. C. Hillman, Eds.), pp. 481–503. Unwin Hyman, London.

Hawkes, K. (1987). How much food do foragers need? In *Food and Evolution* (M. Harris and E. B. Ross, Eds.), pp. 341–355. Temple Univ. Press, Philadelphia.

Hawkes, K., and J. O'Connell (1992). On optimal foraging models and subsistence transitions. *Curr. Anthropol.* **33,** 63–66.

Hawkes, K., K. Hill, and J. O'Connell (1982). Why hunters gather: Optimal foraging and the Ache of eastern Paraguay. *Am. Ethnol.* **9,** 379–398.

Hawkes, K., H. Kaplan, K. Hill, and A. M. Hurtado (1987). Aché at the settlement: Contrasts between farming and foraging. *Human Ecol.* **15,** 133–161.

Hawkes, K., J. F. O'Connell, and L. Rogers (1997). The behavioral ecology of modern hunter-gatherers, and human evolution. *Trends Ecol. Evol.* **12,** 29–32.

Hayden, B. (1975). The carrying capacity dilemma: An alternate approach. *Soc. Am. Archaeol. Mem.* **30,** 11–21.

Hayden, B. (1981). Subsistence and ecological adaptations of modern hunter/gatherers. In *Omnivorous Primates* (R. S. O. Harding and G. Teleki, Eds.), pp. 344–421. Columbia Univ. Press, New York.

Hayden, B. (1990). Nimrods, piscators, pluckers, and planters: The emergence of food production. *J. Anthropol. Archaeol.* **9,** 31–69.

Hayden, B. (1992). Models of domestication. In *Transitions to Agriculture in Prehistory* (A. B. Gebauer and T. D. Price, Eds.), pp. 11–19. Prehistory Press, Madison, WI.

Hayden, B. (1995). A new overview of domestication. In *Last Hunters–First Farmers: New Perspectives on the Prehistoric Transition to Agriculture* (T. D. Price and A. B. Gebauer, Eds.), pp. 273–299. School of American Research Press, Sante Fe, NM.

Haynes, C. V. (1991). Archaeological and paleohydrological evidence for a terminal Pleistocene drought in North America and its bearing on Pleistocene extinction. *Quat. Res.* **35,** 438–450.

Headland, T. N. (1987). The wild yam question: How well could independent hunter-gatherers live in a tropical rainforest ecosystem? *Human Ecol.* **15,** 463–491.

Headland, T. N. and R. C. Bailey (1991). Introduction: Have hunter–gatherers ever lived in tropical rain forest independently of agriculture? *Human Ecol.* **19,** 115–285.

Heiser, C. B., Jr. (1976). Peppers. *Capsicum* (Solanaceae). In *Evolution of Crop Plants* (N. W. Simmonds, Eds.), pp.265–268. Longman, London.

Henry, D. O. (1989). *From Foraging to Agriculture:The Levant at the End of the Ice Age.* Univ. of Pennsylvania Press, Philadelphia.

Heppner, J. B. (1991). Faunal regions and the diversity of Lepidoptera. *Trop. Lepidoptera* **2**(Suppl. 1).

Herrera, L., M. Cardale de Schrimpff, and W. Bray (1982/1983). El hombre y su medio ambiente en Calima (altos Río Calima y Río Grande, Cordillera Occidental). *Rev. Colombiana Antropol.* Bogotá **24,** 381–424.

Herrera, L., W. Bray, M. Cardale de Schrimpff, and P. Botero (1992). Nuevas fechas de radiocarbono para el Precerámico en la Cordillera Occidental de Colombia. In *Archaeology and Environment in Latin America* (O. R. Ortiz-Troncoso and T. van Der Hammen, Eds.), pp. 145–164. Universiteit van Amsterdam, Amsterdam.

Hester, T. R., G. Ligabue, J. D. Eaton, H. J. Shafer, and R. E. W. Adams (1982). Archaeology at Colha, Belize: The 1981 season. In *Archaeology at Colha, Belize: The 1981 Interim Report* (T. R. Hester, H. J. Shafer, and J. D. Eaton, Eds.), pp. 1–10. Center for Archaeological Research, University of Texas at San Antonio, and Centro Studi e Richerche Ligabue, Venezia.

Hester, T. R., H. Iceland, D. Hudler, R. Brewington, H. J. Shafer, and J. Lohse. (1993). New evidence on the preceramic era in northern Belize: A preliminary overview. *Newslett. Friends: Texas Archaeol. Res. Lab.* **1,** 19–23.

Heusser, L. E., and N. J. Shackleton (1994). Tropical climatic variation on the Pacific slopes of the Ecuadorian Andes based on a 25,000–year pollen record from deep-sea sediment core Tri 163–31B. *Quat. Res.* **42,** 222–225.

Hewlett, B. S., J. M. H. Van de Koppel, and L. L. Cavalli-Sforza (1986). Exploration and mating range of Aka Pygmies of the central African Republic. In *African Pygmies* (L. L. Cavalli-Sforza, Ed.), pp. 65–79. Academic Press, Orlando.

Hill, B. (1972–1974) A new chronology of the Valdivia ceramic complex from the coastal zone of Guayas Province, Ecuador. *Ñawpa-Pacha* **10–12,** 1–32.

Hill, K., and K. Hawkes (1983). Neotropical hunting among the Aché of eastern Paraguay. In *Adaptive Responses of Native Amazonians* (R. Hames and W. Vickers, Eds.), pp. 139–188. Academic Press, New York.

Hill, K., K. Hawkes, M. Hurtado, and H. Kaplan (1984). Seasonal variance in the diet of Aché hunter–gatherers in Eastern Paraguay. *Human Ecol.* **12,** 101–135

Hill, K., H. Kaplan, K. Hawkes, and A. M. Hurtado (1987). Foraging decisions among Aché hunter–gatherers: New data and implications for optimal foraging models. *Ethol. and Sociobiol.* **8,** 1–36.

Hillman, G. C. (1996). Late Pleistocene changes in wild plant-foods available to hunter-gatherers of the northern Fertile Crescent: Possible preludes to cereal cultivation. In *The Origins and Spread of*

Agriculture and Pastoralism in Eurasia (D. R. Harris, Ed.), pp. 159–203. Smithsonian Institution Press, Washington, DC.

Hillman, G. C., and M. S. Davies (1990). Measured domestication rates in crops of wild type wheats and barley and their archaeological implications. *J. World Prehistory* **4**, 157–222.

Hillman, G. C., and M. S. Davies (1992). Domestication rate in wild wheats and barley under primitive cultivation: Preliminary results and archaeological implications of field measurements of selection coefficient. In *Préhistoire de l'Agriculture: Nouvelles Approches Expérimentales et Ethnographiques* (P. C. Anderson, Ed.), Monographie du CRA No. 6, pp. 114–158. Centre National de la Recherche Scientifique, Paris.

Hillman, G. C., S. M. Colledge, and D. R. Harris (1989). Plant-food economy during the Epipalaeolithic period at Tell Abu Hureyra, Syria: Dietary diversity, seasonality, and modes of exploitation. In *Foraging and Farming: The Evolution of Plant Exploitation* (D.R. Harris and G. Hillman, Eds.), pp. 240–268. Unwin Hyman, London.

Hladik, A., and E. Dounias (1993). Wild yams of the African rain forest as potential food resources. In *Tropical Forests, People and Food* (C.M. Hladik, A. Hladik, O.F. Linares, H. Pagezy, A. Semple, and M. Hadley (Eds.), pp. 163–176. UNESCO/Parthenon Publishing Group, Paris.

Hladik, A., E. G. Leigh, Jr., and F. Bourlière (1993). Food production and nutritional value of wild and semi-domesticated species-background. In *Tropical Forests, People and Food* (C. M. Hladik, A. Hladik, O. F. Linares, H. Pagezy, A. Semple, and M. Hadley, Eds.), pp. 127–138. UNESCO/Parthenon Publishing Group, Paris.

Hodell, D. A., J. H. Curtis, G. A. Jones, A. Higuera-Gundy, M. Brenner, M. W. Binford, and K. T. Dorsey (1991). Reconstruction of Caribbean climate change over the past 10,500 years. *Nature* **352**, 790–792.

Hodell, D. A., J. H. Curtis, and M. Brenner (1995). Possible role of climate in the collapse of Classic Maya civilization. *Nature* **375**, 391–394.

Holden, C. (1986). Regrowing a dry tropical forest. *Science* **234**, 809–810.

Holmberg, A. R. (1969). *Nomads of the Longbow*. Smithsonian Institution Press, Washington, DC.

Hoopes, J. W. (1995). Interaction in hunting and gathering societies as a context for the emergence of pottery in the Central American Isthmus. In *The Emergence of Pottery: Technology and Innovation in Ancient Societies* (W. K Barnett and J. W. Hoopes, Eds.), pp. 185–198. Smithsonian Institution Press, Washington, DC.

Howe, H. F. (1982). Fruit production and animal activity in two tropical trees. In *The Ecology of a Tropical Forest: Seasonal Rhythms and Long-Term Changes* (E. G.Leigh, Jr., A. S. Rand, and D. M. Windsor, Eds.), pp.189–200. Smithsonian Institution Press, Washington, DC.

Howe, H. F. (1985). Gomphothere fruits: A critique. *Am. Nat.* **125**, 853–865.

Hubbell, S. P. (1979). Tree dispersion, abundance and diversity in a tropical dry forest. *Science* **203**, 1299–1308.

Hubbell, S. P., and R. Foster (1983). Diversity of canopy trees in a Neotropical forest and implications for conservation. In *Tropical Rain Forest: Ecology and Management* (S. Sutton, T. Whitmore, and S. Chadwick, Eds.), pp. 25–41. Blackwell, Oxford, UK.

Huber, O. (1995). Vegetation. In *Flora of the Venezuela Guiana Vol.1 Introduction* (J. A. Steyermark, P. E. Berry, and B. K. Holst, Eds.), pp. 97–160. Timber Press, Portland, OR.

Hurtado, A. M., and K. Hill (1987). Early dry season subsistence ecology of Cuiva (Hiwi) foragers of Venezuela. *Human Ecol.* **15**, 163–187.

Hurtado, A. M., K. Hawkes, K. Hill, and H. Kaplan (1985). Female subsistence strategies among Aché hunter-gatherers of Eastern Paraguay. *Human Ecol.* **13**, 1–28.

Iceland, H., T. R. Hester, H. J. Shafer, and D. Hudler (1995). The Colha preceramic project: A status report. *Newslett. Texas Archaeol. Res. Lab.* **3**, 11–15.

Iltis, H. H. (1983). From teosinte to maize: The catastrophic sexual transmutation. *Science* **222**, 886–894.

Iltis, H. H. (1987). Maize evolution and agricultural origins. In *Grass Systematics and Evolution* (T. R. Soderstrom, K. W. Hilu, C. S. Campbell, and M. E. Barkworth, Eds.), pp. 195–213. Smithsonian Institution Press, Washington, DC.

Iltis, H. H., and J. F. Doebley (1980). Taxonomy of *Zea* (Gramineae). II: Subspecific categories in the *Zea mays* complex and a generic synopsis. *Am. J. Bot.* **67,** 994–1004.

Imbrie, J., and Imbrie, K. P. (1979). *Ice Ages: Solving the Mystery.* Enslow, Short Hills, NY.

Islebe, G. A., H. Hooghiemstra, and K. van der Borg (1995). A cooling event during the Younger Dryas chron in Costa Rica. *Paleogeogr. Paleoclimatol. Paleoecology* **117,** 73–80.

Islebe, G. A., H. Hoohiemstra, M. Brenner, J. H., Curtis, and D. A. Hodell (1996). A Holocene vegetation history from lowland Guatemala. *Holocene* **6,** 265–271.

Jackson, I. J. (1977). *Climate, Water and Agriculture in the Tropics.* Longman, New York.

Janzen, D. H. (1969). Seed-eaters versus seed size, number, toxicity and dispersal. *Evolution* **23,** 1–27.

Janzen, D. H. (1983). Plants: Species accounts. In *Costa Rican Natural History* (D. Janzen, Ed.), pp. 184–185. Univ. of Chicago Press, Chicago.

Janzen, D. H. (1986.) *Guanacaste National Park: Tropical Ecological and Cultural Restoration.* Tinker Foundation, New York.

Janzen, D. H., and P. S. Martin (1982). Neotropical anachronisms: Fruits the gomphotheres ate. *Science* **215,** 19–27.

Janzen, D. H., and D. E. Wilson (1983). Mammals: Introduction. In *Costa Rican Natural History* (D. H. Janzen, Ed.), pp. 426–442. Univ. of Chicago Press, Chicago.

Jennings, D. L. (1995). Cassava. In *Evolution of Crop Plants* (J. Smartt and N. W. Simmonds,. Eds.), pp. 128–132. Longman, Essex, UK.

Jochim, M. A. (1976). *Hunter–Gatherer Subsistence and Settlement, a Predictive Model.* Academic Press, New York.

Jochim, M. A. (1988). Optimal foraging and the division of labor. *Am. Anthropol.* **90,** 130–136.

Johns, T. (1990). *With Bitter Herbs They Shall Eat It. Chemical Ecology and the Origins of Human Diet and Medicine.* Univ. of Arizona Press, Tucson.

Johnson, A. (1983). Machiguenga gardens. In *Adaptive Responses of Native Amazonians* (R. B. Hames and W. T. Vickers, Eds.), pp. 29–63. Academic Press, New York.

Johnson, A., and A. Baksh (1987). Ecological and structural influences on the proportions of wild foods in the diets of two Machiguenga communities. In *Food and Evolution* (M. Harris and E. B. Ross, Eds.), pp. 387–405. Temple Univ. Press, Philadelphia.

Johnson, A., and C. A. Behrens (1982). Nutritional criteria in Machiguenga food production decisions: A linear-programming analysis. *Human Ecol.* **10,** 167–189.

Jones, J. G. (1988). Middle to Late Preceramic (6000–3000 B.P.) subsistence patterns on the central coast of Peru: The coprolite evidence. Unpublished master's thesis, Department of Anthropology, Texas A&M University, College Station.

Jones, J. G. (1991). Pollen evidence of prehistoric forest modification and Maya cultivation in Belize. Unpublished Ph.D. dissertation, Department of Anthropology, Texas A&M University, College Station.

Jones, J. G. (1994). Pollen evidence for early settlement and agriculture in Northern Belize. *Palynology* **18,** 205–211.

Jones, J. G. (1996). The first farmers of the Maya lowlands: Palynological evidence for Pre-Maya subsistence and agriculture. Paper presented at the Ninth International Palynological Congress, Houston, TX.

Jones, R. (1969). Fire-stick farming. *Aust. Nat. History* **16,** 224–228.

Jones, R. (1980). Hunters in the Australian coastal savanna. In *Human Ecology in Savanna Environments* (D. R. Harris, Ed.), pp. 107–146. Academic Press, London.

Jones, R., and B. Meehan (1989). Plant foods of the Gidjingali: Ethnographic and archaeological perspectives from northern Australia on tuber and seed exploitation. In *Foraging and Farming: The Evolution of Plant Exploitation* (D. R. Harris and G. C. Hillman, Eds.), pp. 120–135. Unwin Hyman, London.

Junk, W. J. (1984). Ecology of the *várzea,* floodplain of Amazonian whitewater rivers. In *The Amazon: Limnology and Landscape Ecology of a Mighty Tropical River and Its Basin* (H. Sioli, Ed.), pp. 215–243. Junk, Dordrecht.

Kahn, F. (1993). Amazonian palms: Food resources for the management of forest ecosystems. In *Tropical Forests, People and Food* (C. M. Hladik, A. Hladik, O. F. Linares, H. Pagezy, A. Semple, and M. Hadley, Eds.), pp. 153–162. UNESCO/Parthenon Publishing Group, Paris.

Kami, J., V. B. Velásquez, D. G. Debouck, and P. Gepts (1995). Identification of presumed ancestral DNA sequences of phaseolin in *Phaseolus vulgaris*. *Proc. Natl. Acad. Sci. USA* **92**, 1101–1104.

Kaplan, H. and K. Hill (1992). The evolutionary ecology of food acquisition. In *Evolutionary Ecology and Human Behavior* (E. A. Smith and B. Winterhalder, Eds.), pp. 167–201. Aldine. New York.

Kaplan, L., and L. Kaplan (1988). *Phaseolus* in archaeology. In *Genetic Resources of Phaseolus Beans,* (P. Gepts, Eds.), pp. 125–142. Kluwer, Dordrecht.

Kaplan, L., and C. E. Smith, Jr. (1988) Carbonized plant remains from the Calima region, Valle del Cauca, Colombia. *Pro Calima* **5**, 43–44.

Keegan, W. F. (1986). The optimal foraging analysis of horticultural production. *Am. Anthropol.* **88**, 92–107.

Keeley, L. R. (1995). Protoagricultural practices among hunter–gatherers: A cross-cultural survey. In *Last Hunters–First Farmers: New Perspectives on the Prehistoric Transition to Agriculture* (T. D. Price and A. B. Gebauer, Eds.), pp. 243–272. School of American Research Press, Santa Fe, NM.

Kelly, T. C. (1993). Preceramic projectile-point typology in Belize. *Ancient Mesoamerica* **4**, 205–227.

Kennett, D. J., and B. Voorhies (1996). Oxygen isotopic analysis of archaeological shells to detect seasonal use of wetlands on the southern Pacific coast of Mexico. *J. Archaeol. Sci.* **23**, 689–704.

Khairallah, M. M., B. B. Sears, and M. W. Adams (1992). Mitochondrial restriction fragment length polymorphisms in wild *Phaseolus vulgaris* L.: Insights on the domestication of the common bean. *Theor. Appl. Genet.* **84**, 915–922.

Kirkby, A. V. T. (1973). The use of land and water resources in the past and present valley of Oaxaca, Mexico (Vol. 1). *Mem. Univ. Michigan Mus. of Anthropol.,* Ann Arbor, MI.

Klein, R. G. (1992). The archaeology of modern human origins. *Evol. Anthropol.* **1**, 5–14.

Klein, R. G. (1995). Anatomy, behavior, and modern human origins. *J. World Prehistory* **9**, 167–198.

Koenig, R. L., S. P. Singh, and P. Gepts (1990). Novel phaseolin types in wild and cultivated common bean (*Phaseolus vulgaris,* Fabaceae). *Econ. Bot.* **44**, 50–60.

Koinange, E. M. K., and P. Gepts (1992). Hybrid weakness in wild *Phaseolus vulgaris* L. *J. Heredity* **83**, 135–139.

Krapovickas, A. (1969). The origin, variability, and spread of the groundnut (*Arachis hypogaea*). In *The Domestication and Exploitation of Plants and Animals* (P. J. Ucko and G. W. Dimbleby. Eds.), pp. 427–441. Aldine, Chicago.

Krebs, J. R., and N. B. Davies (Eds.) (1993). *An Introduction to Behavioural Ecology,* 3rd. ed. Blackwell, Oxford, UK.

Kuhry, P. (1988). *Paleobotanical–Paleocological Studies of Tropical High Andean Peatbog Sections (Cordillera Oriental, Colombia).* Cramer, Berlin.

Kutzbach, J. E., and W. F. Ruddiman (1993). Model description, external forcing, and surface boundary conditions. In *Global Climates since the Last Glacial Maximum* (H. E. Wright, Jr., J. E. Kutzbach, T. Webb, III, W. F. Ruddiman, F. A. Street Perrott, and P. J. Bartlein, Eds.), pp. 12–23. Univ. of Minnesota Press, Minneapolis.

Kutzbach, J. E., and T. Webb, III (1993). Conceptual basis for understanding late-Quaternary climates. In *Global Climates since the Last Glacial Maximum* (H. E. Wright, Jr., J. E. Kutzbach, T. Webb, III, W. F. Ruddiman, F. A. Street Perrott, and P. J. Bartlein, Eds.), pp. 5–11. Univ. of Minnesota Press, Minneapolis.

Kutzbach, J. E., P. J. Guetter, P. J. Behling, and R. Selin (1993). Simulated climatic changes: Results of the COHMAP climate-model experiments. In *Global Climates since the Last Glacial Maximum* (H. E. Wright, Jr., J. E. Kutzbach, T. Webb, III, W. F. Ruddiman, F. A. Street Perrott, and P. J. Bartlein Eds.), pp. 24–93. Univ. of Minnesota Press, Minneapolis.

Lackey, J. A., and W. G. D'Arcy (1980). *Phaseolus. Ann. Missouri Bot. Garden,* **67**, 746–750.

Ladefoged, T. N. (1995). The evolutionary ecology of Rotuman political integration. *J. Anthropol. Archaeol.* **14**, 341–358.

Lancaster, P. A., J. S. Ingram, M. Y. Lim, and D. G. Coursey (1982). Traditional cassava-based foods: Survey of processing techniques. *Econ. Bot.* **36**, 12–45.

Lathrap, D. W. (1970). *The Upper Amazon.* Praeger, New York.

Lathrap, D. W. (1973a). Gifts of the cayman: Some thoughts on the subsistence basis of Chavin. In *Variation in Anthropology: Essays in Honor of John McGregor* (D. W. Lathrap and J. Douglas, Eds.), pp. 91–105. Illinois Archaeological Survey, Urbana.

Lathrap, D. W. (1973b). The antiquity and importance of long distance trade relationships in the moist tropics of pre-Colombian South America. *World Archaeol.* **5**, 170–186.

Lathrap, D. W. (1974). The moist tropics, the arid lands, and the appearance of great art styles in the New World. In *Art and Environment in Native North America* (M. E. King and I. Traylor, Jr., Eds.), Spec. Publ. No. 7, pp. 115–158. The Museum of Texas Tech Univ., Lubbock.

Lathrap, D. W. (1976). Radiation: The application to cultural development of a model from biological evolution. In *The Measures of Man* (E. Giles and J. S. Friedlaender, Eds.), pp. 494–532. Peabody Museum Press, Cambridge.

Lathrap, D. W. (1977a). Our father the cayman, our mother the gourd: Spinden revisited, or a unitary model for the emergence of agriculture in the New World. In *Origins of Agriculture* (C. A. Reed, Ed.), pp. 713–751. The Hague, Mouton, The Netherlands.

Lathrap, D. W. (1977b). Review of *The early Mesoamerican village. Science* **195**, 1319–1321.

Lathrap, D. W. (1984). Review of *The Origins of Agriculture: An Evolutionary Perspective. Econ. Geogr.* **60**, 339–344.

Lathrap, D. W. (1987). The introduction of maize in prehistoric Eastern North America: The view from Amazonia and the Santa Elena Peninsula. In *Emergent Horticultural Economies of the Eastern Woodlands,* (W. F. Keegan, Ed.), Occasional Paper No. 7, pp. 345–371. Southern Illinois University, Carbondale.

Lathrap, D. W., D. Collier, and H. Chandra (1975). *Ancient Ecuador: Culture, Clay, and Creativity, 3000–300 B.C.* Field Museum of Natural History, Chicago.

Lathrap, D. W., J. Marcos, and J. Zeidler (1977). Real Alto: An ancient ceremonial center. *Archaelogy* **30**, 2–13.

Layton, R., R. Foley, and E. Williams (1991). The transition between hunting and gathering and the specialized husbandry of resources (A socioecological approach). *Curr. Anthropol.* **32**, 255–274.

Ledru, M.-P. (1993). Late Quaternary environmental and climatic changes in Central Brazil. *Quat. Res.* **39**, 90–98.

Ledru, M.-P., P. I. Soares Braga, F. Soubiès, M. Fournier, L. Martin, K. Suguio, and B. Turcq (1996). The last 50,000 years in the neotropics (Southern Brazil): Evolution of vegetation and climate. *Palaeogeogr. Palaeoclimatol. Palaeoecol.* **123**, 239–257.

Lee, J. A. (1984). Cotton as a world crop. In *Cotton* (R. J. Kohel and C. F. Lewis, Eds.), Agronomy Series No. 24, pp. 1–25. American Society of Agronomy/Crop Science Society of America/Soil Science Society of America, Madison, WI.

Lee, R., and I. DeVore (Eds.) (1968). *Man The Hunter.* Aldine, Chicago.

Leigh, E. G., Jr., A. S. Rand, and D. M. Windsor (Eds.) (1982). *The Ecology of a Tropical Forest: Seasonal Rhythms and Long-Term Changes.* Smithsonian Institution Press, Washington, DC.

Leigh, E. G., Jr., D. M. Windsor, A. S. Rand, and R. B. Foster (1990). The impact of the "El Niño" drought of 1982–83 on a Panamanian semideciduous forest. In *Global Ecological Consequences of the 1982–83 El Niño-Southern Oscillation* (P. W. Glynn, Ed.), pp. 473–486. Elsevier Scientific, Amsterdam.

Lemon, R. R. H., and C. S. Churcher (1961). Pleistocene geology and paleontology of the Talara region, northwest Peru. *Am. J. Sci.* **259**, 410–429.

León, J. (1964). *Plantas alimenticias Andinas,* Boletín Técnico No. 6. Instituto Interamericano de Ciencias Agrícolas Zona Andina, Lima.

León, J. (1994). Plant genetic resources of the New World. In *Neglected Crops. 1492 from a Different Perspective,* (J. E. H. Bermejo and J. León, Eds.), pp. 3–33. Food and Agriculture Organization of the United Nations, Rome.

Levington, J. S. (1982). *Marine Ecology*. Prentice Hall, Englewood Cliffs, NJ.

Levi-Strauss, C. (1950). The use of wild plants in tropical South America. In *Handbook of South American Indians* (J. H. Steward, Ed.), Bull. No. 143, pp. 465–486. Smithsonian Institution Bureau of American Ethnology, Washington, DC.

Leyden, B. (1984). Guatemalan forest synthesis after Pleistocene aridity. *Proc. Natl. Acad. Sci. USA* **81,** 4856–4859.

Leyden, B. (1985). Late Quaternary aridity and Holocene moisture fluctuations in the Lake Valencia Basin, Venezuela. *Ecology* **66,** 1279–1295.

Leyden, B. (1987). Man and climate in the Maya lowlands. *Quat. Res.* **28,** 407–414.

Leyden, B. (1995). Evidence of the Younger Dryas in Central America. *Quat. Sci. Rev.* **14,** 833–839.

Leyden, B., M. Brenner, D. A. Hodell, and J. H. Curtis (1993). Late Pleistocene climate in the Central American lowlands. *Climate Change Continental Isotopic Rec.* Geophys. Monogr. **78,** 165–178.

Leyden, B., M. Brenner, D. A. Hodell, and J. H. Curtis (1994). Orbital and internal forcing of climate on the Yucatan peninsula for the past ca. 36 ka. *Paleogeogr. Paleoclimatol. Paleoecol.* **109,** 193–210.

Linares, O. F. (1976). "Garden hunting" in the American tropics. *Human Ecol.* **4,** 331–349.

Linares, O. F. (1979). What is lower Central American archaeology? *Annu. Rev. Anthropol.* **8,** 21–43.

Lippi, R. D. (1983). La Ponga and the Machalilla phase of coastal Ecuador. Unpublished Ph.D dissertation, Department of Anthropology, University of Wisconsin, Madison.

Lippi, R. D., R. Mck. Bird, and D. M. Stemper (1984). Maize recovered at La Ponga, an early Ecuadorian site. *Am. Antiquity* **49,** 118–124.

Lira Saade, R., and S. Montes Hernández (1994). Cucurbits (*Cucurbita spp.*). In *Neglected Crops. 1492 from a Different Perspective* (J. E. H. Bermejo and J. León, Eds.), pp. 63–77. Food and Agriculture Organization of the United Nations, Rome.

Lister, A. M. and A. V. Sher (1995). Ice cores and mammoth extinction. *Nature* **378,** 23–24.

Liu, K-b., and P. A. Colinvaux (1985). Forest changes in the Amazon Basin during the last glacial maximum. *Nature* **318,** 556–557.

Liu, K-b., and P. A. Colinvaux (1988). A 5200-year history of Amazon rain forest. *J. Biogeogr.* **15,** 231–248.

Livingstone, D. A. (1975). Late Quaternary climatic change in Africa. *Ann. Rev. Ecol. Systematics* **6,** 249–280.

Loaiza-Figueroa, F., K. Ritland, J. A. Laborde Cancino, and S. D. Tanksley. (1989). Patterns of genetic variation of the genus *Capsicum* (Solanaceae) in Mexico. *Plant Systematics Evol.* **165,** 159–188.

López Castaño, C. E. (1993). La edad y el ambiente precerámico en el Magdalena Medio:Resultados de laboratorio del sitio Peñones de Bogotá. *Boletín Arqueología, Fundación Investigaciones Arqueológicas Nacionales* **8,** 13–25.

López Castaño, C. E. (1995a). Dispersión de puntas de proyectil bifaciales en la cuenca media del Río Magdalena. In *Ambito y Ocupaciones Tempranas de la América Tropical* (I. Cavelier and S. Mora, Eds.), pp. 73–82. Instituto Colombiano de Antropología, Fundación Erigaie, Sante Fe de Bogotá.

López Castaño, C. E. (1995b). Preceramic hunter–gatherers in the tropical lowlands of the Middle Magdalena Valley (Colombia, South America). Paper presented at the 60th Annual Meeting of the Society for American Archaeology, Minneapolis.

Lothrop, S. K. (1961). Early migrations to Central and South America: An anthropological problem in light of other sciences. *J. R. Anthropol. Inst.* **91,** 97–123.

Lott, E. J., S. H. Bullock, and J. A. Solís Magallanes (1987). Floristic diversity and structure of upland and Arroyo forests of Coastal Jalisco. *Biotropica* **19,** 228–235.

Lovejoy, T. E., and R. O. Bierregaard, Jr. (1990). Central Amazonian forests and the minimum critical size of ecosystems project. In *Four Neotropical Rainforests* (A. H. Gentry, Ed.), pp. 60–74. Yale Univ. Press, New Haven, CT.

Loy, T. H. (1994). Methods in the analysis of starch residues on prehistoric stone tools. In *Tropical Archaeobotany:Applications and New Developments* (J. G. Hather, Ed.), pp. 86–114. Routledge, London.

Lumbraras, L. G. (1974). *The Peoples and Cultures of Ancient Peru*. Smithsonian Institution Press, Washington, DC.

Lynch, T. F. (Ed.) (1980). *Guitarrero Cave: Early Man in the Andes.* Academic Press, New York.

Lynch, T. F. (1983). The Paleo-Indians. In *Ancient South Americans* (J. D. Jennings, Ed.), pp. 86–137. Freeman, San Francisco.

Lynch, T. F. (1990). Glacial-age man in South America? A critical review. *Am. Antiquity* **55,** 12–36.

MacArthur, R. H. (1972). *Geographical Ecology: Patterns in the Distribution of Species.* Harper & Row, New York.

MacNeish, R. S. (1967). A summary of the subsistence. In *The Prehistory of the Tehuacán Valley, Vol. 1. Environment and Subsistence* (D. S. Byers, Ed.), pp. 290–309. Univ. of Texas Press, Austin.

MacNeish, R. S. (1979). The early man remains from Pikimachay cave, Ayacucho basin, highland Peru. In *Pre-Llano Cultures of the Americas: Paradoxes and Possibilities* (R. L. Humphrey and D. Stanford, Eds.), pp. 1–47. The Anthropological Society of Washington, Washington, DC.

MacNeish, R. S. (1991). *The Origins of Agriculture and Settled Life.* Univ. of Oklahoma Press, Norman.

MacNeish, R. S., and A. Nelken-Terner (1983). *Final Annual Report of the Belize Archaic Archaeological Reconnaissance.* Center for Archaeological Studies, Boston Univ., Boston.

MacNeish, R. S, S. J. K. Wilkerson, and A. Nelken-Terner (1980). *First Annual Report of the Belize Archaic Archaeological Reconnaissance.* Robert S. Peabody Foundation for Archaeology, Andover, MA.

MacNeish, R. S., R. K. Vierra, A. Nelken-Terner, R. Lurie, and A. Cook (1981). *Second Annual Report of the Ayacucho Archaeological-Botanical Project,* Vol. 9. Robert S. Peabody Foundation for Archaeology, Andover, MA.

Madsen, D. B. (1993). Testing diet breadth models: Examining adaptive change in the late prehistoric Great Basin. *J. Archaeol. Sci.* **20,** 321–329.

Mahdeem, H. (1994). Custard apples (*Annona* spp.). In *Neglected Crops. 1492 from a Different Perspective* (J. E. H. Bermejo and J. León, Eds.), pp. 85–92. Food and Agriculture Organization of the United Nations, Rome.

Malcolm, J. R. (1990). Estimation of mammalian densities in continuous forest north of Manaus. In *Four Neotropical Rainforests* (A. H. Gentry, Ed.), pp. 339–358. Yale Univ. Press, New Haven, CT.

Maloney, B.K. (1994). The prospects and problems of using palynology to trace the origins of tropical agriculture: The case of Southeast Asia. In *Tropical Archaeobotany: Applications and New Developments* (J. G. Hather Ed.), pp. 139–171. Routledge, London.

Malpass, M. A.(1983). The preceramic occupations of the Casma valley, Peru. In *Investigations of the Andean Past: Papers from the First Annual Northeast Conference on Andean Archaeology and Ethnohistory,* pp. 1–20. Latin American Studies Program, Cornell University, Ithaca, NY.

Mangelsdorf, P. C. (1953). Review of *Agricultural Origins and Dispersals. Am. Antiquity* **19,** 87–90.

Mangelsdorf, P. C. (1974). *Corn: Its Origins, Evolution, and Improvement.* Harvard Univ. Press, Cambridge, MA.

Mangelsdorf, P. C. (1986). The origin of corn. *Sci. Am.* **254,** 80–86.

Mangelsdorf, P. C., and R. G. Reeves (1939). The origin of Indian corn and its relatives. *Texas Agric. Exp. Stat. Bull.* **574.**

Mangelsdorf, P. C., R. S. MacNeish, and G. R. Willey (1964). Origins of agriculture in Middle America. In *Handbook of Middle American Indians: Natural Environment and Early Cultures,* Vol. 1 (R. C. West, Ed.), pp. 427–445. Univ. of Texas Press, Austin.

Maquet A., A. Gutierrez, and D. G. Debouck (1990). Further biochemical evidence for the existences of two gene pools in lima beans. *Ann. Rep. Bean Improvement Coop.* **33,** 128–129.

Marcos, J. G. (1978). The ceremonial precinct at Real Alto: Organization of time and space in Valdivia society. Unpublished Ph.D, thesis, Department of Anthropology, University of Illinois, Urbana.

Marcos, J. G., D. W. Lathrap, and J. A. Zeidler (1976). Ancient Ecuador revisited. *Field Mus. Nat. History Bull.* **47**(16), 3–8.

Marengo, J. A., L. M. Druyan, and S. Hastenrath (1993). Observational and modelling studies of Amazonia interannual climate variability. *Climate Change* **23,** 267–286.

Mares, M. A. (1992). Neotropical mammals and the myth of Amazonian biodiversity. *Science* **255,** 976–979.

Markgraf, V. (1993). Climatic history of Central and South America since 18,000 yr B.P.: Comparison of pollen records and model simulations. In *Global Climates since the Last Glacial Maximum* (H. E. Wright, Jr., J. E. Kutzbach, T. Webb, III, W. F. Ruddiman, F. A. Street Perrott, and P. J. Bartlein, Eds.), pp. 357–385. Univ. of Minnesota Press, Minneapolis.

Martin, F. W., A. Jones, and R. M. Ruberté. (1974). A wild *Ipomoea* species closely related to the sweet potato. *Economic Botany* **28**, 287–292.

Martin, L., M. Fournier, P. Mourguiart, A. Sifeddine, and B. Turcq (1993). Southern oscillation signal in South American palaeoclimatic data of the last 7000 years. *Quat. Res.* **39**, 338–346.

Martin, P. S. (1964). Paleoclimatology and a tropical pollen profile. Report of the VIth International Congress on Quaternary (Warsaw 1961), Vol II: Palaeo-Climatological Section, pp. 319–323. Lodz, Poland.

Martyn, D. (1992). *Climates of the World* (P. Senn, Trans.), Developments in Atmospheric Science 18. Polish Scientific, Warszawa.

Matsutani, A. (1972). Spodographic analysis of ash from the Kotosh site. In *Andes 4: Excavations at Kotosh, Peru, 1963 and 1966* (S. Izumi and K. Terada, Eds.), pp. 319–326. Univ. of Tokyo Press, Tokyo.

McCorriston, J., and F. Hole (1991). The ecology of seasonal stress and the origins of agriculture in the Near East. *Am. Anthropol.* **93**, 46–69.

McKey, D., and S. Beckerman (1993). Chemical ecology, plant evolution and traditional manioc cultivation systems. In *Tropical Forests, People and Food* (C. M. Hladik, A. Hladik, O. F. Linares, H. Pagezy, A. Semple, and M. Hadley. Eds.), pp. 83–112. UNESCO/Parthenon Publishing Group, Paris.

McLeod, M. J., S. I. Guttman, and W. H. Eshbaugh (1982). Early evolution of the chili peppers (*Capsicum*). *Econ. Bot.* **36**, 361–368.

Meggers, B. J. (1954). Environmental limitations on the development of culture. *Am. Anthropol.* **56**, 801–824.

Meggers, B. J. (1971). *Amazonia: Man and Culture in a Counterfeit Paradise*. Aldine, Chicago.

Meggers, B. J. (1982). Archaeological and ethnographic evidence compatible with the model of forest fragmentation. In *Biological Diversification in the Tropics*. (G. T. Prance, Ed.), pp. 483–496. Columbia Univ. Press, New York.

Meggers, B. J. (1984). The indigenous peoples of Amazonia, their cultures, land use patterns and effect on the landscape and biota. In *The Amazon: Limnology and Landscape Ecology of a Mighty Tropical River and Its Basin* (H. Sioli, Ed.), pp. 627–648. Junk, Dordrecht.

Meggers, B. J. (1987). The early history of man in Amazonia. In *Biogeography and Quaternary History in Tropical America* (T. C. Whitmore and G. T. Prance Eds.), pp. 151–212. Clarendon, Oxford.

Meggers, B. J., and C. Evans (1957). *Archaeological Investigations at the Mouth of the Amazon,* Bureau of American Ethnology, Bull. 167. Smithsonian Institution Press, Washington, DC.

Meggers, B. J., C. Evans, and E. Estrada (1965). *The Early Formative Period of Coastal Ecuador: The Valdivia and Machalilla Phases,* Smithsonian Contributions to Anthropology, Vol. I. Smithsonian Institution Press, Washington, DC.

Mellars, P. (1996). *The Neanderthal Legacy: An Archaeological Perspective from Western Europe.* Princeton University Press, Princeton.

Meltzer, D. J. (1997). Monte Verde and the Pleistocene peopling of the Americas. *Science* **276**, 754–755.

Meltzer, D. J., and B. Smith (1987). Paleoindian and early Archaic subsistence strategies in eastern North America. In *Foraging, Collecting, and Harvesting: Archaic Period Subsistence and Settlement in the Eastern Woodlands* (S. W. Neusius, Ed.), pp. 3–31. Center for Archaeological Investigations, Southern Illinois University, Carbondale.

Meltzer, D. J., J. M. Adovasio, and T. D. Dillehay (1994). On a Pleistocene human occupation at Pedra Furada, Brazil. *Antiquity* **68**, 695–714.

Merrick, L. C. (1990). Systematics and evolution of a domesticated squash, *Cucurbita argyrosperma,* and its wild and weedy relatives. In *Biology and Utilization of the Cucurbitaceae* (D. M. Bates, R. W. Robinson, and C. Jeffrey, Eds.), pp. 77–95. Cornell Univ. Press, Ithaca, NY.

Merrick, L. C. (1995). Squashes, pumpkins, and gourds. In *Evolution of Crop Plants* (J. Smartt and N. W. Simmonds, Eds.), pp. 97–105. Longman, Essex, UK.

Messer, E. (1978). Zapotec plant knowledge: Classification, uses and communication about plants in Mitla, Oaxaca, Mexico. In *Memoirs of the Museum of Anthropology*, Vol. 10, Part 2. Univ. of Michigan Press, Ann Arbor.

Metcalfe, S. E., F. A. Street-Perrott, R. B. Brown, P. E. Hales, R. A. Perrott, and F. M. Steininger (1989). Late Holocene human impact on lake basins in Central Mexico. *Geoarchaeology* **4**, 119–141.

Miller, N. F. (1992). The origins of plant cultivation in the Near East. In *The Origins of Agriculture: An International Perspective* (C. W. Cowan and P. J. Watson, Eds.), pp. 39–58. Smithsonian Institution Press, Washington, DC.

Milton, K. (1984). Protein and carbohydrate resources of the Maku indians of northwestern Amazonia. *Am. Anthropol.* **86**, 7–27.

Miranda, F. (1947). Estudios sobre la vegetación de Mexico-V rasgos de la vegetación en la cuenca del Río de las Balsas. *Rev. de la Sociedad Mexicana de Historia Nat.* **8**, 95–139.

Mithen, S. (1996). *The Prehistory of the Mind*. Thames and Hudson, London.

Monsalve, J. G. (1985). A pollen core from the Hacienda Lusitana. *Pro Calima* **4**, 40–44.

Moore, A., and G. C. Hillman (1992). The Pleistocene to Holocene transition and human economy in southwest Asia: The impact of the Younger Dryas. *Am. Antiquity* **57**, 482–494.

Mora, S. C., L. F. Herrera, I. Cavelier, and C. Rodríguez (1991). *Cultivars, Anthropic Soils and Stability.* University of Pittsburgh Latin American Archaeology Report No.2. University of Pittsburgh, Department of Anthropology, Pittsburgh.

Moran, E. (1983). Mobility as a negative factor in human adaptability: The case of the South American tropical forest populations. In *Rethinking Human Adaptation: Biological and Cultural Models* (R. Dyson-Hudson and M. A. Little, Eds.), pp. 117–135. Westview, Boulder, CO..

Moran, E. (1993). *Through Amazonian Eyes: The Human Ecology of Amazonian Populations*. Univ. of Iowa Press, Iowa City.

Mora-Urpi, J. (1994). Peach-palm (*Bactris gasipaes*). In *Neglected Crops. 1492 from a Different Perspecticve* (J. E. H. Bermejo and J León, eds.), pp. 211–221. Food and Agriculture Organization of the United Nations, Rome.

Morcote, Riós G. (1994). *Trabajo de Grado: Estudio Paleoetnobotánico en un Yacimiento Precerámico del Medio Río Caquetá (Amazonia Colombiana)*. Universidad Nacional de Colombia, Colombia.

Morera, J. A. (1994). Sapote (*Pouteria sapota*). In *Neglected Crops. 1492 from a Different Perspective* (J. E. H. Bermejo and J. León Eds.), pp. 103–109. Food and Agriculture Organization of the United Nations, Rome.

Morishidi, M. (1996). Genetic variability in *Carica papaya* and related species. Ph.D. dissertation, Department of Horticulture, University of Hawaii at Manoa, Honolulu.

Moseley, M. E. (1975). *The Maritime Foundations of Andean Civilization*. Cummings, Menlo Park, CA.

Moseley, M. E. (1992). Maritime foundations and multilinear evolution: Retrospect and prospect. *Andean Past* **3**, 5–42.

Mulholland, S. C., and C. Prior (1993). AMS radiocarbon dating of phytoliths. In *Current Research in Phytolith Analysis: Applications in Archaeology and Paleoecology* (D. M. Pearsall and D. R. Piperno, Eds.), MASCA Research Papers in Science and Archaeology, Vol. 10, pp. 21–23. The University Museum of Archaeology and Anthropology, Philadelphia.

Murphy, P. G., and A. E. Lugo (1986). Ecology of a tropical dry forest. *Ann. Rev. Ecol. Systematics* **17**, 67–88.

Murphy, P. G., and A. E. Lugo (1995). Dry forests of central America and the Caribbean. In *Seasonally Dry Tropical Forests* (S. H. Bullock, H. A. Mooney and E. Medina Eds.), pp. 9–34. Cambridge Univ. Press, Cambridge, UK.

Nassar, N. M. A. (1978). Conservation of the genetic resources of Cassava (*Manihot esculenta*): Determination of wild species localities with emphasis on probable origin. *Econ. Bot.* **32**, 311–320.

Nassar, N. M. A. (1980). Attempts to hydridize wild *Manihot* species with Cassava. *Econ. Bot.* **34**, 13–15.

National Academy of Sciences (NAS) (1975). *Underexploited Tropical Plants with Promising Economic Value*. NAS, Washington, DC.

National Academy of Sciences (NAS) (1979). *Tropical Legumes: Resources for the Future*. NAS, Washington, DC.

Nee, M. (1990). The domestication of *Cucurbita* (Cucurbitaceae). *Econ. Bot.* **44**(3, Suppl.), 56–58.

Neto, G. G., V. C. M. S. Guarim, and G. T. Prance (1994). Structure and floristic composition of the trees of an area of cerrado near Cuiabá, Mato Grosso, Brazil. *Kew Bull.* **49,** 499–509.

Nicholls, N. (1989). How old is Enso? *Climate Change* **14,** 111–115.

Nienhuis, J., J. Tivang, and P. Skroch. (1995). Genetic relationships among cultivars and landraces of lima bean (*Phaseolus lunatus* L.) as measured by RAPD markers. *J. Am. Soc. Hort. Sci.* **120,** 300–306.

Nishiyama, I. (1971). Evolution and domestication of the sweet potato. *Bot. Mag. Tokyo* **84,** 377–387.

Nix, H. A. (1983). Climate of tropical savannas. In *Ecosystems of the World 13: Tropical Savannas* (F. Bourliere, Ed.), pp. 37–62. Elsevier, Amsterdam.

Noda, H., C. R. Bueno, and D. F. Silva Filho (1994). Guinea arrowroot (*Calathea allouia*). In *Neglected Crops. 1492 from a Different Perspective* (J. E. H. Bermejo and J. León, Eds.), pp. 239–244. Food and Agriculture Organization of the United Nations, Rome.

Norr, L. (1989). Nutritional consequences of prehistoric subsistence strategies in lower Central America. Unpublished Ph.D. dissertation, Department of Anthropology, University of Illinois, Urbana.

Norr, L. (1995). Interpreting dietary maize from bone stable isotopes in the American tropics: The state of the art. In *Archaeology in the Lowland American Tropics: Current Analytical Methods and Recent Applications* (P. W. Stahl, Ed.), pp. 198–223. Cambridge Univ. Press, Cambridge, UK.

Oberhänsli, H., P. Heinze, L. Diester-Haass and G. Wefer (1990). Upwelling off Peru during the last 430,000 yr and its relationship to the bottom-water environment, as deduced from coarse grain-size distributions and analyses of benthic foraminifers at holes 679D, 680B, and 681B, Leg 112. In *Proceedings of the Ocean Drilling Program, Scientific Results,* Vol. 112 (E. Seuss, R. van Huene, *et al.,* Eds.), pp. 369–390. Texas A&M Univ., College Station.

O'Brien, M. J., and T. D. Holland (1990) Variation, selection, and the archaeological record. In *Archaeological Method and Theory* (M. B. Schiffer, Ed.), Vol. 2, pp. 31–79. Univ. of Arizona Press, Tucson.

O'Connell, J., and K. Hawkes (1981). Alyawara plant use and optimal foraging theory. In *Hunter–Gatherer Foraging Strategies* (B. Winterhalder and E. A. Smith Eds.), pp. 99–125. Univ. of Chicago Press, Chicago.

O'Connell, J., and K. Hawkes (1984). Food choice and foraging sites among the Alyawara. *J. Anthropol. Res.* **40,** 504–535.

O'Hara, S. L., F. A. Street-Perrott, and T. P. Burt (1993). Accelerated soil erosion around a Mexican highland lake caused by prehispanic agriculture. *Nature* **362,** 48–51.

Oliver, J. R., and C. S. Alexander (1990). The Pleistocene peoples of western Venezuela: The terrace sequence of Río Pedregal and new discoveries in Paraguaná. Paper presented at the First World Summit on the Peopling of the Americas (Center for the study of the first Americans-University of Maine), Orono, MA.

Onwueme, I. C. (1978). *The Tropical Tuber Crops. Yams, Cassava, Sweet Potato, and Cocoyams*. Wiley, Chichester, UK.

Opler, P. A., G. W. Frankie, and H. G. Baker (1980). Comparative phenological studies of treelet and shrub species in tropical wet and dry forests in the lowlands of Costa Rica. *J. Ecol.* **68,** 167–188.

Orjeda, G., R. Freyre, and M. Iwanaga (1990). Production of 2n pollen in diploid *Ipomoea trifida,* a putative wild ancestor of sweet potato. *J. Heredity* **81,** 462–467.

Osborn, A. J. (1977). Strand loopers, mermaids and other fairy tales: Ecological determinants of marine resource utilization—The Peruvian case. In *For Theory Building in Archaeology* (L. R. Binford, Ed.), pp.157–205. Academic Press, New York.

Owen-Smith, N. (1987). Pleistocene extinctions: The pivotal role of megaherbivores. *Paleobiology* **13,** 351–362.

Oyuela-Caycedo, A. (1993). Sedentism, Food Production, and Pottery Origins in the Tropics: The Case of San Jacinto 1, Colombia. Unpublished Ph.D. dissertation, University of Pittsburgh, Pittsburgh.

Oyuela-Caycedo, A. (1995). Rocks versus clay: The evolution of pottery technology in the case of San Jacinto 1, Colombia. In *The Emergence of Pottery: Technology and Innovation in Ancient Societies* (W. K. Barnett and J. W. Hoopes, Eds.), pp. 133–144. Smithsonian Institution Press, Washington, DC.

Oyuela-Caycedo, A. (1996). The study of collector variability in the transition to sedentary food producers in northern Colombia. *J. World Prehistory* **10**, 49–93.

Paine, R. R., and A. Freter (1996). Environmental degradation and the Classic Maya collapse at Copan, Honduras (A.D. 600–1250). *Ancient Mesoamerica* **7**, 37–47.

Parsons, J. J. (1969). Ridged fields in the Río Guayas valley, Ecuador. *Am. Antiquity* **34**, 76–80

Parsons, M. H.(1970). Preceramic subsistence on the Peruvian coast. *Am. Antiquity* **35**, 292–304.

Patiño, V. M. (1958). Importáncia de los frutales en la alimentación y en la vida e costumbres de los pueblos Americanas de la parte equinoccial. *Ann. Congr. Int. Am.* **35**, 169–175.

Pearsall, D. M. (1977). Early movements of maize between Mesoamerica and South America. *J. Steward Anthropol. Soc.* **9**, 41–75.

Pearsall, D. M. (1979). The application of ethnobotanical techniques to the problem of subsistence in the Ecuadorian Formative. Unpublished Ph.D. dissertation, Department of Anthropology, University of Illinois, Urbana. University Microfilms, Ann Arbor, MI.

Pearsall, D. M. (1983). Evaluating the stability of subsistence strategies by use of paleoethnobotanical data. *J. Ethnobiol.* **3**, 121–137.

Pearsall, D. M. (1987). Evidence for prehistoric maize cultivation on raised fields at Peñón del Río, Guayas, Ecuador. In *Pre-Hispanic Agricultural Fields in the Andean Region* (W. M. Denevan, K. Mathewson, and G. Knapp, Eds.), Vol. 359(ii), pp. 279–295. British Archaeological Reports International, Oxford, UK.

Pearsall, D. M. (1988). An overview of Formative period subsistence in Ecuador: Palaeoethnobotanical data and perspectives. In *Diet and Subsistence: Current Archaeological Perspectives* (B. V. Kennedy and G. M. LeMoine, Eds.), Proceedings of the Nineteenth Annual Chacmool Conference, pp. 149–164. The University of Calgary Archaeological Association, Calgary, Canada.

Pearsall, D. M. (1989a) Adaptation of prehistoric hunter–gatherers to the High Andes: The changing role of plant resources. In *Foraging and Farming: The Evolution of Plant Exploitation* (D.R. Harris and G. C. Hillman, Eds.), pp. 318–332. Unwin Hyman, London.

Pearsall, D. M. (1989b). *Paleoethnobotany: A Handbook of Procedures.* Academic Press, San Diego.

Pearsall, D. M. (1992). The origins of plant cultivation in South America. In *The Origins of Agriculture: An International Perspective* (C. W. Cowan and P. J. Watson, Eds.), pp. 173–205. Smithsonian Institution Press, Washington, DC.

Pearsall, D. M. (1994a). Macrobotanical analysis. In *Regional Archaeology in Northern Manabí, Ecuador, Vol. 1. Environment, Cultural Chronology, and Prehistoric Subsistence in the Jama River Valley* (J. A. Zeidler and D. M. Pearsall, Eds.), University of Pittsburgh Memoirs in Latin American Archaeology No. 8., pp. 149–159. Department of Anthropology, University of Pittsburgh, Pittsburgh.

Pearsall, D. M. (1994b). Phytolith analysis. In *Regional Archaeology in Northern Manabí, Ecuador, Vol. 1. Environment, Cultural Chronology, and Prehistoric Subsistence in the Jama River Valley* (J. A. Zeidler and D. M. Pearsall, Eds.), University of Pittsburgh Memoirs in Latin American Archaeology No. 8., pp. 161–174. Department of Anthropology, University of Pittsburgh, Pittsburgh.

Pearsall, D. M. (1995a). Domestication and agriculture in the New World tropics. In *Last Hunters–First Farmers: New Perspectives on the Prehistoric Transition to Agriculture* (T. D. Price and A. B. Gebauer, Eds.), pp. 157–192. School of American Research Press, Santa Fe, NM.

Pearsall, D. M. (1995b). "Doing" paleoethnobotany in the tropical lowlands: Adaptation and innovation in methodology. In *Archaeology in the Lowland American Tropics. Current Analytical Methods and Recent Applications* (P. W. Stahl, Ed.), pp. 113–129. Cambridge Univ. Press, Cambridge, UK.

Pearsall, D. M. (1996). Reconstructing subsistence in the lowland tropics: A case study from the Jama River valley, Manabí, Ecuador. In *Case Studies in Environmental Archeology* (E. Reitz, S. Scudder, and L. Newsom, Eds.), pp. 223–254. Plenum, New York.

Pearsall, D. M. (1997). Agricultural evolution and the emergence of Formative societies in Ecuador. In *Development of Agriculture and Emergence of Formative Civilizations in Pacific Central and South America* (M. Blake, Ed.), in press. Washington State Univ. Press, Seattle.

Pearsall, D. M. (1998). Subsistence in the Ecuadorian Formative: Overview and comparison to the central Andes. In *Dumbarton Oaks Conference on the Ecuadorian Formative* (J. S. Raymond and R. Burger, Eds.), in press. Dumbarton Oaks, Washington, DC.

Pearsall, D. M., and D. R. Piperno (1990). Antiquity of maize cultivation in Ecuador: Summary and reevaluation of the evidence. *Am. Antiquity* **55,** 324–337.

Pearsall, D. M., and D. R. Piperno (Eds.) (1993). *Current Research in Phytolith Analysis: Applications in Archaeology and Paleoecology.* MASCA Research Papers in Science and Archaeology, Vol. 10. The University Museum of Archaeology and Anthropology, Philadelphia.

Pearsall, D. M., and P. Stahl (1996). Multidisciplinary perspectives on environment and subsistence change in the Jama river valley, Manabí, Ecuador. Paper presented at the 61st Annual Meeting of the Society for American Archaeology, New Orleans.

Pearsall, D. M., and J. A. Zeidler (1994). Regional environment, cultural chronology, and prehistoric subsistence in Northern Manabí. In *Regional Archaeology in Northern Manabí, Ecuador, Vol. 1, Environment, Cultural Chronology, and Prehistoric Subsistence in the Jama River Valley,* (J. A. Zeidler and D. M. Pearsall, Eds.) University of Pittsburgh Memoirs in Latin American Archaeology, No. 8, pp. 201–216. Department of Anthropology, University of Pittsburgh Press, Pittsburgh.

Percival, A. E., and R. J. Kohel (1990). Distribution, collection, and evaluation of *Gossypium. Adv. Agron.* **44,** 225–256.

Percy, R. G., and J. F. Wendel (1990). Allozyme evidence for the origin and diversification of *Gossypium barbadense* L. *Theor. Appl. Genet.* **79,** 529–542.

Perlman, S. M. (1980). An optimum diet model, coastal variability, and hunter–gatherer behavior. In *Advances in Archaeological Method and Theory,* (M. B. Schiffer. Ed.), Vol. 3, pp. 257–310. Academic Press, New York.

Philander, S. G. (1990). *El Niño, La Niña, and the Southern Oscillation.* International Geophysics Series. Academic Press, San Diego.

Phillips, L. L. (1976). Cotton. *Gossypium* (Malvaceae). In *Evolution of Crop Plants* (N. W. Simmonds, Ed.), pp. 196–200. Longman, London.

Pickersgill, B. (1984). Migrations of chili peppers, *Capsicum* spp., in the Americas. In *Pre-Columbian Plant Migrations* (D. Stone, Ed.), Papers of the Peabody Museum of Archaeology and Ethnology, pp. 105–123. Harvard Univ. Press, Cambridge, MA.

Pickersgill, B., and C. B. Heiser, Jr. (1977). Origins and distribution of plants domesticated in the New World tropics. In *Origins of Agriculture* (C. A. Reed, Ed.), pp. 803–835. Mouton, The Hague, The Netherlands.

Piperno, D. R. (1985a). Phytolith analysis and tropical paleoecology: Production and taxonomic significance of siliceous forms in New World plant domesticates and wild species. *Rev. Paleobot. Palynol.* **45,** 185–228.

Piperno, D. R. (1985b). Phytolith records from prehistoric raised fields in the Calima region, Colombia. *Pro Calima* **4,** 37–40.

Piperno, D. R. (1985c). Phytolithic analysis of geological sediments from Panama. *Antiquity* **59,** 13–19.

Piperno, D. R. (1988a). *Phytolith analysis: An Archaeological and Geological Perspective.* Academic Press, San Diego.

Piperno, D. R. (1988b). Primer informe sobre los Fitolitos de las plantas del OGSE-80 y la evidéncia del cultivo de maiz en el Ecuador. In *La Historia Temprana de la Península de Santa Elena, Ecuador: Cultura las Vegas.* (K. E. Stothert, Ed.), Miscelánea Antropológica Ecuatoriana. Serie Monográfica 10, pp. 203–214. Museos del Banco Central de Ecuador, Guayaquil.

Piperno, D. R. (1989a). The occurrence of phytoliths in the reproductive structures of selected tropical angiosperms and their significance in tropical paleoecology, paleoethnobotany, and systematics. *Rev. Palaeobot. Palynol.* **61,** 147–173.

Piperno, D. R. (1989b). Non-affluent foragers: Resource availability, seasonal shortages, and the emergence of agriculture in Panamanian tropical forests. In *Foraging and Farming: The Evolution of Plant Exploitation* (D. R. Harris and G. C. Hillman, Eds.), pp. 538–554. Unwin Hyman, London.

Piperno, D. R. (1990a). Fitolitos, arqueología y cambios prehistóricos de la vegetación en un lote de cincuenta hectareas de la Isla de Barro Colorado. In *Ecología de un Bosque Tropical* (E. G. Leigh, Jr., A. S. Rand, and D. M. Windsor, Eds.), pp. 153–156. Smithsonian Tropical Research Institute, Balboa, Panama.

Piperno, D. R. (1990b). Aboriginal agriculture and land usage in the Amazon basin, Ecuador. *J. Archaeol. Sci.* **17,** 665–677.

Piperno, D. R. (1991). The status of phytolith analysis in the American tropics. *J. World Prehistory* **5,** 155–191.

Piperno, D. R. (1993). Phytolith and charcoal records from deep lake cores in the American Tropics. In *Current Research in Phytolith Analysis: Applications in Archaeology and Paleoecology* (D. M. Pearsall and D. R. Piperno, Eds.), MASCA Research Papers in Science and Archaeology, Vol.10, pp. 58–71. The University Museum of Archaeology and Anthropology, Philadelphia.

Piperno, D. R. (1994). Phytolith and charcoal evidence for prehistoric slash and burn agriculture in the Darien rainforest of Panama. *Holocene* **4,** 321– 325.

Piperno, D. R. (1995a). Plant microfossils and their application in the New World tropics. In *Archaeology in the Lowland American Tropics: Current Analytical Methods and Recent Applications* (P. W. Stahl, Ed.), pp. 130–153. Cambridge Univ. Press, Cambridge, UK.

Piperno, D. R. (1995b). Late Pleistocene/early Holocene human ecology of central America. Paper presented at the 60th annual meeting of the Society for American Archaeology, Minneapolis.

Piperno, D. R. (1997a). The origins and development of food production in Pacific Panama. In *Pacific Latin America in Prehistory: The Evolution of Archaic and Formative Cultures* (M. Blake, Ed.), in press. Washington State Univ. Press, Olympia.

Piperno, D. R. (1997b). Phytoliths and microscopic charcoal from Leg 155: A vegetational and fire history of the Amazon Basin during the last 75,000 years. In *Proceedings of the Ocean Drilling Program* (R. Flood, D. Piper, L. Peterson, and A. Klaus, Eds.), in press Vol. 155. Texas A&M Univ., College Station.

Piperno, D. R. (1997c). Phytoliths from the site of Peña Roja. Report to the Fundación Erigaie.

Piperno, D. R., and P. Becker (1996). A vegetational history of a site in the central Amazon Basin derived from phytolith and charcoal records from natural soils. *Quat. Res.* **45,** 202–209.

Piperno, D. R., and I. Holst (1996a). The presence of ancient starch grains on stone grinding tools from archaeological sites in the neotropics. (Manuscript on file at the Smithsonian Tropical Research Institute.)

Piperno, D. R., and I. Holst (1996b). Diagnostic characters of phytoliths from the rinds of wild and modern *Cucurbita* spp., and their presence in early Holocene sites from the tropics. Unpublished manuscript.

Piperno, D. R., and I. Holst (1997). The presence of starch grains on prehistoric stone tools from the humid tropics: Indications of early tuber use and agriculture in Panama. *J. Archaeol. Sci.,* in press.

Piperno, D. R., and D. M. Pearsall (1993). Phytoliths in the reproductive structures of maize and teosinte: Implications for the study of maize evolution. *J. Archaeol. Sci.* **20,** 337–362.

Piperno, D. R., and D. M. Pearsall (1997). The silica bodies of tropical American grasses: Morphology, taxonomy and implications for grass systematics and fossil phytolith identification. *Smithsonian Contrib. Bot.,* in press.

Piperno, D. R., K. Husum Clary, R. G. Cooke, A. J. Ranere, and D. W. Weiland (1985). Preceramic maize from central Panama: Evidence from phytoliths and pollen. *Am. Anthropol.* **87,** 871–878.

Piperno, D. R., M. B. Bush, and P. A. Colinvaux (1990). Paleoenvironments and human occupation in Late-Glacial Panama. *Quat. Res.* **33,** 108–116.

Piperno, D. R., M. B. Bush, and P. A. Colinvaux (1991a). Paleoecological perspectives on human adaptation in Central Panama. I. The Pleistocene. *Geoarchaeology* **6,** 210–226.

Piperno, D. R., M. B. Bush, and P. A. Colinvaux (1991b). Paleoecological perspectives on human adaptation in Central Panama. II. The Holocene. *Geoarchaeology* **6**, 227–250.

Piperno, D. R., M. B. Bush, and P. A. Colinvaux (1992). Patterns of articulation of culture and the plant world in prehistoric Panama: 11,500 B.P.–3000 B.P. In *Archaeology and Environment in Latin America* (O. R. Ortiz-Troncoso and T. Van der Hammen, Eds.), pp. 109–127. Universiteit van Amsterdam, Amsterdam.

Plucknett, D. L. (1976). Edible aroids. In *Evolution of Crop Plants* (N. W. Simmonds, Ed.), pp. 10–12. Longman, London.

Pohl, M. D., P. R. Bloom, and K. O. Pope (1990). Interpretation of wetland farming in Northern Belize: Excavations at San Antonio, Río Hondo. In *Ancient Maya Wetland Agriculture in Northern Belize: Excavations on Albion Island* (M. Pohl, Ed.), pp. 187–278. Westview, Boulder, CO.

Pohl, M. D., K. O. Pope, J. G. Jones, J. S. Jacob, D. R. Piperno, S. de France, D. L. Lentz., J. A. Gifford., F. Valdez, Jr., M. E. Danforth, and J. K. Josserand (1996). Early agriculture in the Maya lowlands. *Latin Am. Antiquity* **7**, 355–372.

Pope, K. O., and B. H. Dahlin (1989). Ancient Maya wetland agriculture: New insights from ecological and remote sensing research. *J. Field Archaeol.* **16**, 87–106.

Popper, V. S. (1982). Análisis general de las muestras. In *Precerámico Peruano. Los Gavilanes. Mar, desierto, y oasis en la historia del hombre* (D. Bonavia, Ed.), pp. 148–156. Corporación Financiera de Desarrollo S. A. (COFIDE). Oficina de Asuntos Culturales: Instituto Arqueológico Alemán, Comisión de Arqueología General y Comparada. Editorial Ausonia-Talleres Gráficos S. A., Lima.

Pozorski, S. G. (1983). Changing subsistence priorities and early settlement patterns on the north coast of Peru. *J. Ethnobiol.* **3**, 15–38.

Pozorski, S. G., and T. Pozorski (1979). An early subsistence exchange system in the Moche valley, Peru. *J. Field Archeol.* **6**, 413–432.

Pozorski, S. G., and T. Pozorski (1987). *Early Settlement and Subsistence in the Casma Valley, Peru.* Univ. of Iowa Press, Iowa City.

Pozorski, S. G., and T. Pozorski (1990). Reexamining the critical Preceramic/Ceramic period transition: New data from coastal Peru. *Am. Anthropol.* **92**, 481–491.

Pozorski, S. G., and T. Pozorski (1991). The impact of radiocarbon dates on the maritime hypothesis: Response to Quilter. *Am. Anthropol.* **93**, 454–455.

Prance, G. T. (Ed.) (1982). *Biological Diversification in the Tropics.* Columbia Univ. Press, New York.

Prance, G. T. (1984). The pejibaye, *Guilielma gasipaes* (HBK) Bailey, and the papaya, *Carica papaya* L. In *Pre-Columbian Plant Migrations* (D. Stone, Ed.), Papers of the Peabody Museum of Archaeology and Ethnology, pp. 85–104. Harvard Univ. Press, Cambridge, MA.

Prance, G. T. (1986). Etnobotânica de algumas tribus Amazônicas. In *Sums Etnológica Brasileira* (B. Ribiero, Ed.), Vol. 1. Voces: Petrópolis.

Prance, G. T. (1987). Vegetation. In *Biogeography and Quaternary History in Tropical America* (T. C. Whitmore and G. T. Prance, Eds.), pp. 28–45. Clarendon, Oxford, UK.

Price, T. D. (1995). Social inequality at the origins of agriculture. In *Foundations of Social Inequality* (T. D. Price and G. M. Feinman, Eds.), pp. 129–151. Plenum, New York.

Price, T. D., and A. B. Gebauer (1992). The final frontier: Foragers to farmers in southern Scandinavia. In *Transitions to Agriculture in Prehistory* (A. B. Gebauer and T. D. Price, Eds.), pp. 97–116. Prehistory Press, Madison, WI.

Price, T. D., and A. B. Gebauer (1995). New perspectives on the transition to agriculture. In *Last Hunters-First Farmers: New Perspectives on the Prehistoric Transition to Agriculture* (T. D. Price and A. B. Gebauer, Eds.), pp. 3–19. School of American Research Press, Santa Fe, NM.

Puleston, D. E., and O. S. Puleston (1971). An ecological approach to the origins of Maya civilization. *Archaeology* **24**, 330–337.

Purseglove, J. W. (1968). *Tropical Crops. Dicotyledons.* Wiley, New York.

Purseglove, J. W. (1972). *Tropical Crops. Monocotyledons.* Longman, London.

Putzer H. (1984). The geological evolution of the Amazon Basin and its mineral resources. In *The Amazon: Limnology and Landscape Ecology of a Mighty Tropical River and Its Basin* (H. Sioli, Ed.), pp. 15–46. Junk, Dordrect.

Quilter, J., B. Ojeda E., D. M. Pearsall, D. H. Sandweiss, J. G. Jones, and E. S. Wing (1991). Subsistence economy of El Paraíso, an early Peruvian site. *Science* **251,** 277–283.

Quinn, W. H. (1992). A study of Southern Oscillation-related climatic activity for A.D. 622–1900 incorporating Nile River flood data. In *El Niño: Historical and Paleoclimatic Aspects of the Southern Oscillation* (H. F. Diaz and V. Markgraf, Eds.), pp. 119–149. Cambridge Univ. Press, Cambridge, UK.

Quinn, W. H., and V. T Neal (1992). The historical record of El Niño events. In *Climate Since* A.D. 1500 (R. S. Bradley and P. D. Jones, Eds.), pp. 623–648. Routledge, London.

Quinn, W. H., V. T. Neal, and S. E. Antunez de Mayolo (1987). El Niño occurrences over the past four and a half centuries. *J. Geophys. Res.* **92,** 14,449–14,461.

Ranere, A. J. (1975). Toolmaking and tool use among the preceramic peoples of Panama. In *Lithic Technology: Making and Using Stone Tools* (E. Swanson, Ed.), pp. 173–209. Mouton, The Hague, The Netherlands.

Ranere, A. J. (1980a). Human movement into tropical America at the end of the Pleistocene. In *Anthropological Papers in Memory of Earl H. Swanson* (L. B. Harten, C. N. Warren, and D. R. Tuohy, Eds.), pp. 41–47. Idaho Museum of Natural History, Pocatello.

Ranere, A. J. (1980b). Stone tools and their interpretation. In *Adaptive Radiations in Prehistoric Panama* (O. F. Linares and A. J. Ranere, Eds.), Peabody Museum Monographs No. 5, pp. 118–137. Harvard Univ. Press, Cambridge, MA.

Ranere, A. J. (1980c). Preceramic shelters in the Talamancan Range. In *Adaptive Radiations in Prehistoric Panama* (O. F. Linares and A. J. Ranere, Eds.), Peabody Museum Monographs No. 5, pp. 16–43. Harvard Univ. Press, Cambridge, MA.

Ranere, A. J. (1992). Implements of change in the Holocene environments of Panama. In *Archaeology and Environment in Latin America* (O. R. Ortiz-Troncoso and T. Van der Hammen, Eds.), pp. 25–44. Universiteit van Amsterdam, Amsterdam.

Ranere, A. J., and R. G. Cooke (1991). Paleoindian occupation in the Central American Tropics. In *Clovis: Origins and Adaptations* (R. Bonnichsen and K. L. Turnmire, Eds.), pp. 237–253. Center for the Study of the First Americans, Oregon State Univ., Corvallis.

Ranere, A. J., and R. G. Cooke (1995). Evidéncias de ocupación humana en Panamá a postrimerías del pleistoceno y a comienzos del holoceno. In *Ambito y Ocupaciones Tempranas de la America Tropical* (I. Cavelier and S. Mora, Eds.), pp. 5–26. Instituto Colombiana de Antropología, Fundacion Erigaie, Sante Fe' de Bogota.

Ranere, A. J., and R. G. Cooke (1996). Stone tools and cultural boundaries in prehistoric Panama: An initial assessment. In *Paths to Central American Prehistory* (F. W. Lange, Ed.), pp. 49–77. Univ. Press of Colorado, Niwot.

Ranere, A. J., and P. Hansell (1978). Early subsistence patterns along the Pacific Coast of Central Panama. In *Prehistoric Coastal Adaptations* (B. L. Stark and B. Voorhies, Eds.), pp. 43–59. Academic Press, New York.

Raymond, J. S. (1981). Maritime foundations of Andean civilization: A reconsideration of the evidence. *Am. Antiqity* **46,** 806–821.

Raymond, J. S. (1988). Subsistence patterns during the early Formative in the Valdivia valley, Ecuador. In *Diet and Subsistence: Current Archaeological Perspectives.* Proceedings of the Nineteenth Annual Chacmool Conference (B. V. Kennedy and G. M. Lemoine, Eds.), pp. 159–163. The University of Calgary Archaeological Association, Calgary, Canada.

Raymond, J. S. (1993). Ceremonialism in the Early Formative of Ecuador. *Senri Ethnol. Stud.* **37,** 25–43.

Raymond, J. S., J. Marcos, and D. W. Lathrap (1980). Evidence of Formative settlement in the Guayas Basin, Ecuador. *Curr. Anthropol.* **21,** 700–701.

Raz, R., P. Puigdomenech, and J. A. Martínez-Izquierdo (1991). A new family of repetitive nucleotide sequences is restricted to the genus *Zea. Gene* **105,** 151–158.

Reading, A. J., R. D. Thompson, and A. C. Millington (1995). *Humid Tropical Environments.* Blackwell, Oxford, UK.

Reed, C. A. (1977). Origins of agriculture: Discussion and some conclusions. In *Origins of Agriculture* (C. A. Reed, Ed.), pp. 879–943. Mouton, The Hague, The Netherlands.

Redding, R. (1988). A general explanation of subsistence change: From hunting and gathering to food production. *J. Anthropol. Archaeol.* **7,** 56–97.

Redford, K. H. (1993). Hunting in Neotropical forests: A subsidy from nature. In *Tropical Forests, People and Food* (C. M. Hladik, A. Hladik, O. F. Linares, H. Pagezy, A. Semple, and M. Hadley, Eds.), pp. 227– 246. UNESCO/Parthenon Publishing Group, Paris.

Reichel-Dolmatoff, G. (1957). Momíl: A formative sequence from the Sinú valley, Colombia. *Am. Antiquity* **22,** 226–234.

Reichel-Dolmatoff, G. (1985). *Monsú: Un Sitio Arqueológico.* Banco Popular, Bogotá, Colombia.

Reichert, E. T. (1913). *The Differentiation and Specificity of Starches in Relation to Genera, Species, etc.* Carnegie Institution of Washington, Washington, DC.

Reitz, E. J. (1994). Environmental change at Ostra base camp: A Peruvian Preceramic site. Manuscript on file, University of Missouri, Columbia.

Renvoize, B. S. (1972). The area of origin of *Manihot esculenta* as a crop plant—A review of the evidence. *Econ. Bot.* **26,** 352–360.

Richards, P. W. (1952). *The Tropical Rain Forest: An Ecological Study.* Cambridge Univ. Press, Cambridge, UK.

Richards, P. W. (1996). *The Tropical Rain Forest: An Ecological Study, 2nd ed.* Cambridge Univ. Press, Cambridge, UK.

Richardson, J. B., III (1978). Early man on the Peruvian north coast, early maritime exploitation and the Pleistocene and Holocene environment. In *Early Man in America from a Circum-Pacific Perspective* (A. Bryan, Ed.), Occasional Papers No. 1, pp. 274–289. Department of Anthropology, University of Alberta, Edmonton, Canada.

Richardson, J. B., III (1983). The Chira beach ridges, sea level change, and the origins of maritime economies on the Peruvian coast. *Ann. Carnegie Mus.* **52,** 265–276.

Richardson, J. B., III (1995). Mangroves of the Peruvian north coast: Climate and cultural adaptations in the Holocene. Paper presented at the 60th Annual Meeting of the Society for American Archaeology, Minneapolis.

Rind, D., and D. Peteet (1985). Terrrestrial conditions at the Last Glacial Maximum and CLIMAP sea-surface temperature estimates: Are they consistent? *Quat. Res.* **24,** 1–23.

Rindos, D. (1984). *The Origins of Agriculture: An Evolutionary Perspective.* Academic Press, Orlando.

Robinson, J. G., and K. H. Redford (1989). Body size, diet, and population variation in Neotropical forest mammal species: Predictors of local extinction. *Adv. in Neotropical Mammalogy,* 567–594.

Rodgers, J .C., III, and S. P. Horn (1996). Modern pollen spectra from Costa Rica. *Palaeogeogr. Palaeoclimatol. Palaeoecol.* **124,** 53–71.

Rodríguez, C. (1995). Sites with early ceramics in the Caribbean littoral of Colombia: A discussion of periodization and typologies. In *The Emergence of Pottery: Technology and Innovation in Ancient Societies* (W. K. Barnett and J. W. Hoopes, Eds.), pp. 145–156. Smithsonian Institution Press, Washington, DC.

Rogers, D. J. (1965). Some botanical and ethnological considerations of *Manihot esculenta. Econ. Bot.* **19,** 369–377.

Rogers, D. J., and S. G. Appan (1973). *Flora Neotropica. Monograph No. 13, Manihot Manihotoides.* Hafner, New York.

Rogers, D. J., and H. S. Fleming (1973). A monograph of *Manihot esculenta* with an explanation of the taximetrics methods used. *Econ. Bot.* **27,** 1–113.

Rollins, H. B., J. B. Richardson, III, and D. H. Sandweiss (1986). The birth of El Niño: Geoarchaeological evidence and implications. *Geoarchaeology* **1,** 3–15.

Roosevelt, A. C. (1980). *Parmana: Prehistoric Maize and Manioc Subsistence along the Amazon and Orinoco.* Academic Press, New York.

Roosevelt, A. C. (1987). Chiefdoms in the Amazon and Orinoco. In *Chiefdoms in the Americas* (R. D. Drennan and C. A. Uribe, Eds.), pp. 153–185. University Press of America, Lanham, MD.

Roosevelt, A. C. (1989). Resource management in Amazonia before the conquest: Beyond ethnographic projection. In *Resource Management in Amazonia: Indigenous and Folk Strategies* (D. A. Posey and W. Balée, Eds.), Advances in Economic Botany, Vol. 7, pp. 30–62. The New York Botanical Garden, Bronx, NY.

Roosevelt, A. C. (1991). *Moundbuilders of the Amazon: Geophysical Archaeology on Marajó Island, Brazil.* Academic Press, San Diego.

Roosevelt, A. C., R. A. Housley, M. I. Da Silveira, S. Maranca, and R. Johnson (1991). Eighth millennium pottery from a prehistoric shell midden in the Brazilian Amazon. *Science* **254,** 1621–1624.

Roosevelt A. C., M. L. da Costa, C. L. Machado, M. Michab, N. Mercier, H. Valladas, J. Feathers, W. Barnett, M. I. da Silveira, A. Henderson, J. Silva, B. Chernoff, D. S. Reese, J. A. Holman, N. Toth, and K. Schick (1996). Paleoindian cave dwellers in the Amazon: The peopling of the Americas. *Science* **272,** 373–384.

Ross, E. B. (1978). Food taboos, diet, and hunting strategy: The adaptation to animals in Amazon cultural ecology. *Curr. Anthropol.* **19,** 1–36.

Rossen, J., T. D. Dillehay, and D. Ugent (1996). Ancient cultigens or modern intrusions?: Evaluating plant remains in an Andean case study. *J. Archaeol. Sci.* **23,** 391–407.

Rouse, I., and J. M. Cruxent (1963). *Venezuelan Archaeology.* Yale Univ. Press, New Haven, CT.

Rowe, J. H. (1967). Form and meaning in Chavin art. In *Peruvian Archaeology. Selected Readings* (J. H. Rowe and D. Menzel, Eds.), pp. 72–103. Peek, Palo Alto, CA.

Ruddle, K., D. Johnson, P. K. Townsend, and J. D. Rees (1978) *Palm Sago. A Tropical Starch from Marginal Lands.* Univ. Press of Hawaii, Honolulu.

Rue, D. J. (1987). Early agriculture and early postclassic Maya occupation in western Honduras. *Nature* **326,** 285–286.

Rue, D. J. (1988). Archaic middle American agriculture and settlement: Recent pollen data from Honduras. *J. Field Archaeol.* **16,** 177–184.

Rumney, G. R. (1968). *Climatology and the World's Climates.* MacMillan, London.

Russell, K. W. (1988). *After Eden: The Behavioral Ecology of Early Food Production in the Near East and North Africa,* British Archaeological Reports, International Series 391. Oxford, UK.

Rust, W. F., and B. W. Leyden (1994). Evidence of maize use at early and middle preclassic La Venta Olmec sites. In *Corn and Culture in the Prehistoric New World* (S. Johannessen and C. A. Hastorf Eds.), pp. 181–202. Westview, Boulder, CO.

Rust, W. F., and R. J. Sharer (1988). Olmec settlement data from La Venta, Tabasco, Mexico. *Science* **242,** 102–104.

Sage, R. F. (1995). Was low atmospheric CO_2 during the Pleistocene a limiting factor for the origin of agriculture? *Global Change Biol.* **1,** 93–106.

Sahlins, M. (1972). *Stone Age Economics.* Aldine, Chicago.

Salati, E., and P. B. Vose (1984). Amazon Basin: A system in equilibrium. *Science* **225,** 129–138.

Salgado López, H. (1989). *Medio Ambiente y Asentamientos Humanos Prehispánicos en el Calima Medio.* Instituto Vallecaucano de Investigaciones Científicas, Cali.

Salgado López, H. (1995). El precerámico en el cañon del Río Calima, Cordillera Occidental. In *Ambito y Ocupaciones Tempranas de la América Tropical* (I. Cavelier and S. Mora, Eds.), pp. 91–98. Instituto Colombiano de Antropología, Fundación Erigaie, Santa Fe de Bogotá.

Salick, J. (1995). Toward an integration of evolutionary ecology and economic botany: Personal perspectives on plant/people interactions. *Ann. Missouri Bot. Garden* **82,** 25–33.

Sampaio, E. V. S. B. (1995). Overview of the Brazilian caatinga. In *Seasonally Dry Tropical Forests* (S. H. Bullock, H. A. Mooney, and E. Medina, Eds.), pp. 35–63. Cambridge Univ. Press, Cambridge, UK.

Sanchez, P. (1981). Blowing in the wind: Deforestation and long-range Implications. *Soils Humid Tropics Stud. Third World Soc.* **14,** 347–410.

Sandweiss, D. H. (1986). The beach ridges at Santa, Peru: El Niño, uplift, and prehistory. *Geoarchaeology* **1,** 17–28.

Sandweiss, D. H., J. B. Richardson, III, E. J. Reitz, H. B. Rollins, and K. A. Maasch (1996). Geoarchaeological evidence from Peru for a 5000 years B.P. onset of El Niño. *Science* **273,** 1531–1533.

Sarma, A. V. N. (1974). Holocene paleoecology of South Coastal Ecuador. *Proc. Am. Philos. Soc.* **118**, 93–134.

Sarmiento, G. (1975). The dry plant formations of South America and their floristic connections. *J. Biogeogr.* **2**, 233–251.

Sarmiento, G., and M. Monasterio (1975). A critical consideration of the environmental conditions associated with occurrence of savanna ecosystems in tropical America. In *Tropical Ecological Systems: Trends in Terrestrial and Aquatic Research* (F. B. Golley and E. Medina, Eds.), pp. 223–250. Springer-Verlag, Heidelberg.

Sauer, C. O. (1936). American agricultural origins: A consideration of nature and agriculture. In *Essays in Anthropology in Honor of A. L. Kroeber* (R. H. Lowie, Ed.), pp. 279–298. Univ. of California Press, Berkeley.

Sauer, C. O. (1944). A geographic sketch of early man in America. *Geogr. Rev.* **34**, 529–573.

Sauer, C. O. (1947). Early relations of man to plants. *Geogr. Rev.* **37**, 1–25.

Sauer, C. O. (1950). Cultivated plants of South and Central America. In *Handbook of South American Indians* (J. Steward, Ed.), Bureau of American Ethnology Bull. No. 143, Vol. 6, pp. 487–543. U.S. Government Printing Office, Washington, DC.

Sauer, C. O. (1952). *Agricultural Origins and Dispersals.* American Geographical Society, New York.

Sauer, C. O. (1958). Man in the ecology of tropical America. *Proc. Ninth Pacific Sci. Congr.* Bangkok **20**, 104–110.

Sauer, C. O. (1966). *The Early Spanish Main.* Univ. of California Press, Berkeley.

Sauer, J. D. (1964). Revision of *Canavalia. Brittonia* **16**, 106–181.

Sauer, J. D. (1993). *Historical Geography of Crop Plants: A Select Roster.* CRC Press, London.

Schmitz, P. I. (1987) Prehistoric hunters and gatherers of Brazil. *J. World Prehistory* **1**, 53–125.

Schubert, C. (1988). Climatic changes during the Last Glacial Maximum in northern South America and the Caribbean: A review. *Interciencia* **13**, 128–137.

Schüle, W. (1992). Vegetation, megaherbivores, man and climate in the Quaternary and the genesis of closed forests. In *Tropical Forests in Transition* (J. G. Goldammer, Ed.), pp. 45–76. Birkhäuser-Verlag, Basel.

Schultes, R. (1989). Seje: An oil-rich palm for domestication. *Elaeis* (Palm Oil Research Institute of Malaysia, Kuala Lumpur) **1**, 126–131.

Schwabe, G. H. (1969). Towards an ecological characterisation of the South American continent. In *Biogeography and Ecology in South America* (E. J. Fittkan, J. Illies, H. Klinge, G. H. Schwabe and A. Alvarado, Eds.), Monographiae Biologicae, Vol. 18, pp. 113–136. Junk, The Hague, The Netherlands.

Schwarz, F. A., and J. S. Raymond (1996). Formative settlement patterns in the Valdivia Valley, SW Coastal Ecuador. *J. Field Archaeol.* **23**, 205–224.

Schwerdtfeger, W. (Ed.) (1976). *World Survey of Climatology Vol. 12. Climates of Central and South America.* Elsevier, Amsterdam.

Scora, R. W. and B. Bergh. (1990). The origins and taxonomy of avocado (*Persea americana*) Mill. Lauraceae. In *International Symposium on the Culture of Subtropical and Tropical Fruits and Crops, Vol. 1* (J. C. Robinson, Ed.), Acta Horticulturae 275, pp. 387–393. International Society for Horticultural Science, Nelspruit, South Africa.

Shackleton, N. J. (1987). Oxygen isotopes, ice volume, and sea level. *Quat. Sci. Rev.* **6**, 183–190.

Shackleton, N. J., M. A. Hall, J. Line, and C. Shuxi (1983). Carbon isotope data in core V19–30 confirm reduced carbon dioxide concentration in the ice age atmosphere. *Nature* **306**, 319–322.

Sharer, R. J., and D. C. Grove (1989). *Regional Perspectives on the Olmec.* Cambridge Univ. Press, Cambridge, UK.

Sheets, P. D., and McKee, B. R. (Eds.) (1994). *Archaeology, Volcanism, and Remote Sensing in the Arenal Region, Costa Rica.* Texas Univ. Press, Austin.

Shiotani, I., and T. Kawase (1989). Genomic structure of the sweet potato and hexaploids in *Ipomoea trifida* (H.B.K.) Don. *Jpn. J. Breed.* **39**, 57–66.

Simmonds, N. W. (Ed.) (1976). *Evolution of Crop Plants.* Longman, London.

Simms, S. R. (1987). *Behavioral Ecology and Hunter–Gatherer Foraging.* British Archaeological Reports International Series 381. Oxford, UK.

Simpson, B. B. (1975). Glacial climates in the eastern tropical South Pacific. *Nature* **253,** 34–36.

Singh, A. K. (1995). Groundnut. In *Evolution of Crop Plants* (J. Smartt and N. W. Simmonds, Eds.), pp. 246–250. Longman, Essex, UK.

Singh, S. P., P. Gepts, and D. G. Debouck (1991). Races of common bean (*Phaseolus vulgaris,* Fabaceae). *Econ. Bot.* **45,** 379–396.

Sioli, H. (1984). The Amazon and its main affluents: Hydrography, morphology of the river courses, and river types. In *The Amazon: Limnology and Landscape Ecology of a Mighty Tropical River and Its Basin* (H. Sioli, Ed.), pp. 127–166. Junk, Dordrecht.

Slobodkin, L. B., and H. L. Sanders (1969). On the contribution of environmental predictability to species diversity. *Brookhaven Symp. Biol.* **22,** 82–95

Smartt, J., and N. W. Simmonds (Eds.) (1995). *Evolution of Crop Plants, 2nd ed.* Longman, Singapore.

Smith, B. D. (1987). The independent domestication of indigenous seed-bearing plants in Eastern North America. In *Emergent Horticultural Economies of the Eastern Woodlands* (W. F. Keegan, Ed.), Vol. 7, 3–48. Southern Illinois Univ. Press, Carbondale.

Smith, B. D. (1992). Prehistoric plant husbandry in eastern North America. In *The Origins of Agriculture: An International Perspective* (C. W. Cowan and P. J. Watson, Eds.), pp. 101–120. Smithsonian Institution Press, Washington, DC.

Smith, B. D. (1995a). *The Emergence of Agriculture.* Scientific American Library, New York.

Smith, B. D. (1995b). The origins of agriculture in the Americas. *Evol. Anthropol.* **3,** 174–184.

Smith, B. D. (1997). The initial domestication of *Cucurbita pepo* in the Americas 10,000 years ago. *Science* **276,** 932–934.

Smith, C. E., Jr. (1967). Plant remains. In *The Prehistory of the Tehuacán Valley, Vol. I. Environment and Subsistence* (D. S. Byers, Ed.), pp. 220–255. Univ. of Texas Press, Austin.

Smith, C. E., Jr. (1980). Plant remains from the Chiriquí sites and ancient vegetational patterns. In *Adaptive Radiations in Prehistoric Panama* (O.F. Linares and A.J. Ranere, Eds.), Peabody Museum Monographs No. 5, pp. 146–174. Harvard Univ. Press, Cambridge, MA.

Smith, C. E., Jr. (1986). Preceramic plant remains from Guilá Naquitz. In *Guilá Naquitz: Archaic Foraging and Early Agriculture in Archaic Mexico* (K. Flannery, Ed.), pp. 265–274. Academic Press, Orlando.

Smith, C. E., Jr. (1988). Floral remains. In *La Galgada. Peru. A Preceramic Culture in Transition* (T. Grieder, A. B. Mendoza, C. E. Smith, and R. M. Malina, Eds.), pp. 125–151. Univ. of Texas Press, Austin.

Smith, E. A. (1981). The application of optimal foraging theory to the analysis of hunter–gatherer group size. In *Hunter–Gatherer Foraging Strategies* (B. Winterhalder and E. A. Smith, Eds.), pp. 84–108. Univ. of Chicago Press, Chicago.

Smith, E. A. (1983). Anthropological applications of optimal foraging theory: A critical review. *Curr. Anthropol.* **24,** 625–651.

Smith, E. A., and B. Winterhalder (1992). Natural selection and decision making: Some fundamental principles. In *Evolutionary Ecology and Human Behavior* (E. A. Smith and B. Winterhalder, Eds.), pp. 25–60. Aldine, New York.

Smith, N. J. H., J. T. Williams, D. L. Plucknett, and J. P. Talbot (1992). *Tropical Forests and Their Crops.* Comstock, Ithaca, NY.

Smith, P. E. L., and T. C. Young (1972). The evolution of agriculture and culture in greater Mesopotamia: A trial model. In *Population Growth: Anthropological Implications* (B. Spooner, Eds.), pp. 1–59. MIT Press, Cambridge.

Smythe, N. (1970). Relationships between fruiting seasons and seed dispersal methods in a Neotropical forest. *Am. Nat.* **104,** 25–35.

Smythe, N., W. E. Glanz, and E. G. Leigh, Jr. (1982). Population regulation in some terrestrial frugivores. In *The Ecology of a Tropical Forest: Seasonal Rhythms and Long-Term Changes* (E. G.

Leigh, Jr., A. S. Rand, and D. M. Windsor., Eds.), pp. 227–238. Smithsonian Institution Press, Washington, DC.

Snarskis, M. J. (1979). Turrialba: A Paleo-Indian quarry and workshop site in eastern Costa Rica. *Am. Antiquity* **44**, 125–138.

Sombroek, W. G. (1984). Soils of the Amazon region. In *The Amazon: Limnology and Landscape Ecology of a Mighty Tropical River and Its Basin* (H. Sioli, Ed.), pp. 521–536. Junk, Dordrecht.

Sonnante, G., T. Stockton, R. O. Nodari, V. L. Becerra Velásquez, and P. Gepts (1994). Evolution of genetic diversity during the domestication of common-bean (*Phaseolus vulgaris* L.) *Theor. Appl. Genet.* **89**, 629–635.

Speth, J. D. (1987). Early hominid subsistence strategies in seasonal habitats. *J. Archaeol. Sci.* **14**, 13–29.

Speth, J. D., and K. A. Spielmann (1983). Energy source, protein metabolism, and hunter-gatherer subsistence strategies. *J. Anthropol. Archaeol.* **2**, 1–31.

Spinden, H. J. (1917). The origin and distribution of agriculture in America. Paper presented at the XIX International Congress of Americanists, 1915, Washington, DC.

Stahl, P. W. (1995) (Ed.) *Archaeology in the Lowland American Tropics: Current Analytical Methods and Recent Applications.* Cambridge University Press, Cambridge, UK.

Stahl, P. W. (1998). Formative subsistence and the Ecuadorian archaeofaunal record. In *Dumbarton Oaks Conference on the Ecuadorian Formative* (J. S. Raymond and R. Burger, Eds.), in press. Dumbarton Oaks, Washington, DC.

Stalker, H. T. (1990). A morphological appraisal of wild species in section *Arachis* of peanuts. *Peanut Sci.* **17**, 117–122.

Stalker, H. T., and J. P. Moss (1987). Speciation, cytogenetics, and utilization of *Arachis* species. *Adv. Agron.* **41**, 1–40.

Staller, J. E. (1992). El sitio Valdivia tardío de La Emerenciana en la costa sur del Ecuador y su significación del desarrollo de complejidad en la costa oeste de Sud America. *Cuadernos de História y Arqueología* **46–47**, 14–37.

Staller, J. E. (1994). Late Valdivia occupation in the southern coastal El Oro Province, Ecuador: Excavations at the early Formative period (3500–1500 B.C.) site of La Emerenciana. Unpublished Ph.D. dissertation, Department of Anthropology, Southern Methodist University, Dallas.

Stanford, D. (1991). Clovis: origins and adaptations: An introductory perspective. In *Clovis: Origins and Adaptations* (R. Bonnichsen and K. L. Turnmire, Eds.), pp. 1–13. Center for the Study of the First Americans, Oregon State Univ., Corvallis.

Stark, B. L. (1986). Origins of food production in the New World. In *American Archaeology Past and Future: A Celebration of the Society for American Archaeology 1935–1985* (D. J. Meltzer, D. D. Fowler and J. A. Sabloff, Eds.), pp. 277–322. Smithsonian Institution Press, Washington, DC.

Stearman, A. M. (1991). Making a living in the tropical forest: Yuquí foragers in the Bolivian Amazon. *Human Ecol.* **19**, 245–260.

Stephens, S. G. (1973). Geographical distribution of cultivated cottons relative to probable centers of domestication in the New World. In *Genes, Enzymes, and Populations* (A. M. Srb, Ed.), pp. 239–254. Plenum, New York.

Stephens, S. G. and M. E. Moseley (1974). Early domesticated cottons from archaeological sites in central coastal Peru. *Am. Antiquity* **39**, 109–122.

Steward, J. H. (1948). Culture areas of the tropical forests. In *Handbook of South American Indians, Vol. 3. The Tropical Forest Tribes* (J. H. Steward, Ed.), pp. 883–899. Smithsonian Institution Press, Washington, DC.

Steward, J. H. (1949). South American cultures: An interpretive summary. In *Handbook of South American Indians, Vol. 5. The Comparative Ethnology of South American Indians* (J. H. Steward Ed.), pp. 669–772. Smithsonian Institution Press, Washington, DC.

Storey, W. B., B. Bergh, and G. A. Zentmyer (1986). The origin, indigenous range, and dissemination of the avocado. In *Yearbook of the California Avocado Society for the Year 1986, pp. 127–133.* California Avocado Society. Saticoy, CA.

Stothert, K. E. (1985). The preceramic Las Vegas culture of coastal Ecuador. *Am. Antiquity* **50,** 613–637.

Stothert, K. E. (Ed.) (1988). *La Prehistoria Temprana de la Península de Santa Elena, Ecuador: Cultura Las Vegas.* Miscelánea Antropológica Ecuatoriana. Serie Monográfica 10. Museos del Banco Central del Ecuador, Guayaquil

Stothert, K. E. (1992). Early economies of coastal Ecuador and the foundations of Andean civilization. *Andean Past* **3,** 43–54.

Street-Perrott, F. A., R. A. Perrott, and D. D. Harkness (1989). Anthropogenic soil erosion around Lake Patzcuaro, Michoacan, Mexico, during the Preclassic and late Postclassic–hispanic periods. *Am. Antiquity* **54,** 759–765.

Sturtevant, W. C. (1969). History and ethnography of some West Indian starches. In *The Domestication and Exploitation of Plants and Animals* (P. J. Ucko and G. W. Dimbleby, Eds.), pp. 177–199. Aldine, Chicago.

Talbert, L. E., J. F. Doebley, S. Larson, and V. L. Chandler (1990). *Tripsacum andersoni* is natural hydrid involving *Zea* and *Tripsacum*: Molecular evidence. *Am. J. Bot.* **77,** 722–726.

Teltser, P. A. (Ed.) (1995). *Evolutionary Archaeology: Methodological Issues.* Univ. of Arizona Press, Tucson.

Terborgh, J. (1986). Community aspects of frugivory in tropical forests. In *Frugivores and Seed Dispersal* (A. Estrada and T. H. Fleming, Eds.), pp. 371–384. Junk, Dordrecht.

Terborgh, J. (1990) An overview of research at Cocha Cashu biological station. *Four Neotropical Rainforests* (A. H. Gentry, Ed.), pp. 48–59. Yale Univ. Press, New Haven, CT.

Terborgh, J., and S. J. Wright (1994). Effects of mammalian herbivores on plant recruitment in two Neotropical forests. *Ecology* **75,** 1829–1833.

Terborgh, J. H., J. W. Fitzpatrick, and L. H. Emmons (1984). An annotated check-list of bird and mammal species of Cocha Cashu Biological Station, Manu National Park. Peru. *Fieldiana (Zool.)* **21,** 1–29.

Thompson, L. G., E. Mosley-Thompson, M. E. Davis, P. N. Lin, K. A. Henderson, J. Cole-Dai, J. F. Bolzan, and K. B. Liu (1995). Late glacial stage and Holocene tropical ice core records from Huascarán, Peru. *Science* **269,** 46–50.

Trewartha, C. (1961). *The Earth's Problem Climates.* Univ. of Wisconsin Press, Madison.

Turnbull, C. M. (1986). Survival factors among Mbuti and other hunters of the Equatorial rain forest. In *African Pygmies* (L. Cavalli-Sforza, Ed.), pp. 103–123. Academic Press, Orlando.

Turner, B. L., II, and P. D. Harrison (Eds.) (1983). *Pulltrouser Swamp: Ancient Maya Habitat, Agriculture, and Settlement in Northern Belize.* Univ. of Texas Press, Austin.

Ubelaker, D. H. (1984). Prehistoric human biology of Ecuador: Possible temporal trends and cultural correlations. In *Paleopathology at the Origins of Agriculture* (M. N. Cohen and G. J. Armelagos, Eds.), pp. 491–511. Academic Press, Orlando.

Ubelaker, D. H. (1988). Human remains from OGSE-46, La Libertad, Guayas Province, Ecuador. *J. Washington Acad. Sci.* **78,** 3–16.

Ubelaker, D. H. (1995). Biological research with archaeologically recovered human remains from Ecuador: Methodological issues. In *Archaeology in the Lowland American Tropics* (P. W. Stahl, Ed.), pp. 181–197. Cambridge Univ. Press, Cambridge, UK.

Ubelaker, D. H. (1998). Health issues in the early Formative of Ecuador: Skeletal biology of Real Alto. In *Dumbarton Oaks Conference on the Ecuadorian Formative* (J. S. Raymond and R. Burger, Eds.), in press. Dumbarton Oaks, Washington, DC.

Ugent, D., S. Pozorski, and T. Pozorski (1982). Archaeological potato tuber remains from the Casma Valley of Peru. *Econ. Bot.* **36,** 182–192.

Ugent, D., S. Pozorski, and T. Pozorski (1984). New evidence for ancient cultivation of *Canna edulis* in Peru. *Econ. Bot.* **38,** 417–432.

Ugent, D., S. Pozorski, and T. Pozorski (1986). Archaeological manioc (*Manihot*) from Coastal Peru. *Econ. Bot.* **40,** 78–102.

Uhl, C., J. B. Kauffman, and D. L. Cummings (1988). Fire in the Venezuelan Amazon 2: Environmental conditions necessary for forest fires in the evergreen rainforest of Venezuela. *Oikos* **53,** 176–184.

Ulijaszek, S. J., and S. P. Poraituk (1993). Making sago in Papua, New Guinea: Is it worth the effort?. In *Tropical Forests, People and Food* (C. M. Hladik, A. Hladik, O. F. Linares, H. Pagezy, A. Semple, and M. Hadley, Eds.), pp. 271–280. UNESCO/Parthenon Publishing Group, Paris.

Umlauf, M. (1988). Paleoethnobotanical investigations at the Initial Period site of Cardal, Peru. Unpublished master's thesis, Department of Anthropology, University of Missouri, Columbia.

Umlauf, M. (1993). Phytolith evidence for Initial Period maize at Cardal, central coast of Peru. In *Current Research in Phytolith Analysis: Applications in Archaeology and Paleoecology* (D. M. Pearsall and D. R. Piperno, Eds.), MASCA Research Papers in Science and Archaeology, Vol. 10, pp. 125–129. The University Museum of Archaeology and Anthropology, Philadelphia.

Unger-Hamilton, R. (1989). The Epi-paleolithic southern Levant and the origins of cultivation. *Curr. Anthropol.* **30**, 88–103.

Van der Hammen, T. (1974). The Pleistocene changes of vegetation and climate in tropical South America. *J. Biogeogr.* **1**, 3–26.

Van der Hammen, T., and M. L. Absy (1994). Amazonia during the last glacial. *Paleogeogr. Paleoclimatol. Paleoecol.* **109**, 247–261.

Van der Hammen, T., and E. González (1960). Upper Pleistocene and Holocene climate and vegetation of the 'Sabana de Bogotá' (Colombia, South America). *Leidse Geol. Mededelingen* **25**, 261–315.

Van der Hammen, T., and H. Hooghiemstra (1995). The El Abra Stadial, a Younger Dryas equivalent in Colombia. *Quat. Sci. Rev.* **14**, 841–851.

Van der Hammen, T., J. F. Duivenvorden, J. M. Lips, L. E. Urrego, and N. Espejo (1991). El cuaternario tardio del área del medio Caquetá (Amazonia Colombiana). *Colombia Amazonica* **5**, 63–90.

Van der Merwe, N. J., J. A. Lee-Thorp, and J. S. Raymond (1993). Light, stable isotopes and the subsistence base of Formative cultures at Valdivia, Ecuador. In *Prehistoric Human Bone: Archaeology at the Molecular Level* (J.B. Lambert and G. Grupe, Eds.), pp. 63–97. Springer-Verlag, Berlin.

van Schaik, C. P., J. W. Terborgh, and S. J. Wright (1993). The phenology of tropical forests: Adaptive significance and consequences for primary consumers. *Annu. Rev. Ecol. Systematics* **24**, 353–377.

Van Wambeke, A. (1992). *Soils of the Tropics: Properties and Appraisal.* Mc Graw-Hill, Ithaca, NY.

Vargas, J., J. Tohme, and D. G. Debouck (1990). Common bean domestication in the southern Andes. *Ann. Rep. Bean Improvement Coop.* **33**, 104–105.

Vaughn, H. H., E. S. Deevey, Jr., and S. E. Garrett-Jones. (1985). Pollen stratigraphy of two cores from the Peten Lake District, with an appendix on two deep-water cores. In *Prehistoric Lowland Maya Environment and Subsistence Economy* (M. Pohl, Ed.), pp. 73–89. Harvard Univ. Press, Cambridge, MA.

Vaupel, J. W. (1988). Inherited frailty and longevity. *Demography* **25**, 277–287.

Veintimilla, C. (n.d.). Analysis of past vegetation in the Jama River Valley, Manabí Province. Master's thesis in progress, Department of Anthropology, University of Missouri, Columbia.

Villalba, R. (1994). Tree-ring and glacial evidence for the Medieval Warm Epoch and the Little Ice Age in southern South America. *Climatic Change* **26**, 183–197.

Vincent, A. (1985). Plant foods in savanna environments: A preliminary report of tubers eaten by the Hadza of northern Tanzania. *World Archaeol.* **17**, 131–148.

Vincentini, K. R. C. F. (1993). Análise palinológica de uma vereda em Cromínia (Go). Thesis Univ. Brasília.

Vradenburg, J. A. (1992). Analysis of the human skeletal remains from the Late (1150–800 B.C.) Initial Period Site of Cardal, Lurin Valley, Peru. Unpublished Master's thesis, Department of Anthropology, University of Missouri, Columbia.

Vradenburg, J. A., R. A. Benfer, Jr., and L. Sattenspiel (1997). Evaluating archaeological hypotheses of population growth and decline on the central coast of Peru. In *Integrating Archaeological Demography: Multidisciplinary Approaches to Prehistoric Population* (R. R. Paine, Ed.), Center for Archaeological Investigations, Southern Illinois University at Carbondale, Carbondale, Occasional Paper No. 24, pp. 150–172.

Wagner, P. L. (1964). Natural vegetation of Middle America. In *Handbook of Middle American Indians, Vol 1.* (R. C. West, Ed.), pp. 216–264. Univ. of Texas Press, Austin.

Wagner, P. L. (1977). The concept of environmental determinism in cultural evolution. In *Origins of Agriculture* (C. A. Reed, Ed.), pp.49–74. Mouton, The Hague, The Netherlands.

Walsh, R. P. D. (1996). Climate. In *The Tropical Rain Forest: An Ecological Study.* 2nd ed. by P. W. Richards, pp. 159–205. Cambridge Univ. Press, Cambridge, UK.

Watson, P. J. (1991). Origins of food production in western Asia and eastern North America: A consideration of interdisciplinary research in anthropology and archaeology. In *Quaternary Landscapes* (L. C. K. Shane and E. J. Cushing, Eds.), pp. 1–37. Univ. of Minnesota Press, Minneapolis.

Watson, P. J. (1995). Explaining the transition to agriculture. In *Last Hunters–First Farmers: New Perspectives on the Prehistoric Transition to Agriculture* (T. D. Price and A. B. Gebauer, Eds.), pp. 21–38. School of American Research Press, Santa Fe, NM.

Watts, W. A., and J. P. Bradbury (1982). Paleoecological studies at Lake Patzcuaro on the west-central Mexican plateau and at Chalco in the Basin of Mexico. *Quat. Res.* **17,** 56–70.

Webb, S. D. (1997). The great American faunal interchange. In *Central America: A Natural and Cultural History* (A. G. Coates, Ed.), Yale University Press, New Haven, CT, in press.

Weiland, D. (1984). Prehistoric settlement patterns in the Santa María drainage of Panama: A preliminary analysis. In *Recent Developments in Isthmian Archaeology* (F. W. Lange, Ed.), British Archaeological Reports International Series 212, pp. 31–54. Oxford, UK.

Weir, G., and P. Dering (1986). The lomas of Paloma: Human-environmental relationships in a central Peruvian fog oasis—Archaeobotany and palynology. In *Andean Archaeology. Papers in Memory of Clifford Evans* (R. Matos M., S. A. Turpin, and H. H. Eling, Jr., Eds.), Monographs XXVII. Institute of Archaeology, University of California, Los Angeles.

Weir, G. H., R. A. Benfer, Jr., and J. G. Jones (1988). Preceramic to Early Formative subsistence on the central coast. In *Economic Prehistory of the Central Andes* (E. S. Wing and J. C. Wheeler, Eds.), BAR International Series 427, pp. 56–94, Oxford, UK.

Wells, L. E. (1987). An alluvial record of El Niño events from northern coastal Peru. *J. Geophys. Res.* **92**(C13), 14,463-14,470.

Wells, L. E. (1990). Holocene history of the El Niño phenomenon as recorded in flood sediments of northern coastal Peru. *Geology* **18,** 1134–1137.

Wendel, J. F. and V. A. Albert (1992). Phylogenetics of the cotton genus (*Gossypium*): Character-state weighted parsimony analysis of chloroplast-DNA restriction site data and its systematic and biogeographic implications. *Systematic Bot.* **17,** 115–143

Wendel, J. F., C. L. Brubacker, and A. E. Percival (1992). Genetic diversity in *Gossypium hirsutum* and the origin of upland cotton. *Am. J. Bot.* **79,** 1291–1310.

Whitaker, T. W. (1968). Ecological aspects of the cultivated *Cucurbita. Hort. Sci.* **3,** 9–11.

Whitaker, T. W., and W. P. Bemis (1964). Evolution in the genus *Cucurbita. Evolution* **18,** 553–559.

Whitaker, T. W., and W. P. Bemis (1976). Cucurbits. In *Evolution of Crop Plants* (N. W. Simmonds, Ed.), pp. 64–69. Longman, London.

Whitledge, T. E. (1978). Regeneration of nitrogen by zooplankton and fish in the northwest Africa and Peru upwelling ecosystems. In *Upwelling Ecosystems* (R. Boje and M. Tomczak, Eds.), pp. 90–100. Springer-Verlag, Berlin.

Whitmore, T. C. (1990). *An Introduction to Tropical Rain Forests.* Oxford Univ. Press, Oxford, UK.

Whitmore, T. C., and G. T. Prance (Eds.) (1987). *Biogeography and Quaternary History in Tropical America.* Clarendon, Oxford, UK.

Wijmstra, T. A., and T. Van der Hammen (1966). Palynological data on the history of tropical savannas in northern South America. *Leidse Geol. Mededelingen* **38,** 71–83.

Wilbert, J. (1976). *Manicaria saccifera* and its cultural significance among the Warao Indians of Venezuela. *Bot. Mus. Leaflets* **24,** 275–335.

Willey, G. R., and C. R. McGimsey, III (1954). *The Monagrillo Culture of Panama.* Papers of the Peabody Museum of Archaeology and Ethnology No. 49. Harvard Univ. Press, Cambridge, MA.

Wilson, D. J. (1981). Of maize and men: A critique of the maritime hypothesis of state origins on the coast of Peru. *Am. Anthropol.* **83,** 93–120.

Wilson, H. D., J. Doebley, and M. Duvall (1992). Chloroplast DNA diversity among wild and domesticated members of *Cucurbita* (Cucurbitaceae). *Theor. Appl. Genet.* **84,** 859–865.

Winterhalder, B. (1981). Optimal foraging strategies and hunter gatherer research in anthropology: Theory and models. In *Hunter–Gatherer Foraging Strategies* (B. Winterhalder and E. A. Smith, Eds.), pp. 13–35. Univ. of Chicago Press, Chicago.

Winterhalder, B. (1986), Diet choice, risk, and food sharing in a stochastic environment. *J. Anthropol. Archaeol.* 5, 369–392.

Winterhalder, B. (1990). Open field, common pot: Harvest variability and risk avoidance in agricultural and foraging societies. In *Risk and Uncertainty in Tribal and Peasant Economies* (E. Cashdan, Ed.), pp. 67–87.Westview, Boulder, CO.

Winterhalder, B. (1993). Work, resources and population in foraging societies. *Man* **28,** 321–340.

Winterhalder, B., and C. Goland (1993). On population, foraging efficiency, and plant domestication. *Curr. Anthropol.* **34,** 710–715.

Winterhalder, B., and E. A. Smith (1992). Evolutionary ecology and the social sciences. In *Evolutionary Ecology and Human Behavior* (E. A. Smith and B. Winterhalder, Eds.), pp. 3–23. Aldine, New York.

Winterhalder, B., W. Baillargeon, F. Cappelletto, I. R. Daniel Jr., and C. Prescott (1988). The population ecology of hunter-gatherers and their prey. *J. Anthropol. Archaeol.* **7,** 289–328.

Wobst, H. M. (1974). Boundary conditions for paleolithic social systems: A simulation approach. *Am. Antiquity* **39,** 147–178.

Wobst, H. M. (1978). The archaeoethnology of hunter–gatherers or the tyranny of the ethnographic record in archaeology. *Am. Antiquity* **43,** 303–308.

Wright, H. E., Jr. (1993). Environmental determinism in Near Eastern prehistory. *Curr. Anthropol.* **34,** 458–469.

Wright, K. I. (1994). Ground-stone tools and hunter–gatherer subsistence in Southwest Asia: Implications for the transition to farming. *Am. Antiquity* **59,** 238–263.

Wright, S. J. (1992). Seasonal drought, soil fertility and the species density of tropical forest plant communities. *Trends Ecol. Evol.* **7,** 260–263.

Wright, S. J., and C. P. van Schaik (1994). Light and the phenology of tropical trees. *Am. Nat.* **143,** 192–199.

Wright, S. J., M. E. Gompper, and B. DeLeon (1994). Are large predators keystone species in Neotropical forests? The evidence from Barro Colorado Island. *Oikos* **71,** 279–294.

Wynne, J. C., and T. Halward (1989). Cytogenetics and genetics of *Arachis*. *Crit. Rev. Plant Sci.* **8,** 189–220.

Yesner, D. R. (1987). Life in the "Garden of Eden": Causes and consequences of the adoption of marine diets by human societies. In *Food and Evolution* (M. Harris and E. B. Ross, Eds.), pp. 285–309. Temple Univ. Press, Philadelphia.

Yost, J. A., and P. M. Kelley (1983). Shotguns, blowguns, and spears: The analysis of technological efficiency. In *Adaptive Responses of Native Amazonians* (R. Hames and W. Vickers, Eds.), pp.189–224. Academic Press, New York.

Zeidler, J. A. (1984). Social Space in Valdivia Society: Community Patterning and Domestic Structure at Real Alto, 3000–2000 B.C. Unpublished Ph.D. dissertation, Department of Anthropology, University of Illinois, Urbana.

Zeidler, J. A. (1986). La evolución local de asentamientos Formativos en el litoral Ecuatoriano: El caso de Real Alto. In *Arqueología de la Costa Ecuatoriana: Nuevos Enfoques* (J. G. Marcos, Ed.), pp. 85–127. Biblioteca Ecuatoriana de Arqueología, Guayaquil.

Zeidler, J. A. (1991). Maritime exchange in the Early Formative period of coastal Ecuador: Geopolitical origins of uneven development. *Res. Econ. Anthropol.* **13,** 247–268.

Zeidler, J. A. (1994). Archaeological testing in the Middle Jama Valley. In *Regional Archaeology in Northern Manabí, Ecuador, Vol. 1. Environment, Cultural Chronology, and Prehistoric Subsistence in the*

Jama River Valley (J. A. Zeidler and D. M. Pearsall, Eds.), University of Pittsburgh Memoirs in Latin American Archaeology No. 8, pp. 71–98. Department of Anthropology, University of Pittsburgh, Pittsburgh.

Zeidler, J. A., and D. M. Pearsall (Eds.) (1994). *Regional Archaeology in Northern Manabi, Ecuador. Vol. 1: Environment, Cultural Chronology, and Prehistoric Subsistence in the Jama River Valley,* University of Pittsburgh Memoirs in Latin American Archaeology, No. 8. Department of Anthropology, University of Pittsburgh, Pittsburgh.

Zeidler, J. A., C. E. Buck, and C. D. Litton (1997). The integration of archaeological phase information and radiocarbon results from the Jama River Valley, Ecuador: A Bayesian approach. *Latin Am. Antiquity,* in press.

Zeitlin, R. N. (1984). A summary report on three seasons of field investigations into the Archaic period prehistory of lowland Belize. *Am. Anthropol.* **86,** 358–369.

Zohary, D. (1989). Domestication of the southwest Asian neolithic crop assemblage of cereals, pulses, and flax: The evidence from the living plants. In *Foraging and Farming: The Evolution of Plant Exploitation* (D. R. Harris and G. C. Hillman, Eds.), pp. 359–373. Unwin Hyman, London.

Zohary, D. (1992). Domestication of the Neolithic Near Eastern crop assemblage. In *Préhistoire de L'Agriculture: Nouvelles Approches Expérimentales et Ethnographiques* (P. C. Anderson, Ed.), Monographie du CRA No. 6. Centre National de la Recherche Scientifique, Paris.

Zohary, D., and M. Hopf (1988). *Domestication of Plants in the Old World.* Oxford Univ. Press, Oxford, UK.

Zuta, S., T. Rivera and A. Bustamante (1978). Hydrologic aspects of the main upwelling areas off Peru. In *Upwelling Ecosystems* (R. Boje and M. Tomczak, Eds.), pp. 235–257. Springer-Verlag, Berlin.

Index of Common and Scientific Plant Names

babassu, 156, 158
Bactris gasipaes, 56, 58, 156, 158
bamboo, 103
barley, 3, 9, 86, 102
batana, 155, 156
bean, common, 132, 134–7, 139, 163, 249, 257, 262, 275–6, 293, 319
bean, jack , 82, 130, 132–4, 139, 164, 229, 248, 275
bean, scarlet runner, 139
beans, 1, 5, 19, 22, 23, 24, 30, 132–9, 220, 233, 239, 308, 309, 316
bean, lima, 23, 32, 132, 134, 137–9, 164, 271, 275–6, 319
beans, snap, 23, 132
Begonia, 271
Bertholletia excelsa, 56, 205
Bombacaceae, 259
Bombacopsis, 206
Borreria, 301, 303
Brazil nut, 56, 205
Bromeliaceae, 156
Brosimum, 300–2
Brosimum alicastrum, 157
Brysonima, 291, 293
Brysonima crassifolia, 291
Byrsonima crispa, 205
Bunchosia, 205
Bunchosia armeniaca, 275–6
Burseraceae, 298

cacao, 260, 263
cactus, 44, 89, 94, 99, 100, 103, 172, 183, 234
Calathea, 35, 59, 111, 115, 163, 197–9, 213, 220, 227
Calathea allouia, 29, 35, 113, 115, 163, 165, 187, 192, 197, 203–6, 213, 216, 221, 227, 244, 248, 287
Calathea latifolia, 115, 197
Calathea macrosepala, 115
Calocarpum mammosum, 157
Calycophysum, 216
Campomanesia zanthocarpa, 84,
Canavalia, 132–4, 139, 248, 250, 252, 257, 275
Canavalia brasilensis, 133
Canavalia ensiformis, 130, 132–3, 165
Canavalia piperi, 133
Canavalia plagiosperma, 132–3, 164

Canna, 111, 118–20, 163, 250, 251, 252
Canna coccinea, 118
Canna edulis, 110, 112, 118–9, 128, 163, 165, 248, 274
Canna indica, 118
Canna paniculata, 118
Cannaceae, 111
Capsicum, 110, 152–3
Capsicum annuum, 152–4, 163, 165
Capsicum baccatum, 152–3, 164
Capsicum chacoense, 154
Capsicum chinense, 152–3, 264
Capsicum eximium, 154
Capsicum frutescens, 152–4, 163, 165
Capsicum praetemlisum, 154
Capsicum pubescens, 152–4, 164
Carica papaya, 156
Carica pubescens, 156
Caricaceae, 156
Caryocar, 200, 205
cashew, 156
Casimiroa sinensis, 84
cassava, 110, 120 (also see manioc)
Cecropia, 259, 290
Celtis, 205
Cheno-Ams, 301, 302, 303
Chenopodium, 204, 248, 265
Chenopodium quinoa, 5, 206
chick pea, 9
chiles (chile peppers), 5, 29, 233, 260, 263, 275–6, 300, 301
chirimoya, 156
Chrysophyllum ganocarpum, 84
ciruela de fraile, 157, 275
Cladium, 301
coca, 249, 253, 265–6, 316
cocoyam, 116, 117, 118
Colocarpum mammosum, 155–8
Colocarpum sapota, 157
Colocasia esculenta, 116
Compositae, 261, 299, 302 (also see Asteraceae)
conifer, 234
corn, 32, 308 (also see maize)
corn, purple dye, 215
corn, sweet, 215
corozo, 156
cotton, 27, 82, 139; 147–52, 164, 208, 229, 233, 249, 250, 252, 254, 263, 265, 271, 273–7, 300, 301, 316
coyol, 58, 156, 217–8, 233
Crescentia cujete, 140, 186

Orbigyna martiana, 156
Orbigyna phalerata, 156, 158

Quercus, 94, 172, 201, 290, 295
quinoa, 5, 25, 204, 206, 208, 248, 265–6

Pachyrrhizus, 120, 128, 274
palm, 33, 35, 43, 57–8, 62, 75, 86, 155, 200,
    203, 205, 220, 230, 249, 250, 252, 262,
    291, 293, 316
palm, American oil, 58, 291
palm, peach, 56, 58, 156, 158, 164
palm, sago, 57, 59
Palmae, 205, 249, 293 (also see Arecaceae)
papaya, 155
papaya de monte, 156
peanut, 32, 110, 128–32, 139, 164, 204, 206,
    208–9, 249, 265, 275, 316
pehibaye, 58
Persea, 200, 202, 205
Persea americana, 157, 200, 264, 274,
    293
Phaseolus, 129, 132, 134–9, 233, 271,
    283
Phaseolus acutifolius, 139
Phaseolus coccineus, 139
Phaseolus lunatus, 134, 164, 165, 275
Phaseolus vulgaris, 134, 163, 249, 275, 293
Philodendron sellam, 84
pine, 44
pineapple, 156
Plantago, 200
Poaceae, 177, 217, 259, 294, 301, 303 (also see
    Gramineae)
Podocarpus, 94, 97, 172
Polymnia, 120
Polymnia sonchifolia, 128
popcorn, 215
popcorn, Argentine, 215, 310
potato, 25, 274
potato, sweet, See sweet potato
potato, white, 1, 276
Pouteria caimito, 157, 264
Pouteria obovata, 157
Pouteria lucuma, 157
Pouteria sapota, 157
prickly pear, 99, 100, 234
Prosopis, 157, 228
Prosopis juliflora, 100
Psidium, 264
Psidium guajava, 157, 158, 249, 274
pumpkin, summer, 143

Rajania, 117
Rajania cordata, 117
ramón, 157, 300–302
Rapanea, 94, 172
Rheedia brazilense, 84
Rhizomatosae, 131
Rhizophora, 303

sapodilla, 157
Sapotaceae, 155, 157, 205, 217, 249, 252,
    293
sapote, 155–8, 205, 249, 293
sapote, black, 157
Scirpus/Cyperus, 250
sedge, 99, 103, 250, 252
seje, 155, 156
Setaria, 160, 233
Solanum tuberosum, 274
soursop, 249, 252, 274
Spondias, 206, 291, 293, 303
Spondias mombin, 156
Spondias purpurea, 156
squash, 1,5, 9, 23, 24, 29, 30, 35, 139–47,
    163–4, 182, 204, 206, 222, 229, 233, 237,
    244, 245, 248, 252, 262, 265, 271, 274,
    287, 292, 314, 316, 317, 326
squash, summer, 143
Sterculiaceae, 157
sumpweed, 1
sunflower, 1
sweet potato, 1, 116, 117, 120, 122, 126–8,
    163–4, 220, 274, 276, 292, 293, 297, 312,
    316
sweet sop, 156
Symplocus, 94, 172

teosinte, 16, 22, 79, 86, 102, 105, 107, 160,
    161, 163, 182, 225–6, 238
teosinte, Balsas, 215, 221
Terminalia, 298
Theobroma bicolor, 264
Theobroma cacao, 157

Subject Index

Page numbers followed by *f*, *n*, and *t*, denote figures, notes, and tables, respectively.

Abeja, 251*f*, 263–265
El Abra, 169
Acacia, 99
Acerola, 157*t*
Aché Indians, 58, 68, 72
  food sharing and, 241
  hunting and, 62–63
  meat and, 71
  returns to labor and, 85, 87–88
  seasonality and, 70–71
  self-sufficiency and, 77–78
Achira
  at Chorrera sites, 258
  at coastal Peruvian sites, 271, 274*t*, 278
  at early Valdivia sites, 250
  at Formative period sites, 248*t*
  at Jama Valley, 252
  in Machalilla culture, 257
  phytogeography, 110–112, 115, 118–120,
    128, 163, 165*f*
Acicular phytoliths, 221, 224
Acorns, 234
Acrocomia mexicana (coyol), 58, 156*t*,
  217–218
Acrocomia spp., 200
Acrocomia vinifera, 217
Adaptive radiation, 24–25
Adze, at San Isidro, 200
Aerial roots, 42
Aeropuerto, 265

Africa
  bottle gourd and, 140
  cotton and, 149
  as origin of agriculture, 23
  tuber densities in, 59–60
  ungulate biomass, 67
Agave spp., 100
Agouti, 62, 63, 64*f*, 66*t*, 76
Agouti paca, 65*f*
Agricultural Origins and Dispersals, 18–19
Agriculture, *see also* Food production
  adaptive radiation and, 24–25
  in Belize, 299–308
  at Caverna de Pedra Pintada, 281–282
  crop origins and, 1
  defined, 6–7
  distinguished from food production, 8
  early debates on, 2–4
  early scholars of, 18–26
  elites and, 307
  influence of El Niño on, 279–280
  at Lake Ayauch¹, 260–261
  Neotropical compared to Near Eastern,
    3–4
  in Panama, 295, 296
  riparian, 260–261
  symbiotic relationship with foragers,
    77–78
  Valdivia culture and, 246, 250, 252
  wetland, 260–261, 304, 306–307

Babassu, 156*t*, 158
Bactris gasipaes (peach palm), 56, 58, 156*t*, 158, 164
Bajos, 304
Balsas Valley, 29, 79, 160, 162
Bari period, 310
Barley
  Mediterranean climate and, 102
  rate of domestication, 9
  returns to labor, 86
Barra, 309
Batana, 155, 156*t*, 158
Beach ridges, 269–270
Beans, *see* Common bean; Jack bean; Lima bean; *Phaseolus* spp.
Begonia spp., 271
Behavior
  in development of food production, 28, 324
  risk avoidance and, 239–241
Behavioral ecology, 11, 16
Belize
  Formative period, 299–308
  plant remains at, 292–293*t*
  pollen and phytolith record, 305*f*
Bertholletia excelsa (Brazil nut), 56, 205*t*
Biface axes, 307
Bifacial stone technology, 218; *see also* Lithic assemblages
  Paijan culture and, 228
  at Río Santa Maria sites, 210, 212, 213, 214
Bighorn sheep, 85*t*, 89
Biodiversity
  food availability and, 56
  intermediate disturbance hypothesis, 74–75
  in tropics, 42–43
Birds, as game, 67, 219
Bitter manioc, 124–126, 128
  at Caverna de Pedra Pintada, 282
  in Colombia, 286
Black sapote, 157*t*
Blowdowns, 74–75
Blow guns, 62
Bone, human, *see* Skeletal data
Bottle gourd, 24
  in Africa, 140
  at Caverna de Pedra Pintada, 282
  at coastal Peruvian sites, 271, 275
  in early horticulture, 217
  at Formative period sites, 249*t*
  in house gardens, 141
  on Obelisk Tello, 110

  in Panama, 314
  at Peña Roja, 203–204
  phytogeography, 139–142
  phytoliths, 35, 197*n*
  at preceramic sites, 141, 204*t*
  seeds of, 140
  at Vegas sites, 186, 187, 196–197, 244
Brazil
  early food production in, 29
  origin of manioc and, 123
Brazil nut, 56, 205*t*
Breadboard metates, 296
"Broad spectrum revolution," 325
Brocket deer, 62, 65*f*, 66*t*
Brosimum alicastrum (ramón), 157*t*, 300, 302
Budares
  at Caverna de Pedra Pintada, 282
  in Colombia, 286
  in Formative period Venezuela, 283, 284
Buffalo, 85*t*
Bunchosia armeniaca (ciruela de fraile), 157*t*, 275*t*, 276
Bundle burials, 198
Burials, *see also* Cemeteries
  "elite," 296
  at Vegas sites, 198, 199*f*
Burning, *see also* Swidden cultivation
  at Calima Valley, 202–203
  in development of food production, 312, 316, 324
  in early Holocene, 217
  at Lake Ayauchi occupation, 259–260
  at Lake Geral, 281
  in seasonal forests, 51, 52
  susceptibility of tropical trees, 180
  at La Yeguada, 175–179
Bushmen, maximum supportable population, 68
Butternut squash, 141*f*
Buttresses, 42
Byrsonima (nance), 157*t*, 205*t*, 291, 293*t*

Caatinga, 43, 80
Caballo Muerto, 277
Cactus, 79, 99, 100
Caiman, 67, 73
Calathea allouia, *see* Leren
Calathea spp., 59, 111, 296
Calima Valley, 204–205*t*, 248–249*t*, 251*f*, 262
Calocarpum mammosum, 155, 157*t*, 158